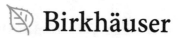

Modeling and Simulation in Science, Engineering and Technology

More information about this series at http://www.springer.com/series/4960

Nicola Bellomo • Pierre Degond • Eitan Tadmor
Editors

Active Particles, Volume 2

Advances in Theory, Models, and Applications

 Birkhäuser

Editors

Nicola Bellomo
Department of Mathematical Sciences
Politecnico di Torino
Torino, Italy

Pierre Degond
Department of Mathematics
Imperial College London
London, UK

Eitan Tadmor
CSCAMM, CSIC Bldg 406
University of Maryland
College Park, USA

ISSN 2164-3679 ISSN 2164-3725 (electronic)
Modeling and Simulation in Science, Engineering and Technology
ISBN 978-3-030-20299-6 ISBN 978-3-030-20297-2 (eBook)
https://doi.org/10.1007/978-3-030-20297-2

Mathematics Subject Classification (2010): 35K55, 35Q92, 35Q70, 37N40, 60H10, 49J15, 74A25, 70F45, 76N10, 82D99, 91A26, 92D25, 91C20, 82B21, 35Q91, 49J45

This book is published under the imprint Birkhäuser, www.birkhauser-science.com by the registered company Springer Nature Switzerland AG.
The registered company address is: Gewerbestrasse 11, 6330 Cham, Switzerland

Preface

This edited book, with title "Active Particles, Volume 2 Advances in Theory, Models, and Applications" collects eight surveys on active matter, a follow-up to Volume 1 under the same title. It blends together contributions which indicate the diversity of the subject matter in theory and applications. These contributions discuss different aspects of active matter at different scales of organization, modeled by agent-based, kinetic, and hydrodynamic descriptions. The study of different models involves different mathematical tools—from the analysis of nonlinear partial differential equations, kinetic theory, and statistical and stochastic dynamics to network theory, mean field approximations, control theory, and flocking analysis. Simulations involve particle dynamics and finite-volume methods and spectral and finite-element methods. The content provides surveys of recent results with a look ahead toward research perspectives. Hence, this book is a timely outlet in providing the scientific community with an up-to-date overview of the current research conducted by leading experts in this field.

The book covers a broad range of applications, including *biological network formation and network theory* in Chapters "Kinetic and Moment Models for Cell Motion in Fiber Structures", "Kinetic Models for Pattern Formation in Animal Aggregations: A Symmetry and Bifurcation Approach", and "High-Resolution Positivity and Asymptotic Preserving Numerical Methods for Chemotaxis and Related Models"; *social systems* in Chapters "Kinetic Models for Pattern Formation in Animal Aggregations: A Symmetry and Bifurcation Approach", "Aggregation-Diffusion Equations: Dynamics, Asymptotics, and Singular Limits", "Control Strategies for the Dynamics of Large Particle Systems", "Kinetic Equations and Self-organized Band Formations", and "A Stochastic-Statistical Residential Burglary Model with Finite Size Effects"; *control theory of sparse systems* in Chapter "Kinetic Equations and Self-organized Band Formations"; *dynamics of swarming and flocking systems* in Chapters "Kinetic Models for Pattern Formation in Animal Aggregations: A Symmetry and Bifurcation Approach", "Kinetic Equations and Self-organized Band Formations", and "Singular Cucker-Smale Dynamics"; and *stochastic particles and mean field approximation* in Chapters "Control Strategies for the Dynamics of Large Particle Systems", "Singular Cucker-Smale Dynamics", and "A Stochastic-Statistical Residential Burglary Model with Finite Size Effects".

The variety of applications and the interdisciplinary use of different mathematical tools reflect the interest of applied mathematicians in modeling, qualitative analysis, and computing of large systems of active particles, which are viewed as living, hence complex, systems. This new frontier of science offers a range of new challenging problems.

The research activity in the field meets an equally productive scientific environment. In particular, we mention the Ki-Net—an NSF Research Network focused on "Kinetic description of emerging challenges in multiscale problems of natural sciences" (www.ki-net.umd.edu). The Ki-Net, through its main three hubs in the Universities of Maryland, Wisconsin, and UT Austin and an inter-linked network of 20+ nodes, fostered a series of activities with the main intellectual focus on development, analysis, computation, and application of quantum dynamics, network dynamics, and kinetic models of biological processes. As such, Ki-Net was a primary outlet for the presentation of the recent activities in the above areas of active matter. Indeed, many of the authors in this special volume were involved in Ki-Net activities (www.ki-net.umd.edu/content/activities), and we use this opportunity to acknowledge the NSF support of Ki-Net grant #1107444 for funding these activities.

Torino, Italy   Nicola Bellomo
London, UK   Pierre Degond
College Park, MD, USA   Eitan Tadmor

Contents

Kinetic and Moment Models for Cell Motion in Fiber Structures

Raul Borsche, Axel Klar, and Florian Schneider

Abstract This review focuses on kinetic and macroscopic models for the migration of cells in fiber structures. Typical applications of cell migration models in such geometries are tumor cell invasion into tissue, or tissue-engineering and the movement of fibroblasts on artificial scaffolds during wound healing.

1 Introduction

This review focuses on kinetic and macroscopic models for the migration of cells in fiber structures. Typical applications of cell migration models in such geometries are tumor cell invasion into tissue, see [24, 27, 28, 38–40], or tissue-engineering and the movement of fibroblasts on artificial scaffolds during wound healing [2, 55].

Continuous approaches to cell migration involve systems of various types of partial differential equations, sometimes coupled to ODEs, and allow to describe the dynamics of cells interacting with their surroundings. Most of them are directly set on the macroscopic scale and involve drift-diffusion-reaction equations for the density of the cells [4, 12, 16, 19–21, 47–49]. Kinetic models for cell migration start on a mesoscopic scale [4, 18, 46]. A parabolic scaling is classically used to deduce the macroscopic description of the cell density from kinetic transport equations for the cell density function depending on time, position, velocity [18, 30–32, 58]. On the macrolevel, this leads to the drift-diffusion-reaction equation for the cell density mentioned above. For surveys and extended reference lists, see, for example, [3, 5, 6].

In this review we consider kinetic equations for cell migration with a focus on moment closure techniques allowing to handle the mesoscopic kinetic transport equation numerically. In the kinetic context, the distribution function for the cells is

R. Borsche · A. Klar (✉) · F. Schneider
TU Kaiserslautern, Kaiserslautern, Germany
e-mail: borsche@mathematik.uni-kl.de; klar@mathematik.uni-kl.de;
schneider@mathematik.uni-kl.de

© Springer Nature Switzerland AG 2019
N. Bellomo et al. (eds.), *Active Particles, Volume 2*, Modeling
and Simulation in Science, Engineering and Technology,
https://doi.org/10.1007/978-3-030-20297-2_1

1

a mesoscopic quantity depending not only on time and position, but also on the cell velocity and potentially other variables. Among other methods, these dependencies can be discretized by moment closures [15, 51, 53, 60, 61, 63], which transform the scalar, but high-dimensional transport equation into a hyperbolic system for moments of the cell density with respect to the velocity variable. Using moment closure approaches one obtains macroscopic systems of equations intermediate between kinetic and scalar macroscopic drift-diffusion equations, see, for example, [8, 34, 45, 46]. Linear moment closure models are in many situations reasonable approximations of the kinetic equation. However, a major drawback of these approximations is that they do not guarantee the positivity of the cell density. Thus, one considers nonlinear closure methods to deal with this problem. Additionally, we also introduce partial-moment closures for the kinetic equations and obtain associated hydrodynamic equations, see [29, 60].

Besides considering the above class of models on multi-dimensional geometries, we consider these models also on network topologies. In this case, the crucial point is to define suitable coupling conditions. In [7–9, 17] coupling conditions for kinetic equations and their moment approximations have been discussed. We refer also to [8, 11, 13, 42], where full- and partial-moment closure equations are studied on a graph and coupling conditions for these models are derived from the kinetic coupling conditions. We note that in the diffusive limit, when the scaling parameter goes to zero, all coupling conditions converge to those of the limit equation, i.e. the conservation of mass through nodes and a continuity condition, compare, for example, [7].

The paper is structured as follows: Sect. 2 states a general class of kinetic equations for cell motion and discusses two specific examples. In the first example we consider a kinetic model describing cell migration on a fiber structure, see [25]. The influence of the fiber structure is described by a haptotaxis term modeling the alignment of the cells in the directions of the fibers. The model is specified by an adequate choice of the turning kernels to describe cell reorientations in response to the interactions with the surrounding tissue. As the focus is on the study of the moment models, no proliferation or decay terms are considered. The second example is a classical kinetic chemotaxis equation, see [18]. It is used to describe cell motion on a network given by a fibrous structure in the second part of the review. Section 3 is dedicated to the derivation of first and higher order moment closures. Numerical simulations for one- and two-dimensional situations are presented in Sect. 4, which also provides a comparison between the different approaches. Section 5 considers the kinetic chemotaxis model and the associated moment equations on a network of fibers. In particular the coupling conditions for the kinetic equations and moment models are presented. Section 6 shows the numerical results for tripod and more general networks. Conclusions and an outlook are given in Sect. 7.

2 Kinetic Models for Cell Motion

In this section we discuss a general class of kinetic equations and special models for cell motion based on haptotaxis and chemotaxis.

2.1 A Class of Kinetic Equations for Cell Motion

In this section we describe the kinetic equation from which we derive approximating moment equations. Let $x \in \mathbb{R}^n$, $n = 1, 2, 3$, $t \in \mathbb{R}^+$, and $v \in V = \mathbb{S}^2$ denote the mechanical variables, i.e., a position vector, the time variable, and a velocity vector, respectively. Thereby assuming constant speed of the cells. We note that the velocity space $V = \mathbb{S}^2$ in three dimensions can be reduced via projection to $V = [-1, 1]$ for one spatial dimension and to $V = B_1(0) = \{v \in \mathbb{R}^2, |v| \leq 1\}$ for two spatial dimensions.

We consider the kinetic equation for the mesoscopic cell density function f

$$\partial_t f + \frac{1}{\epsilon} v \cdot \nabla_x f = (\frac{1}{\epsilon^2}\mathscr{L}_1 + \frac{1}{\epsilon}\mathscr{L}_2)f, \tag{1}$$

where ϵ is a scaling parameter specified later. The turning operators $\mathscr{L}_1$ and $\mathscr{L}_2$ are defined as

$$\mathscr{L}_i f = \int_V \left(k_i(x, v, v')f(v') - k_i(x, v', v)f(v) \right) dv', \quad i \in \{1, 2\}. \tag{2}$$

Both turning operators conserve mass, i.e.

$$\int_V \mathscr{L}_i(f)(v)dv = 0. \tag{3}$$

The first kernel and the combined kernel $k_1 + k_2$ are assumed to be strictly positive and bounded from above:

$$0 < k_{1,min} \leq k_1(x, v', v) \qquad\qquad \leq k_{1,max}, \tag{4}$$

$$0 < k_{min} \quad \leq k_1(x, v', v) + k_2(x, v', v) \leq k_{max}. \tag{5}$$

Additionally, the first kernel satisfies

$$\int_V k_1(x, v', v)dv' = \kappa_1(x)$$

with some known function κ_1. To simplify notations we use in the following

$$\langle f \rangle = \int_V f(v) dv.$$

We assume that there is a probability distribution $F = F(x, v) > 0$, $\langle F \rangle = 1$ that is first-order symmetric,

$$\langle v F \rangle = 0, \tag{6}$$

and fulfills for each $x \in \mathbb{R}^n$ the detailed balance condition

$$k_1(x, v', v) F(x, v) = k_1(x, v, v') F(x, v'). \tag{7}$$

This function will be called the equilibrium distribution of the turning operator. Finally, we will denote the density and mean flux by

$$\rho = \int_V f dv = \langle f \rangle, \quad q = \langle v f \rangle.$$

In the following two subsections we specify the physical situation and introduce two examples for turning kernels which will be considered in the following sections.

2.2 Haptotaxis-Kernels

In this subsection we describe a kinetic equation modeling glioma invasion in a tissue fiber structure. The influence of the fiber structure is modelled via a haptotaxis term describing the tendency of the cells aligning with the fibers. Such equations have been developed in [30, 58]. Following [25] one obtains for $v \in c\mathbb{S}^2$ with the speed $c > 0$, the kinetic equation

$$\partial_t f + v \cdot \nabla_x f = -\lambda_0 (f - F\rho) + \lambda_H \nabla_x Q(x) \cdot (vf - Fq), \tag{8}$$

where λ_0 denotes the part of the turning rate, which is independent of the cell-state. $F = F(x, v)$ represents the normalized directional distribution of tissue fiber. This distribution is in general not isotropic, typical examples are functions F with a bilinear dependence on v. $Q = Q(x)$ is the macroscopic volume fraction of tissue fibers and

$$\lambda_H = \lambda_H(Q(x)) \tag{9}$$

denotes the cell-state dependent part of the turning rate. A non-dimensional form of (8) is

$$\partial_t f + \frac{t_0 c}{x_0} \nabla_x \cdot (vf) = -t_0 \lambda_0 (f - F\rho) + \frac{t_0 c}{x_0} \lambda_H \nabla_x Q \cdot (fv - Fq), \qquad v \in \mathbb{S}^2$$

(10)

with the characteristic time t_0 and the characteristic distance x_0. Identifying the Strouhal number $St = \frac{x_0}{t_0 c}$ and the Knudsen number $Kn = \frac{1}{t_0 \lambda_0}$ as the characteristic parameters, we write the above as

$$\partial_t f + \frac{1}{St} \nabla_x \cdot (vf) = -\frac{1}{Kn} (f - F\rho) + \frac{1}{St} \lambda_H \nabla_x Q \cdot (fv - Fq).$$

(11)

Note that (11) fits into the more general framework (1) by assuming $\eta = \frac{St^2}{Kn}$ to be a constant, identifying the parameter ϵ as $\epsilon = St$ and choosing the kernels k_1 and k_2 as

$$k_1(x, v, v') = \eta F,$$
$$k_2(x, v, v') = -\lambda_H \nabla_x Q \cdot v' F.$$

(12)

As ϵ tends to 0 the macroscopic approximation is

$$\partial_t \rho - \nabla_x \cdot (\nabla_x \cdot (\rho D) - \rho \lambda_H \nabla_x Q D) = 0$$

(13)

with

$$\eta D = \int_V v \otimes v F dv = D_F$$

which is a generally anisotropic drift-diffusion equation.

2.3 Chemotaxis-Kernels

In this subsection we consider a classical kernel describing chemotaxis, see [18]. We will use this example in the second part of the review to describe cell motion on a fiber structure via considering the equations on networks. This means we consider

$$k_1(x, v, v') = \lambda F(x, v) = \frac{\lambda}{|\mathbb{S}^2|},$$

(14)

$$k_2(x, v, v') = \alpha \overline{\nabla m} \cdot vF,$$

with constants λ and α and a limiter chosen, for example, as

$$\bar{x} = \frac{x}{\sqrt{1 + |x|^2}},$$

compare [11, 53]. The corresponding linear reorientation operator is

$$\mathcal{L}_2 f(v) = \alpha \rho v F(v) \overline{\nabla m}$$

and the kinetic equation is given by

$$\partial_t f + \frac{1}{\epsilon} \nabla_x \cdot (vf) = -\frac{\lambda}{\epsilon^2} (f - F\rho) + \frac{1}{\epsilon} \alpha \rho \overline{\nabla m} \cdot vF. \tag{15}$$

This is a flux-limited kinetic chemotaxis equation [18]. The chemoattractant concentration $m(t, x)$ is usually governed by a diffusion equation

$$\partial_t m - D_m \Delta_x m = \gamma \rho - \delta m \tag{16}$$

with a production proportional to the population density ρ with rate γ and an exponential decay with rate δ. As ϵ tends to 0 the macroscopic approximation is a flux-limited Keller-Segel model

$$\partial_t \rho + \nabla_x \cdot \left(\frac{\alpha}{3\lambda} \rho \overline{\nabla_x m} - \frac{1}{3\lambda} \nabla_x \rho \right) = 0. \tag{17}$$

In 1D one obtains for $x \in \mathbb{R}$, velocity $v \in V = [-1, 1]$, $F(v) = 1/2$, $k_1(v, v') = \lambda/2$ and $k_2(v, v') = \frac{1}{2} \alpha v \overline{\partial_x m}$ the scaled equation

$$\begin{cases} \partial_t f + \frac{1}{\epsilon} v \partial_x f = -\frac{\lambda}{\epsilon^2} \left(f - \frac{\rho}{2} \right) + \frac{1}{2\epsilon} \alpha v \overline{\partial_x m} \rho \\ \partial_t m - D_m (\partial_{xx}) m = \gamma \rho - \delta m. \end{cases} \tag{18}$$

3   Moment Models

In this section we give an overview how to derive moment approximations to the general kinetic equation (1).

3.1   Balance Equations

To start with we derive first order moment equations for the density $\rho = \langle f \rangle$ and momentum $q = \langle vf \rangle$. Higher moment approximations can be developed as well, see Sect. 3.5 or [61] for further references. We start with the kinetic equation (1), i.e.,

$$\epsilon^2 \partial_t f + \epsilon v \cdot \nabla_x f = \mathcal{L}_1 f + \epsilon \mathcal{L}_2 f. \tag{19}$$

Multiplication with 1 and v, and integrating with respect to v gives the continuity and momentum equations

$$\epsilon \partial_t \rho + \nabla_x \cdot q = 0, \tag{20}$$

$$\epsilon^2 \partial_t q + \epsilon \nabla_x \cdot P = \langle v \mathcal{L}_1 f \rangle + \epsilon \langle v \mathcal{L}_2 f \rangle. \tag{21}$$

In the momentum equations, the pressure tensor $P := \langle v \otimes vf \rangle$ contains the second moments of f. Since the system is underdetermined, these equations have to be closed by an approximation of P using only ρ and q. This is usually obtained by choosing an ansatz function $f^A(v; \rho, q)$ and defining the approximation

$$P = \langle v \otimes vf \rangle \approx \langle v \otimes vf^A \rangle = P^A.$$

Thus the closed system of equations is

$$\epsilon \partial_t \rho + \nabla_x \cdot q = 0,$$
$$\epsilon^2 \partial_t q + \epsilon \nabla_x \cdot P^A(\rho, q) = \langle v \mathcal{L}_1 f^A(\rho, q) \rangle + \epsilon \langle v \mathcal{L}_2 f^A(\rho, q) \rangle. \tag{22}$$

In the following we consider different ansatz functions and show the resulting closure relations for P^A. In the subsequent derivations it will be useful to consider normalized moments indicated by a hat. For example, normalized momentum and pressure tensor are denoted by

$$\hat{q} := \frac{q}{\rho}, \quad \hat{P} := \frac{P}{\rho}.$$

3.2 Linear ($P_1^{(F)}$-)Closure

One uses the simple linear perturbation ansatz

$$f^A = a(1 + \epsilon v \cdot b) F(v).$$

The multipliers a and b are chosen to fulfill the moment constraints $\langle f^A \rangle = \rho$ and $\langle vf^A \rangle = q$, i.e. they are given by

$$a = \rho,$$

$$\epsilon \langle v \otimes vF \rangle b = \hat{q}.$$

The approximated pressure tensor is

$$P^A = \rho \hat{P}^A(\hat{q}),$$

with

$$\hat{P}^A(\hat{q}) = \frac{\langle v \otimes vf^A \rangle}{\langle f^A \rangle} = D_F + \epsilon \langle v \otimes vv \cdot bF(v) \rangle, \tag{23}$$

where $D_F = \langle v \otimes vF(v) \rangle$ is the pressure tensor of the equilibrium. If F is symmetric, i.e. additionally to $\langle vF(v) \rangle = 0$ we have $\langle v \otimes vvF(v) \rangle = 0$, then the pressure tensor becomes $P^A = \rho D_F$.

Example 1 In the 1D case the linear closure function is

$$f^A = \frac{1}{2}\rho(x, t) + \epsilon \frac{3}{2}vq(x, t).$$

Starting from the 1D kinetic chemotaxis equation this leads to a macroscopic model for chemotaxis as in [13, 42]

$$\begin{cases} \epsilon \partial_t \rho + \partial_x q = 0 \\ \partial_t q + \frac{1}{3\epsilon}\partial_x \rho = -\frac{1}{\epsilon^2}\lambda q + \frac{\alpha}{3\epsilon}\overline{\partial_x m}\rho, \end{cases} \tag{24}$$

which is sometimes called the Cattaneo model for chemotaxis. When $\epsilon \to 0$, the Cattaneo model (24) has the same macroscopic diffusive limit as the kinetic equation (18), i.e. the Keller-Segel equations.

3.3 Nonlinear ($M_1^{(F)}$-)Closure

For this closure we use the approximating function

$$f^A = a \exp(\epsilon v \cdot b)F(v). \tag{25}$$

In contrast to the linear closure discussed in the previous section, the ansatz function f^A is now positive, which leads to several advantages for the resulting approximating equations, see [1, 15, 26]. The computations proceed in a similar way as before. Again, the multipliers a and b are determined from the moment constraints on f^A:

$$(\rho, q) = \left\langle (1, v) f^A \right\rangle = \langle (1, v) a \exp(\epsilon v \cdot b) F(v) \rangle .$$

This gives

$$\hat{q}(b) = \frac{\langle v \exp(\epsilon v \cdot b) F(v) \rangle}{\langle \exp(\epsilon v \cdot b) F(v) \rangle} \quad \text{and} \quad \hat{P}^A(b) = \frac{\langle v \otimes v \exp(\epsilon v \cdot b) F(v) \rangle}{\langle \exp(\epsilon v \cdot b) F(v) \rangle}. \tag{26}$$

Inverting the relation for $\hat{q}(b)$ one obtains $\hat{P}^A(\hat{q})$.

Example 2 In the 1D chemotaxis case the closure using the exponential function leads to

$$\hat{P}^A = h(\hat{q}), \quad h(\hat{q}) = \left(1 - \frac{2}{b}\hat{q}\right), \quad \hat{q} := \frac{q}{\rho} = \coth(b) - \frac{1}{b},$$

where $\lim_{\hat{q} \to 0} h(\hat{q}) = \frac{1}{3}, \hat{q}^2 \le h(\hat{q}) \le 1$. The function h is called the Eddington factor.
The resulting full-moment macroscopic model for chemotaxis reads

$$\begin{cases} \partial_t \rho + \dfrac{1}{\epsilon}\partial_x q & = 0 \\[2mm] \partial_t q + \dfrac{1}{\epsilon}\partial_x \left(\rho h\left(\dfrac{q}{\rho}\right)\right) = -\dfrac{\lambda}{\epsilon^2}q + \dfrac{\alpha}{3\epsilon}\overline{\partial_x m}\rho . \end{cases} \tag{27}$$

The linearization of this model gives again Eq. (24).

3.4 Simplified Nonlinear Closure ($K_1^{(F)}$)

We assume as before $F \ge 0$, $\langle F \rangle = 1$, and $\langle vF \rangle = 0$. This implies that $\mathrm{tr}(< v \otimes vF >) = \mathrm{tr}(D_F) = 1$. Now we want to extend the concept of Kershaw closures [50] for our special situation with not necessarily isotropic function F. We determine the second moment P^A via an interpolation between the free-streaming value $P_\delta = \rho \frac{q \otimes q}{|q|^2}$ for $|q| = \rho$ and the equilibrium solution $P_{eq} = \rho D_F$ for $|q| = 0$ (compare (26) with $b = 0$) and make the ansatz

$$P^A = \rho \hat{P}^A(\hat{q}) := \rho \left(\alpha D_F + (1 - \alpha)\frac{\hat{q} \otimes \hat{q}}{|\hat{q}|^2}\right), \tag{28}$$

where $\alpha = \alpha(\hat{q})$ is given below. To obtain a reasonable model it is crucial to satisfy the so-called realizability conditions [50, 61], i.e. the fact that the moments $\hat{q}$, $\hat{P}^A$ are generated by a non-negative distribution function. In this case we have to ensure that for every $\rho \ge 0$ and $|\hat{q}| \le 1$ we have that [50]

$$\hat{P} - \hat{q} \otimes \hat{q} \geq 0 \quad \text{and} \quad \text{tr}(\hat{P}) = 1.$$

The trace equality immediately follows for all $\alpha \in \mathbb{R}$ since $\text{tr}(D_F) = \text{tr}(\frac{\hat{q} \otimes \hat{q}}{|\hat{q}|^2}) = 1$. Plugging in the definition of P^A gives that

$$\hat{P}^A - \hat{q} \otimes \hat{q} = \alpha D_F + (1 - \alpha - |\hat{q}|^2) \frac{\hat{q} \otimes \hat{q}}{|\hat{q}|^2}$$

is positive semidefinite if $\alpha \geq 0$ and $1 - \alpha \geq |\hat{q}|^2$. We use

$$\alpha = 1 - |\hat{q}|^2, \tag{29}$$

which satisfies both inequalities if $|\hat{q}| \leq 1$. Note that in the special case $D_F = \frac{I}{3}$ the original Kershaw model [50] is recovered.

One can prove that for any distribution $F : \mathbb{S}^2 \mapsto \mathbb{R}^+$ that fulfills $\langle F \rangle = 1$, $\langle Fv \rangle = 0$ and $\langle F(x^\top v)^2 \rangle > 0, \forall x \in \mathbb{S}^2$, the first order moment system (22) together with the Kershaw closure (28) is strictly hyperbolic for all realizable moment vectors (ρ, q), except for $|\hat{q}| = 1$ with $\hat{q}$ parallel to an eigenvector of $\langle Fvv^\top \rangle$. In this case the system matrix still has real eigenvalues but cannot be diagonalized any more.

3.5   Higher-Order Moment Models and Other Angular Bases

Analogously to the first-order moment system, higher-order approximations are conceivable. We give a very brief summary of the general methodology of moment methods. See, for example, [37] for an overview of some of the most common methods.

Let $a(v) = (a_0(v), \dots a_{K-1}(v))$ be the basis of a K-dimensional subspace of $L^2(V)$. The corresponding moments are defined as $u := \langle fa \rangle$. By multiplying the kinetic equation (19) with a and integrating over V we get a system for the moments:

$$\partial_t u + \frac{1}{\epsilon} \nabla_x \cdot \langle va f \rangle = \left\langle (\frac{1}{\epsilon^2} \mathcal{L}_1(f) + \frac{1}{\epsilon} \mathcal{L}_2(f)) a \right\rangle. \tag{30}$$

As in the first-order case, f is approximated by an ansatz function

$$f^A[u](v) \approx f(v)$$

which depends on the moments, such that we get a closed form

$$\partial_t u + \nabla_x \cdot \left\langle va f^A \right\rangle = \frac{1}{\epsilon^2} \left\langle \mathcal{L}_1(f^A) a \right\rangle + \frac{1}{\epsilon} \left\langle \mathcal{L}_2(f^A)) a \right\rangle.$$

One choice for the basis a are spherical harmonics [41, 56]. The classical P_N [59] and M_N [14, 15, 53, 54] methods use the ansatz functions

$$f^A = \alpha_N \cdot a \qquad \text{and} \qquad f^A = \exp(\alpha_N \cdot a),$$

respectively. Analogously to the first-order methods we define the modified $P_N^{(F)}$ and $M_N^{(F)}$ as

$$f^A = (\alpha_N \cdot a) F(v) \qquad \text{and} \qquad f^A = \exp(\alpha_N \cdot a) F(v),$$

respectively, in order to incorporate the equilibrium of the reorientation kernel $F(v)$.

In the following, we discuss a special type of moment basis, the so-called partial moments.

3.6 Half Moments: Partial Moments in One Spatial Dimension

Given a density function $f(t, x, v)$ with $t \in \mathbb{R}^+$, $x \in \mathbb{R}$ and $v \in V = [-1, 1]$, we define the zeroth, first, and second half-moments as

$$(\rho_\pm, q_\pm, P_\pm) := \int_{V_\pm} (1, v, v^2) f dv, \qquad (31)$$

with $V_- := [-1, 0]$, $V_+ := [0, 1]$. This means that we use the angular basis $a = (\mathbb{1}_{V_+}, \mathbb{1}_{V_-}, \mathbb{1}_{V_+} v, \mathbb{1}_{V_-} v)$.

The so-called HP_1 model is given by the ansatz

$$f^A = \begin{cases} a_+(t, x) + b_+(t, x)v & \text{if } v \in V_+, \\ a_-(t, x) + b_-(t, x)v & \text{if } v \in V_-. \end{cases} \implies P_\pm^A = -\frac{1}{6}\rho_\pm \mp q_\pm.$$

As above, the HM_1 model is given by the exponential ansatz

$$f^A = \begin{cases} \exp(a_+(t, x) + b_+(t, x)v) & \text{if } v \in V_+, \\ \exp(a_-(t, x) + b_-(t, x)v) & \text{if } v \in V_-. \end{cases}$$

This gives the normalized first and second half-moments

$$\hat{q}_\pm = \frac{(\pm b_\pm - 1)\exp(\pm b_\pm) + 1}{b_\pm(\exp(\pm b_\pm) - 1)}, \qquad \hat{P}_\pm^A = \frac{(b_\pm^2 \mp 2b_\pm + 2)\exp(\pm b_\pm) - 2}{b_\pm^2(\exp(\pm b_\pm) - 1)}.$$

Example 3 For the chemotaxis problem we get the linear half-moment system as

$$
\begin{cases}
\epsilon \partial_t \rho^\pm + \partial_x q^\pm & = -\dfrac{1}{\epsilon}\lambda\left(\rho^\pm - \dfrac{\rho^+ + \rho^-}{2}\right) \pm \dfrac{1}{4}\alpha\overline{\partial_x m}\left(\rho^+ + \rho^-\right) \\[2mm]
\epsilon \partial_t q^\pm + \partial_x\left(-\dfrac{1}{6}\rho^\pm \pm q^\pm\right) & = -\dfrac{1}{\epsilon}\lambda\left(q^\pm \mp \dfrac{\rho^+ + \rho^-}{4}\right) + \dfrac{1}{6}\alpha\overline{\partial_x m}\left(\rho^+ + \rho^-\right).
\end{cases}
\tag{32}
$$

The linear half-moment model (32) has again the Keller-Segel equations as macroscopic diffusive limit as ϵ goes to 0. The nonlinear closure leads to the half-moment model

$$
\begin{cases}
\epsilon \partial_t \rho^\pm + \partial_x q^\pm & = -\tfrac{1}{\epsilon}\lambda\left(\rho^\pm - \tfrac{\rho^+ + \rho^-}{2}\right) \pm \tfrac{\alpha}{4}\overline{\partial_x m}\left(\rho^+ + \rho^-\right) \\[2mm]
\epsilon \partial_t q^\pm + \partial_x\left(\rho^\pm h^\pm\left(\tfrac{q^\pm}{\rho^\pm}\right)\right) & = -\tfrac{1}{\epsilon}\lambda\left(q^\pm \mp \tfrac{\rho^+ + \rho^-}{4}\right) + \tfrac{\alpha}{6}\overline{\partial_x m}\left(\rho^+ + \rho^-\right),
\end{cases}
\tag{33}
$$

where

$$
h^-(\hat{q}^-) = \hat{q}^-\left(-1 - \frac{2}{b^-}\right) - \frac{1}{b^-}, \qquad h^+(\hat{q}^+) = \hat{q}^+\left(1 - \frac{2}{b^+}\right) + \frac{1}{b^+},
$$

$$
\hat{q}^- := \frac{q^-}{\rho^-} = \frac{1}{\exp(b^-) - 1} - \frac{1}{(b^-)}, \qquad \hat{q}^+ := \frac{q^+}{\rho^+} = \frac{\exp(b^+)}{\exp(b^+) - 1} - \frac{1}{(b^+)}.
$$

We note that an explicit maximum-entropy Eddington factor approximating the above closure is obtained by using the *Kershaw closure* [50]

$$
h^-(\hat{q}^-) = \frac{2}{3}(\hat{q}^-)^2 - \frac{1}{3}\hat{q}^-, \qquad h^+(\hat{q}^+) = \frac{2}{3}(\hat{q}^+)^2 + \frac{1}{3}\hat{q}^+.
\tag{34}
$$

3.7 Quarter Moments: Partial Moments in Two Spatial Dimension

Similarly to the one-dimensional setting, we define quarter-moments of a given density $f(t, \mathbf{x}, \mathbf{v})$, $\mathbf{x} \in \mathbb{R}^2$, $\mathbf{v} = \mathbf{v}(\phi, r)$ with $(\phi, r) \in [0, 2\pi] \times [-1, 1]$

$$
\rho_{\pm\pm'}(t, x) := \int_{V_{\pm\pm'}} f(t, x, v(\phi, r))d\phi dr =: \langle f \rangle_{\pm\pm'},
$$

$$
q_{\pm\pm'}(t, x) := \int_{V_{\pm\pm'}} v(\phi, r) f(t, x, v(\phi, r))d\phi dr =: \begin{pmatrix} q^x_{\pm\pm'}(t, x) \\ q^y_{\pm\pm'}(t, x) \end{pmatrix},
$$

$$P_{\pm\pm'}(t, x) := \int_{V_{\pm\pm'}} v(\phi, r) \otimes v(\phi, r) f(t, x, v(\phi, r)) d\phi dr$$

$$=: \begin{pmatrix} P^{xx}_{\pm\pm'}(t, x) & P^{xy}_{\pm\pm'}(t, x) \\ P^{xy}_{\pm\pm'}(t, x) & P^{yy}_{\pm\pm'}(t, x) \end{pmatrix}$$

with $v(\phi, r) := (\sqrt{1 - r^2} \cos(\phi), \sqrt{1 - r^2} \sin(\phi))^T$ and $V_{++} := [0, \frac{\pi}{2}] \times [-1, 1]$, $V_{-+} := [\frac{\pi}{2}, \pi] \times [-1, 1]$, $V_{--} := [\pi, \frac{3\pi}{2}] \times [-1, 1]$ $V_{+-} := [\frac{3\pi}{2}, 2\pi] \times [-1, 1]$. Here $\pm\pm'$ stands for an arbitrary combination of $+$ and $-$, that means $++, --, +-, -+$. Again, normalized moments can be defined as

$$\hat{q}_{\pm\pm'} := \frac{q_{\pm\pm'}}{\rho_{\pm\pm'}} \quad \text{and} \quad \hat{P}_{\pm\pm'} := \frac{P_{\pm\pm'}}{\rho_{\pm\pm'}},$$

satisfying similar relations as in one dimension. The first moment has to be located within the corresponding quarter sphere, i.e. $\hat{q}_{\pm\pm'} \in v(V_{\pm\pm'})$ [35, 60, 63].

The linear and minimum-entropy closures can be derived as in one dimension using the ansatz functions

$$f^A = \begin{cases} a_{\pm\pm'} + b_{\pm\pm'} \cdot v & \text{if } (\phi, r) \in V_{\pm\pm'} \end{cases}$$

and

$$f^A = \begin{cases} \exp\left(a_{\pm\pm'} + b_{\pm\pm'} \cdot v\right) & \text{if } (\phi, r) \in V_{\pm\pm'}, \end{cases}$$

respectively. The minimum-entropy closure can be obtained again using a suitable two-dimensional table-lookup [35]. This gives the QP_1 and QM_1 models. Its second moment is depicted in Fig. 1.

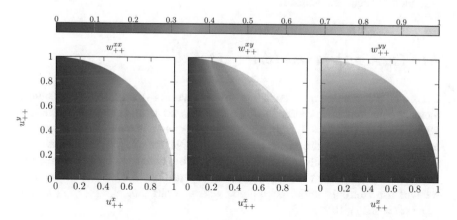

Fig. 1 Quarter-moment minimum-entropy closure

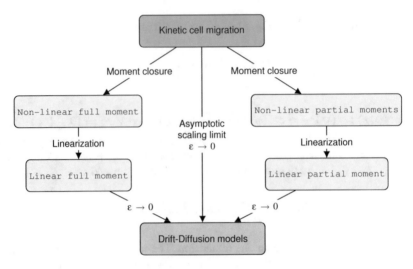

Fig. 2 Hierarchy of cell migration models

3.8 Summary

The moment models and their relations discussed in the previous subsections are summarized in Fig. 2.

4 Numerical Methods and Results

We present numerical examples for the chemotaxis and haptotaxis examples described in Sects. 2.2 and 2.3.

4.1 Chemotaxis in One Dimension

In case of the kinetic chemotaxis equation (1) with (14) and (16), the system of moment equations is discretized using a kinetic scheme on equidistant, structured grids (see, e.g., [35, 37, 43, 62]). The chemoattractant equation (16) is discretized using an implicit finite-difference approximation. We refer to [60] for details.

We compare the results of our simulation to those computed by a scheme with a very fine resolution for the coupled system of equations. Unless otherwise noted we will use isotropic initial conditions, which means $f(0, x, v) = f(0, x) = \rho_{\pm}(0, x) = \pm 2q_{\pm}(0, x)$.

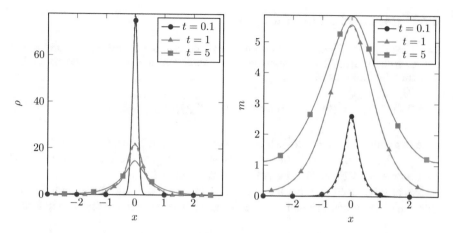

Fig. 3 Half-moment minimum-entropy (solid with marks) and kinetic model (dashed) for the one-spike example

4.1.1 Chemotaxis: One Spike

First, we simulate the effects of chemotaxis on an aggregation of cells in the center of the domain in the absence of a chemoattractant initially. This is modelled by the initial data

$$\rho_\pm(0, x) = \frac{1}{2}\left(100 \exp\left(-100x^2\right) + 10^{-4}\right), \qquad m(0, x) = 0$$

on the domain $x \in [-3, 3]$ with the parameters $D_m = 1$, $\gamma = 1$, $\delta = 1$, $\alpha = 2$ in (14) and (16).

Figure 3 shows that ρ diffuses with decreasing speed, resulting in a "smeared out" version of the initial state. The concentration of the chemoattractant m first increases drastically until it matches ρ, which is caused by the production of the chemoattractant by the cells themselves, and then flattens out as ρ.

When comparing the results obtained by using the different closure relations, one can only observe negligible differences, since the initial condition does not provoke negativity of ρ for the linear closures. In Fig. 3 we show only the Half-moment exponential closure. Most importantly, we see that the macroscopic and kinetic solutions behave very similarly. This shows that the macroscopic model yields very good approximations for this example while being much cheaper computationally: the kinetic reference solution has computation times[1] of 9.9 s and 126.8 s for $\Delta x = 0.1$ and $\Delta x = 0.02$, respectively, with $T = 5$, whereas the macroscopic half-moment solution needs 1.0 s and 3.3 s with the exponential and 0.3 s and 1.3 s with the linear closure.

---

[1]Dual Core 2.6 GHZ, 8 GB RAM.

4.1.2 Chemotaxis: Two Spikes

In this example the (non-isotropic) initial data describes two spikes, that are located symmetrically around the center of the domain and are moving towards it. We consider the initial condition is

$$f(0, x, v) = \frac{100}{0.05\sqrt{\pi}} \left(\exp\left(-10^2((v+1)^2 + (x-1)^2)\right) + \exp\left(-10^2((v-1)^2 + (x+1)^2)\right) \right)$$

together with $m(0, x) = -x^2 + 9$ on the domain $[-3, 3]$ with the parameters $D_m = 0$, $\gamma = 0$, $\delta = 0$, $\alpha = \frac{1}{2}$. In this example we compare the linear and nonlinear half-moment solutions with the full-moment M_1 model.

We observe in Fig. 4 that the two spikes move towards the center, collide, and then oscillate around the center of the domain until they reach a steady state forced

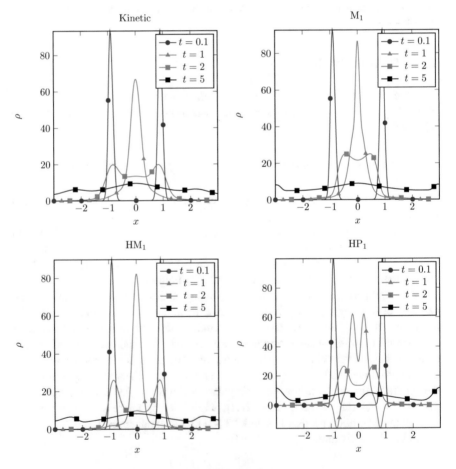

Fig. 4 Comparison of models for the two-spikes example

by the constant chemoattractant concentration. In this case, there are noticeable differences in the closures. In the linear half-moment HP_1 model the density ρ takes negative values on the outer edge of the spikes, but these effects cancel out again after the collision in the center of the domain. The full-moment minimum-entropy M_1 model wrongly predicts a lower speed of propagation when the two spikes hit each other, caused by the well-known "zero-netflux" problem of this model [36, 43, 57, 61]. In contrast, the speed of propagation predicted by the half-moment minimum-entropy HM_1 model is almost exact.

4.2 Haptotaxis and Glioma Invasion in Two Dimensions

In this and the following subsections we consider the haptotaxis example given by Eq. (11) solved with a second-order realizability-preserving scheme. We refer to [25] for details.

All numerical simulations will be done for Eq. (11). We use a quadratic ansatz for the equilibrium fiber distribution

$$F(v) = \frac{3}{4\pi \, \text{trace} \, (D_W)} \left(v^\top D_W v \right).$$

This so-called peanut distribution [30] is a very simple model that relates the fiber distribution to the local water diffusion tensor $D_W \in \mathbb{R}^{3\times3}$, which can be measured by diffusion tensor imaging (DTI) [52]. It has the additional advantage that all the coefficients in the diffusion limit can be computed analytically.

We estimate the volume fraction Q from the local water diffusion tensor D_W using the characteristic length estimate [31]

$$Q(x) = CL(D_W(x)) = 1 - \left(\frac{\text{trace} \, (D_W)}{4\lambda_1} \right)^{\frac{3}{2}}, \tag{35}$$

where λ_1 is the maximum eigenvalue of D_W.

In the moment models we use mass conserving, thermal boundary conditions for the incoming characteristics. This means that outgoing particles are absorbed at the boundary and emitted according to the fiber distribution. For the diffusion approximation the only condition is that there is no flux over the boundary.

Finally, to compare different models we use the pointwise relative difference between two functions $h_1(x), h_2(x)$, i.e. $e_{rel}(h_1(x), h_2(x)) = \frac{|h_1(x) - h_2(x)|}{\|h_2(x)\|_\infty}$.

4.2.1 Glioma Invasion: Abruptly Ending Fiber Strand

This setting models an initially concentrated mass of cells following a white matter tract that abruptly ends. While the latter is not to be expected for a real brain geometry, we use it in order to show some notable effects in the glioma equation. The involved diffusion tensor and volume fraction are both spatially varying. The computational domain is

$$[0, T] \times [0, X]^2 \times c\mathbb{S}^2, \ T = 2, \ X = 3, \ c = \frac{X}{\epsilon T},$$

where the cell speed c is chosen to adjust the parabolic scaling parameter $\epsilon = St = \frac{X}{cT}$ from (19). The coefficients λ_0, λ_H involved in the turning rate are chosen such that $\lambda_0 = c^2 \frac{T}{X^2} = \frac{1}{\epsilon^2 T}$ and $\lambda_H(Q) = \frac{1}{(2+Q)(1+Q)^2}$. This yields $Kn = \frac{X^2}{c^2 T^2}$ and $\eta = 1$.

The fiber geometry is modeled by setting the water diffusion tensor $D_W(x)$ as a function of space. We use a diagonal matrix

$$D_W(x) = \begin{pmatrix} 1 + 5 \exp\left(-\frac{1}{2}\sigma^{-2} \max\left\{0, \ x_1 - \frac{X}{2}, \ |x_2 - \frac{X}{2}| - \frac{1}{10}\right\}\right) & 0 \\ 0 & 1 \end{pmatrix},$$

where only the first eigenvalue varies in space to blend smoothly between a strongly concentrated distribution in x-direction with $D_{00} = 6$ and an isotropic distribution $D_{00} = 1$. For simplicity, the volume fraction $Q(x)$ is then computed as $FA(D_W(x))$.

The initial condition is a square of length $\frac{1}{10}$ centered at $(\frac{1}{2}, \frac{3}{2})$:

$$f(t = 0, x, v) = \frac{1}{4\pi} \begin{cases} 1 & x \in [0.45, 0.55] \times [1.45, 1.55], \\ 10^{-4} & \text{else.} \end{cases}$$

We investigate the different approximations for different values of ϵ and compare the results for $\epsilon \to 0$ with the limit diffusion approximation

$$\partial_t \rho_0 - \nabla_x \cdot (\nabla_x \cdot (\rho_0 D) - \rho_0 \lambda_H \nabla_x Q D) = 0. \tag{36}$$

Figure 5 shows the solution for the diffusion approximation D (Fig. 5a) alongside the first-order Kershaw method $K_1^{(F)}$ (Fig. 5b–f) for $\epsilon \in \{1, 0.5, 0.25, 0.1, 0.01\}$. Additionally, the pointwise relative difference between both models (Fig. 5g–i) for $\epsilon \in \{0, 0.25, 0.1, 0.01\}$ is shown. Far from the diffusion limit at $\epsilon = 1$ the cells travel exactly once through the domain ($St = 1$) and have an expectation of one ($Kn = 1$) velocity jump. Therefore, we see a strongly advection dominated behavior with very little influence from the underlying fiber distribution. As ϵ gets smaller, the Kershaw model becomes more similar to the diffusion approximation. Relative pointwise differences also decrease although even at $\epsilon = 0.01$ some discrepancy in the range of 2–5% remains. We attribute this to inherent differences between the numerical schemes at a not yet fine enough grid. As a reference we show the standard P_5 solution in Fig. 6 at $\epsilon = 0.1$ and $\epsilon = 0.01$. Note that higher-order P_N

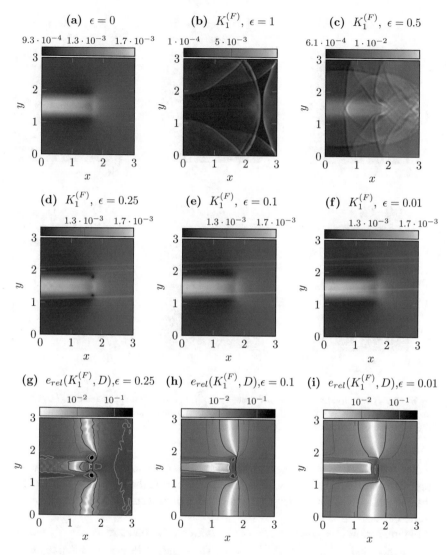

Fig. 5 Comparison between the $K_1^{(F)}$-model and the diffusion approximation as ϵ approaches 0. (**a**) The diffusion approximation. (**b**)–(**f**) $K_1^{(F)}$ for $\epsilon \in \{1, 0.5, 0.25, 0.1, 0.01\}$, respectively. (**g**)–(**i**) The relative difference between the diffusion and the Kershaw model $e_{rel}(K_1, D)$ for $\epsilon \in \{0.25, 0.1, 0.01\}$, respectively. Differences are plotted on a logarithmic scale. Contours are drawn for 0.1 (yellow), 0.05 (green), 0.02 (blue), 0.01 (purple)

models are not shown here as they do not differ significantly from the P_5 solution. From the relative difference to P_5 it is apparent that at $\epsilon = 0.01$ the solution is so close to the diffusion limit and that moment models yield only a marginal improvement. Again, the remaining 2% difference in Fig. 6e can be attributed

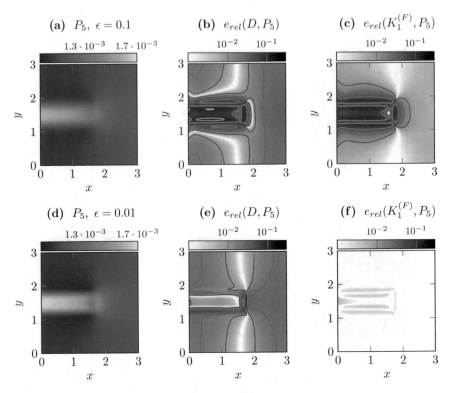

Fig. 6 The P_5 solution and relative difference to $K_1^{(F)}$ and diffusion approximation. Upper row: $\epsilon = 0.1$, lower row: $\epsilon = 0.01$. Differences are plotted on a logarithmic scale. Contours are drawn for 0.1 (yellow), 0.05 (green), 0.02 (blue), 0.01 (purple)

to the differences in the numerical schemes. However, at $\epsilon = 0.1$ the diffusion approximation starts to loose validity and deviates from the P_5 solution up to 10%. The first-order Kershaw model gives a noticeable improvement in this case although a difference of 5% remains.

Figure 7 shows the solution at $\epsilon = 0.1$, both for the standard P_1, P_3, P_5 models and the modified $P_1^{(F)}$, $P_3^{(F)}$, $P_5^{(F)}$ models. The standard P_1 model cannot represent the correct pressure tensor if the distribution is in equilibrium $f(x, v) = F(x, v)$ and thus does not converge to the diffusion limit. This can be observed in Fig. 7a. Here, the modified model $P_1^{(F)}$ that includes F in the ansatz function leads to a great improvement. For the higher moment-orders the difference between standard and modified models becomes less pronounced. This is to be expected since in the special case of $F = v^\top D v$, the P_3 model already contains F.

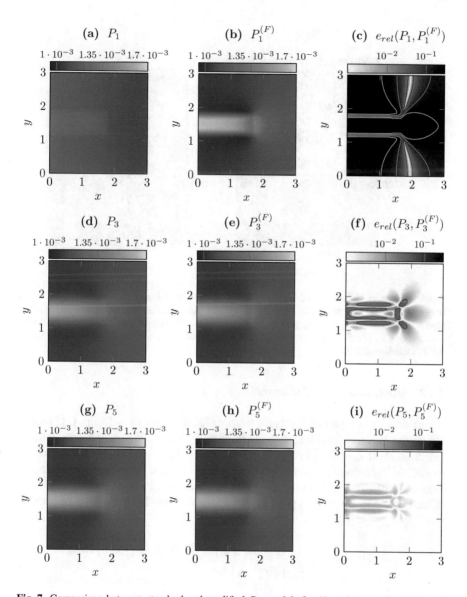

Fig. 7 Comparison between standard and modified P_N models for $N = 1$ (upper row), $N = 3$ (middle), and $N = 5$ (lower row). Color code of contours is the same as in Figs. 5 and 6

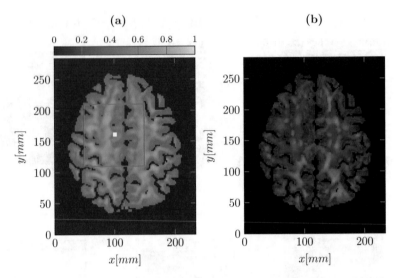

Fig. 8 Estimated volume fraction $Q = CL$ via characteristic length (**a**) and an RGBA coded image of the main axis of D_W (**b**). The color channels RGB encode the x, y, z components of the eigenvector corresponding to the largest eigenvalue of D_W, respectively, while the alpha channel is scaled with CL. The red square indicates the computational domain, while the white square marks the initial cell distribution

4.2.2 Glioma Invasion: 2D Brain Slice

In this numerical experiment we take water diffusion tensors D_W from a DTI scan of the human brain.[2] The tensor field is visualized in Fig. 8. We use the characteristic length estimate $CL(D_W)$ to obtain the volume fractions via (35), which are shown in Fig. 8a. Additionally the main diffusion direction, i.e. the largest eigenvector of D_W, is shown in Fig. 8b as a four-channel color-coded image. The initial tumor mass, marked by the white square in Fig. 8a, is concentrated in a square of length 5 mm

$$f(t = 0, x, v) = \frac{1}{4\pi} \begin{cases} 1 & x \in [98.5, 103.5] \times [158.5, 163.5] \\ 10^{-4} & \text{else} \end{cases}$$

at the center of the spatial domain $[50, 150] \times [110, 210]$ (mm)^2, which is indicated by the red square in Fig. 8a. Moreover, we consider a time span of half a year, which is $T = 1.57 \times 10^7$ s. In Eq. (11) we use the parameters $St = 3.02 \times 10^{-1}$ and the Knudsen number Kn is chosen as $Kn = 6.34 \times 10^{-3}$. Finally we have chosen $\lambda_H(Q) = 2.5 \times 10^1 \frac{1}{(2+Q)(1+Q)^2}$. With this set of parameters the

---

[2]Provided by Carsten Wolters (Institute for Biomagnetism and Biosignal Analysis, WWU Münster).

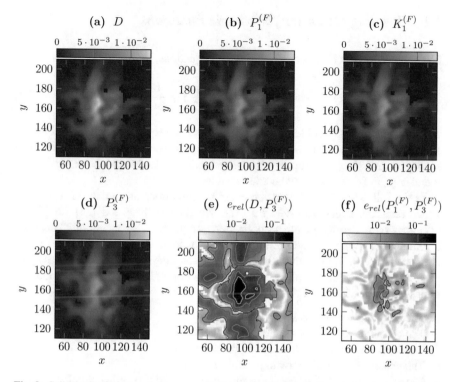

Fig. 9 Cell distribution at end time $t = T$. (**a**) The diffusion approximation D, and (**b**)–(**d**) moment models $P_1^{(F)}$, $K_1^{(F)}$, $P_3^{(F)}$, respectively. (**e**) and (**f**) Relative differences. Color code of contours is the same as in Figs. 5 and 6

characteristic qualitative behavior of glioma cells can be observed, i.e., predominant movement and concentration along white matter tracts. At $\epsilon \approx 0.3$, while the diffusion approximation and moment models are structurally similar, there is a significant difference of up to 20% between them. However, a first-order moment approximation seems to be accurate enough since the difference to the third-order $P_3^{(F)}$ is mostly below 2% (Fig. 9).

5 Cell Motion on Fiber Networks

In this section we develop a hierarchy of coupling conditions for the models proposed in the previous section derived from the chemotaxis kinetic equation (18). We start with the construction of coupling conditions for the kinetic model. These can be used to derive coupling conditions of the remaining models. We refer to [8] and [11] for further details.

5.1 Coupling Condition for Kinetic Equations

The coupled system (18) is composed of two equations, the kinetic problem

$$\partial_t f + \frac{1}{\epsilon} v \partial_x f = -\frac{\lambda}{\epsilon^2} \left(f - \frac{\rho}{2} \right) + \frac{1}{2\epsilon} \alpha v \overline{\partial_x m \rho} \tag{37}$$

and a linear parabolic problem for m. Since these two equations are coupled via the source terms on the right-hand sides, we can develop coupling conditions for both equations separately. Appropriate coupling conditions for the parabolic part can be found, e.g., in [7] and will not be addressed here.

In the following we will develop coupling conditions at a node for a junction which connects edges governed by (37). We refer to [44] and [33] for kinetic equations on networks for other applications. In general, for hyperbolic equations the number of coupling conditions depends on the number of characteristics leaving the node [22, 23]. Thus in the case of (37) the coupling conditions should assign a value to all $f(v)$ with $v \in [0, 1]$. In the construction of the coupling conditions we impose the following conditions:

1. The coupling conditions should be linear.
2. The coupling conditions should be independent of v.
3. The total mass in the system should be conserved.
4. The values of f should remain positive all times.
5. In the case of a 1-to-1 coupling the solution of the coupled problem should coincide with the solution on one continuous edge.
6. In the limit $\epsilon \to 0$ the conditions should converge to the coupling conditions for the Keller-Segel equation, i.e. the continuity of the densities ρ_i.

Especially the first and second condition are chosen to simplify the computations. In the following we will discuss the case of a junction connecting N edges, which are all oriented away from the node. This can be extended to junctions of an arbitrary number of in- and outgoing edges by simple local transformations.

From the first two requirements we can conclude that the coupling conditions are given in the form

$$f^+ = A f^-,$$

where $f_i^+ = f_i(v)$ and $f_i^- = f_i(-v)$ for $v \in [0, 1]$ and $i = 1, \dots, N$. In order to conserve the total mass in the system the matrix $A \in \mathbb{R}^{N \times N}$ has to fulfill

$$\sum_{i=1}^{N} a_{i,j} = 1 \qquad \forall j = 1, \dots, N .$$

The requirement 4 forces all entries of the matrix to be positive $a_{i,j} \geq 0$, $i, j = 1, \dots, N$. In the case of a 1-to-1 coupling, i.e. $N = 2$, we want

$$A = \begin{pmatrix} 0 & 1 \\ 1 & 0 \end{pmatrix}.$$

This we generalize by imposing $a_{i,i} = 0$, $i = 1, \ldots, N$ also for $N > 2$. The equality of the densities in requirement 6 leads to $f_i^+ + f_i^- = f_j^+ + f_j^-$, $i, j = 1, \ldots, N$. Since we can express the f_i^+ in terms of f_i^-, we obtain after some direct computations the following constraint

$$\sum_{j=1}^{N} a_{i,j} = 1 \qquad \forall j = 1, \ldots, N.$$

Applying all the above constraints in the case of a three-way junction $N = 3$, only one free parameter $\alpha \in [0, 1]$ is left in the entries of the matrix A

$$A = \begin{pmatrix} 0 & \alpha & 1 - \alpha \\ 1 - \alpha & 0 & \alpha \\ \alpha & 1 - \alpha & 0 \end{pmatrix}.$$

The only choice in which all edges are treated equally is $\alpha = \frac{1}{2}$. This leads to the very simple set of coupling conditions

$$\begin{bmatrix} f_1^+ \\ f_2^+ \\ f_3^+ \end{bmatrix} = \begin{bmatrix} 0 & 1/2 & 1/2 \\ 1/2 & 0 & 1/2 \\ 1/2 & 1/2 & 0 \end{bmatrix} \begin{bmatrix} f_1^- \\ f_2^- \\ f_3^- \end{bmatrix}. \tag{38}$$

In the cases $N > 3$ some more freedom in the choices of the $a_{i,j}$ is given, but the values $a_{i,j} = \frac{1}{N-1}$ $i \neq j$ and $a_{i,i} = 0$ remain an admissible choice.

5.2 Coupling Condition for the Linear and Nonlinear Half-Moment Model

In this subsection we derive coupling conditions for the half moment system (32) from the kinetic model. As in the previous section we define the quantities $\rho_i^+, \rho_i^-, q_i^+, q_i^-$ for $i = 1, \ldots, N$ on each edge.

Since the kinetic coupling conditions (38) are linear and independent of v we obtain directly the coupling condition for half moment model (32) as

$$\begin{cases} \begin{bmatrix} \rho_1^+ \\ \rho_2^+ \\ \rho_3^+ \end{bmatrix} = \begin{bmatrix} 0 & 1/2 & 1/2 \\ 1/2 & 0 & 1/2 \\ 1/2 & 1/2 & 0 \end{bmatrix} \begin{bmatrix} \rho_1^- \\ \rho_2^- \\ \rho_3^- \end{bmatrix} \\ \begin{bmatrix} q_1^+ \\ q_2^+ \\ q_3^+ \end{bmatrix} = - \begin{bmatrix} 0 & 1/2 & 1/2 \\ 1/2 & 0 & 1/2 \\ 1/2 & 1/2 & 0 \end{bmatrix} \begin{bmatrix} q_1^- \\ q_2^- \\ q_3^- \end{bmatrix}. \end{cases} \tag{39}$$

These are six equations for six outgoing characteristics, which is the correct number of coupling conditions. Note that only the averaging in v and not the exact structure of the closure was used. If the coupling conditions would depend on v, the closure might be relevant.

Recall that all properties of (38) are inherited by the above coupling conditions. For example, the total mass in the system is conserved since

$$\sum_{i=1}^3 q_i = \sum_{i=1}^3 (q_i^+ + q_i^-) = 0.$$

But note that although the coupling conditions maintain the positivity of the densities, this does not necessarily hold for the complete network, as it is, for example, not assured for the linear half-moment model on the edges (32).

5.3 Coupling Condition for the Linear Full Moment Equations

In order to derive coupling conditions for the macroscopic quantities ρ and q we cannot simply average the equations (38) on $[-1, 1]$, since the information is split for positive and negative values of v. From the linear closure function we deduce the following expressions for the half moments

$$f_i^+ = f(v) = \frac{1}{2}\rho_i + \frac{3}{2}v\epsilon q_i , \ f_i^- = f(-v) = \frac{1}{2}\rho_i - \frac{3}{2}v\epsilon q_i,$$

$$v \in [0, 1] , \ i = 1, \ldots, N .$$

Inserting these into (38) we obtain for the case $N = 3$

$$\begin{bmatrix} 2 & -1 & -1 \\ -1 & 2 & -1 \\ -1 & -1 & 2 \end{bmatrix} \begin{bmatrix} \rho_1 \\ \rho_2 \\ \rho_3 \end{bmatrix} + \epsilon\frac{3}{2} \begin{bmatrix} 2 & 1 & 1 \\ 1 & 2 & 1 \\ 1 & 1 & 2 \end{bmatrix} \begin{bmatrix} q_1 \\ q_2 \\ q_3 \end{bmatrix} = 0. \tag{40}$$

Note that in (24) we have only one characteristic moving to the right. Thus for a node connecting three outgoing edges, we have to provide exactly three coupling conditions.

As before all properties of the kinetic coupling conditions transfer to (40), e.g. the classical formulation of the conservation of mass is obtained by summing all three equations.

Remark 1 We note that a more detailed analysis of the situation near the node based on kinetic layers would lead to more accurate conditions, compare [9] for the case of a kinetic problem with the wave equation as limit equation.

Remark 2 To derive coupling conditions for the nonlinear full-moment approximation from the kinetic ones is a challenging topic. In this case one has to take into account the fact that the sign of the eigenvalues might change during the evolution, which leads to a change of the required number of coupling conditions. Such nonlinear problems have been considered for a simpler case deriving coupling conditions for the Burgers equations from a simplified kinetic two-velocity model in [10].

5.4 Coupling Condition for the Keller-Segel Equations

For the Keller-Segel model (36) it is easy to verify, see [8], that the proposed coupling conditions converge for $\epsilon \to 0$ to those introduced in [7]. These conditions are

$$\sum_{i=1}^{N} q_i = 0 \tag{41}$$

$$\rho_i = \rho_j, \qquad i, j = 1, \ldots, N, \ i \neq j,$$

where $q_i = \frac{1}{3\lambda}(\partial_x \rho_i) - \frac{\alpha}{3\lambda}\overline{\partial_x m}\rho_i$. A different possible choice is studied in [13, 42].

6 Numerical Results for Cell Motion on Networks

In this section we investigate the proposed models in several numerical test cases for different values of ϵ. For simplicity we use for the nonlinear half-moment model the explicit Kershaw closure, i.e. $h^{\pm}$ given by (34). In all the considered examples we will use the following values for the parameters $\lambda = \alpha = 1$, $D_m = 1$, $\gamma = 1$ and $\delta = 0.1$. The spatial resolution is $\Delta x = 0.02$ and the time step is chosen according to the CFL condition. At end points where no coupling conditions are imposed zero Neumann boundary conditions are applied. For the kinetic model we discretize the velocity space $V = [-1, 1]$ with $N_v = 50$ cells.

6.1 Numerical Solutions on an Interval

First, we consider the interval $[0, 2]$ and $\Delta x = 0.005$. As initial conditions for the kinetic equation we use

$$f(x, v, 0) = F(v)\rho(x, 0) = \frac{1}{2}\rho(x, 0) \,,$$

with

$$\rho(x, 0) = \begin{cases} 1 & \text{if } 0 \leq x \leq 1 \\ 0 & \text{if } 1 \leq x \leq 2 \,. \end{cases}$$

The remaining initial values for the hydrodynamic equations can be derived from the kinetic initial condition. At $t = 0$ no chemoattractant $m(x, 0) = 0$ is present.

In Figs. 10 and 11 the densities at $t = 0.2$ for the values $\epsilon = 1$, 0.5, 0.1 are shown. For the kinetic model we observe that diffusion depends strongly on the value of ϵ as expected. The closest approximation to the kinetic equations are the half moment models, in particular, the nonlinear half-moment model. For the half moment P_1 model for $\epsilon = 1$ and $\epsilon = 0.5$ one can clearly observe the four waves generated by the advective part. In the full moment P1 model only two waves are used to approximate the kinetic solution. The flux-limited Keller-Segel equation is evolving too fast for large ϵ. In general, the half-moment models are clearly superior to the full moment models and the nonlinear models provide better approximations than the linear models. For small ϵ all models converge to the solution of the Keller-Segel equation.

In the second test we investigate the positivity preserving of the nonlinear moment models which is not guaranteed in the case of the linear models. We use non-equilibrium initial conditions

$$f(x, 0, v) = \begin{cases} a \exp(vb) & \text{if } -1 \leq v \leq 0 \\ 0 & \text{if } 0 \leq v \leq 1 \end{cases} \tag{42}$$

for the kinetic equation (18). a and b are chosen such that the initial data of the half-moment models are given by

$$\rho^-(x, 0) = \int_{-1}^{0} f dv = \rho_0 \,, \qquad q^-(x, 0) = \int_{-1}^{0} v f dv = -0.9 \times \rho_0 \,,$$

$$\rho^+(x, 0) = \int_{0}^{1} f dv = 0 \,, \qquad q^+(x, 0) = \int_{0}^{1} v f dv = 0 \,,$$

$$\tag{43}$$

with

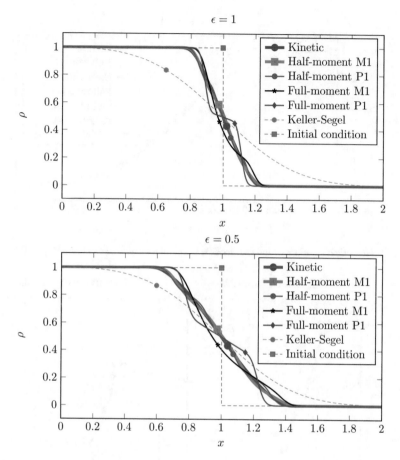

Fig. 10 Numerical solutions of the four models on an interval at time $t = 0.2$ with $\epsilon = 1$ (top) and $\epsilon = 0.5$ (bottom)

$$\rho_0(x) = \begin{cases} 1 & \text{if } 0 \le x \le 1 \\ 10^{-6} & \text{if } 1 \le x \le 2 \end{cases}, \quad m(x, 0) = 0, \ \forall x \in [0, 2] \, .$$

The initial conditions for the *full-moment models* are

$$\rho(x, 0) := \rho^+ + \rho^- = \rho_0, \quad q_\epsilon(x, 0) = -0.9\rho_0 \, .$$

The solutions are plotted in Fig. 12, for $\Delta x = 0.02$ at $t = 0.5$. Even though the initial conditions satisfy the realizability conditions $\rho^\pm \ge |q^\pm|$ (for half-moment models) and $\rho \ge |q|$ (for full-moment models), only the nonlinear full- and half-moment models preserve the positivity of the cell density.

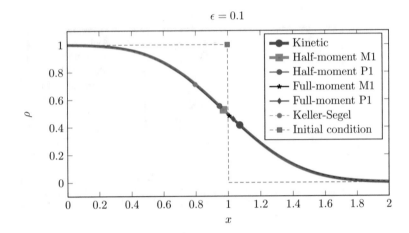

Fig. 11 Numerical solutions of the four models on an interval at time $t = 0.2$ with $\epsilon = 0.1$

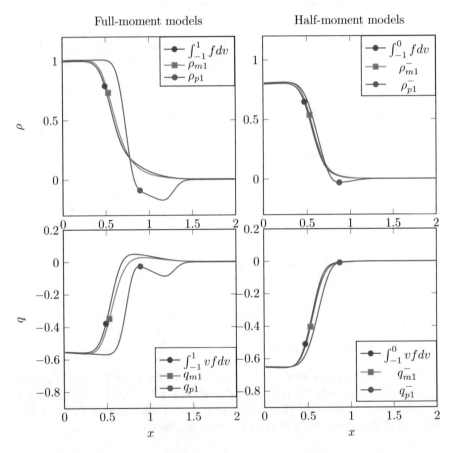

Fig. 12 Linear models (P_1) with negative full or half range densities and positivity preserving nonlinear models (M_1)

6.2 Numerical Solutions on a Tripod Network

This test case focuses on the coupling conditions for a tripod network. We study a junction connecting three outgoing edges. The results of the nonlinear half-moment model with the corresponding coupling conditions are compared with the results of the kinetic, linear half and full moment and Keller-Segel model with coupling conditions described in [8]. On each of the edges we consider the interval [0, 1]. As initial conditions we choose

$$\rho_1(x, 0) = 1, \qquad \rho_2(x, 0) = 4, \qquad \rho_3(x, 0) = 3,$$

which is consistent with the following values for the kinetic equation

$$f_i(x, v, 0) = F(v)\rho_i(x, 0) = \frac{1}{2}\rho_i(x, 0) \qquad i = 1, 2, 3.$$

All other quantities are initially zero.

In Figs. 13, 14 and 15 the densities at $t = 0.2$ for different values of ϵ are shown. The results are similar to those on a single interval. The nonlinear half moment model is again the best approximation to the kinetic result. The nonlinear half-moment solution coincides in all situations with the kinetic solution. As the value of ϵ decreases all solutions approach the solution of the Keller-Segel equation.

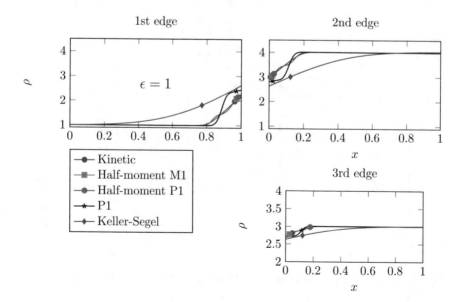

Fig. 13 Numerical solutions on a tripod network at time $t = 0.2$, $\Delta x = 0.02$, $\epsilon = 1$

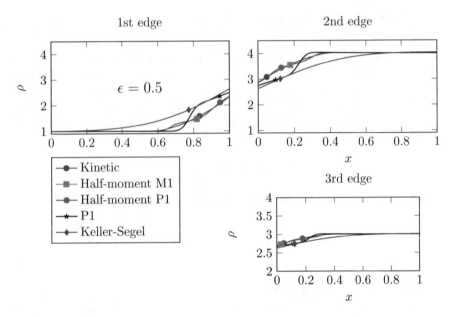

Fig. 14 Numerical solutions on a tripod network at time $t = 0.2$, $\Delta x = 0.02$, $\epsilon = 0.5$

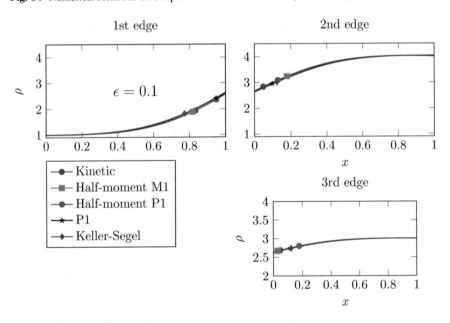

Fig. 15 Numerical solutions on a tripod network at time $t = 0.2$, $\Delta x = 0.02$, $\epsilon = 0.1$

6.3   Numerical Solutions on a Larger Network

In this last test case we consider a larger network of 31 edges and 23 nodes as shown in Fig. 16. The length of the short edges is 0.5, for the longer ones we have 1 and $\sqrt{2}$, respectively. Note that this network does not only contain three way junctions, but also nodes connecting up to five edges. At the open ends of the network Dirichlet boundary conditions are imposed. For the kinetic model we prescribe $f(0, v, t) = \frac{1}{2}$ for $v > 0$. The boundary values for the other models can be derived thereof. As initial conditions all values are set to zero except the density in the edges at the outer boundaries, which is set to 1. For the numerical computation we used 15, 30, and 42 cells for the spatial discretization of the edges, respectively. The velocity space in the kinetic model is resolved with 30 points. The scaling parameter is chosen as $\epsilon = 1$.

In Fig. 16 the density at time $t = 5$ for all four models is shown. There is an inflow from both sides of the network. As observed in the previous tests, the states of the Keller-Segel model propagate faster, such that the network is filled at an earlier time. The solution of the kinetic model and the one of the half moment models almost coincide.

In Fig. 17 the evolution of the total mass in the network up to $T = 30$ is shown. As before, we observe that the Keller-Segel model fills the network much faster than the other three. The values of the half moment and of the kinetic model almost coincide. Concerning the computation times, the least expensive model is the full moment P_1 model with approximately 15% of the computation time of the kinetic model. The Keller-Segel model needs 18%, the half moment P_1 model needs 22%, and the half moment M_1 model needs 23% of the kinetic computation time.

7   Conclusions and Outlook

We investigated the use of first- and higher-order moment closures in comparison with mesoscopic kinetic equations and the diffusion approximation in the context of glioma migration in tissue. In this case, using modified models that explicitly contain the equilibrium distribution given by the fiber structure leads to improved results compared to more classical approaches. Moreover, moment closure models are considered to describe cell migration given by a kinetic equation on a network topology resulting from a fiber structure. In this case additionally coupling conditions for the macroscopic models are derived from conditions for the kinetic models. In the numerical tests we investigated the dynamics for varying values of the scaling parameter. A simulation on a more complicated 2D structure and on larger networks showed the applicability of the methodology to more complicated situations and differences in behavior of macroscopic and kinetic models. In the considered examples the moment models converge as expected to the correct diffusion limit

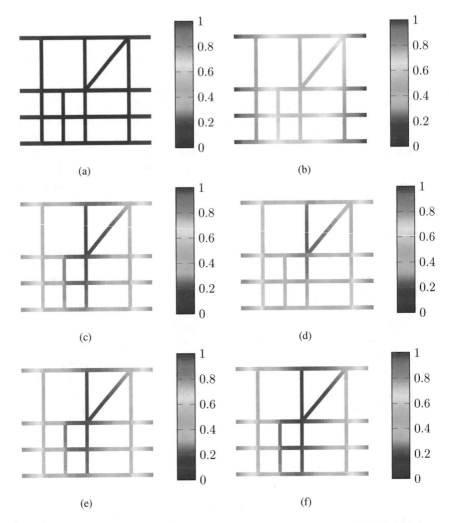

Fig. 16 Comparison of the numerical solutions on a larger network at $t = 5$. (**a**) Initial data. (**b**) Diffusive limit. (**c**) Kinetic. (**d**) Full-moment P1 model. (**e**) Half-moment P1 model. (**f**) Half-moment M1 model

as $\epsilon \to 0$. For larger values of ϵ, in particular nonlinear partial moment models give very accurate results compared to the solution of the full kinetic problem.

Finally, we remark that there are several open issues. Important would be, for example, further analytical investigations of the different equations on the network topologies. Moreover, an extension of the approaches to more complex and clinically relevant situations and associated efficient numerical approaches have to be considered.

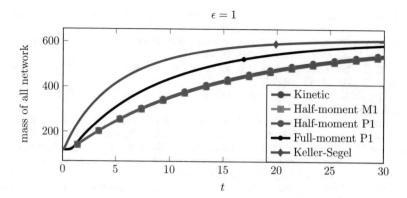

Fig. 17 Total mass over time in the large network

Acknowledgements The second author is supported by DFG grant 1105/27, by BMBF grant 05M16UKB, GlioMaTh and by the DAAD PhD program MIC.

References

1. Anile, A.M., Pennisi, S., Sammartino, M.: A thermodynamical approach to Eddington factors. Journal of Mathematical Physics **32**(2), 544 (1991)
2. B. A. C. Harley H. Kim, M.H.Z.I.V.Y.D.A.L., Gibson, L.J.: Microarchitecture of three-dimensional scaffolds influences cell migration behavior via junction interactions. Biophysical Journal **29**, 4013–4024 (2008)
3. Bellomo, N., Bellouquid, A., Nieto, J., Soler, J.: Complexity and mathematical tools toward the modeling of multicellular growing systems. Mathematical and Computer Modeling **51**, 441–451 (2010)
4. Bellomo, N., Bellouquid, A., Nieto, J., Soler, J.: Multiscale biological tissue models and flux-limited chemotaxis for multicellular growing systems. Mathematical Models and Methods in Applied Sciences **20**(7), 1179–1207 (2010)
5. Bellomo, N., Bellouquid, A., Nieto, J., Soler, J.: On the asymptotic theory from microscopic to macroscopic growing tissue models: an overview with perspectives. Mathematical Models and Methods in Applied Sciences **22**(1), 1130001 (27 pages), (2012)
6. Bellomo, N., Bellouquid, A., Tao, Y., Winkler, M.: Toward a mathematical theory of Keller-Segel models of pattern formation in biological tissues. Math. Models Methods Appl. Sci. **25**(9), 1663–1763 (2015)
7. Borsche, R., Göttlich, S., Klar, A., Schillen, P.: The scalar Keller-Segel model on networks. Math. Models Methods Appl. Sci. **24**(2), 221–247 (2014)
8. Borsche, R., Kall, J., Klar, A., Pham, T.: Kinetic and related macroscopic models for chemotaxis on networks. Mathematical Models and Methods in Applied Sciences **26**(06), 1219–1242 (2016)
9. Borsche, R., Klar, A.: Kinetic layers and coupling conditions for macroscopic equations on networks. SIAM Sci. Computing **40** (2018)
10. Borsche, R., Klar, A.: Kinetic layers and coupling conditions for nonlinear scalar equations on networks. Nonlinearity **31**, 3512–3541 (2018)

11. Borsche, R., Klar, A., Pham, T.H.: Nonlinear flux-limited models for chemotaxis on networks. Networks & Heterogeneous Media **12**(3), 381–401 (2017)
12. Bournaveas, N., Calvez, V.: The one-dimensional Keller-Segel model with fractional diffusion of cells. Nonlinearity **23**(4), 923–935 (2010)
13. Bretti, G., Natalini, R., Ribot, M.: A hyperbolic model of chemotaxis on a network: a numerical study. ESAIM: M2AN **48**(1), 231–258 (2014)
14. Brunner, T.A.: Forms of approximate radiation transport. SAND2002-1778, Sandia National Laboratory (July) (2002)
15. Brunner, T.A., Holloway, J.: One-dimensional Riemann solvers and the maximum entropy closure. Journal of Quantitative Spectroscopy and Radiative Transfer **69**(5), 543–566 (2001)
16. Burger, M., Di Francesco, M., Dolak-Struss, Y.: The Keller-Segel model for chemotaxis with prevention of overcrowding: linear vs. nonlinear diffusion. SIAM J. Math. Anal. **38**(4), 1288–1315 (2006). https://dx.doi.org/10.1137/050637923
17. Camilli, F., Corrias, L.: Parabolic models for chemotaxis on weighted networks. J. Math. Pures Appl. **108**, 459–480 (2017)
18. Chalub, F., Markowich, P., Perthame, B., Schmeiser, C.: Kinetic models for chemotaxis and their drift-diffusion limits. Monatsh. Math. **142**, 123–141 (2004)
19. Chavanis, P.: Jeans type instability for a chemotactic model of cellular aggregation. Eur. Phys. J. B **52**, 433–443 (2006)
20. Chertock, A., Kurganov, A., Wang, X., Wu, Y.: On a chemotaxis model with saturated chemotactic flux. Kinet. Relat. Models **5**(1), 51–95 (2012)
21. Childress, S., Percus, J.: Nonlinear aspects of chemotaxis. Math. Biosci. **56**, 217–237 (1981)
22. Colombo, R.M., Garavello, M.: On the Cauchy problem for the *p*-system at a junction. SIAM J. Math. Anal. **39**(5), 1456–1471 (2008)
23. Colombo, R.M., Guerra, G.: On general balance laws with boundary. J. Differential Equations **248**(5), 1017–1043 (2010)
24. Coons, S.: Anatomy and growth patterns of diffuse gliomas. In: M. Berger, C. Wilson (eds.) The gliomas, pp. 210–225. W.B. Saunders Company, Philadelphia (1999)
25. Corbin, G., Hunt, A., Schneider, F., Klar, A., Surulescu, C.: Higher-order models for glioma invasion: from a two-scale description to effective equations for mass density and momentum. M3AS **28**, 1771–1800 (2018)
26. Coulombel, J., Golse, F., Goudon, T.: Diffusion approximation and entropy-based moment closure for kinetic equations. Asymptotic Analysis **45**(1), 1–34 (2005)
27. D'Abaco, G., Kaye, A.: Integrins: Molecular determinants of glioma invasion. Journal of Clinical Neuroscience **14**, 1041–1048 (2007)
28. Daumas-Duport, C., Varlet, P., Tucker, M., Beuvon, F., Cervera, P., Chodkiewicz, J.: Oligoden-drogliomas. part i: Patterns of growth, histological diagnosis, clinical and imaging correlations: A study of 153 cases. Journal of Neuro-Oncology **34**, 37–59 (1997)
29. Dubroca, B., Klar, A.: Half-moment closure for radiative transfer equations. Journal of Computational Physics **180**, 584–596 (2002)
30. Engwer, C., Hillen, T., Knappitsch, M., Surulescu, C.: Glioma follow white matter tracts: a multiscale DTI-based model. Journal of Mathematical Biology **71**, 551–582 (2015)
31. Engwer, C., Hunt, A., Surulescu, C.: Effective equations for anisotropic glioma spread with proliferation: a multiscale approach. IMA Journal of Mathematical Medicine and Biology **33**, 435–459 (2016)
32. Engwer, C., Knappitsch, M., Surulescu, C.: A multiscale model for glioma spread including cell-tissue interactions and proliferation. Journal of Engineering Mathematics **13**, 443–460 (2016)
33. Fermo, L., Tosin, A.: A fully-discrete-state kinetic theory approach to traffic flow on road networks. Math. Models Methods Appl. Sci. **25**(3), 423–461 (2015)
34. Filbet, F., Laurençot, P., Perthame, B.: Derivation of hyperbolic models for chemosensitive movement. J Math Biol. **50**(2), 189–207 (2005)
35. Frank, M., Dubroca, B., Klar, A.: Partial moment entropy approximation to radiative heat transfer. Journal of Computational Physics **218**(1), 1–18 (2006)

36. Frank, M., Hensel, H., Klar, A.: A fast and accurate moment method for the Fokker-Planck equation and applications to electron radiotherapy. SIAM Journal on Applied Mathematics **67**(2), 582–603 (2007)
37. Garrett, C.K., Hauck, C.: A comparison of moment closures for linear kinetic transport equations: the line source benchmark. Transport Theory and Statistical Physics **42**, 203–235 (2013)
38. Gerstner, E., Chen, P.J., Wen, P., Jain, R., Batchelor, T., Sorensen, G.: Infiltrative patterns of glioblastoma spread detected via diffusion MRI after treatment with cediranib. Neuro-Oncology **12**(5), 466–472 (2010)
39. Giese, A., Kluwe, L., H., M., E., M., Westphal, M.: Migration of human glioma cells on myelin. Neurosurgery **38**, 755–764 (1996)
40. Giese, A., Westphal, M.: Glioma invasion in the central nervous system. Neurosurgery **39**, 235–252 (1996)
41. Gimbutas, Z., Greengard, L.: A fast and stable method for rotating spherical harmonic expansions. Journal of Computational Physics **228**(16), 5621–5627 (2009)
42. Guarguaglini, F.R., Natalini, R.: Global smooth solutions for a hyperbolic chemotaxis model on a network. SIAM J. Math. Anal. **47**(6), 4652–4671 (2015)
43. Hauck, C.D.: High-order entropy-based closures for linear transport in slab geometry. Communications in Mathematical Sciences **9**(1), 187–205 (2011)
44. Herty, M., Moutari, S.: A macro-kinetic hybrid model for traffic flow on road networks. Comput. Methods Appl. Math. **9**(3), 238–252 (2009)
45. Hillen, T.: Hyperbolic models for chemosensitive movement. Mathematical Models and Methods in Applied Sciences **12**(07), 1007–1034 (2002)
46. Hillen, T., Othmer, H.G.: The diffusion limit of transport equations derived from velocity jump processes. Siam Journal on Applied Mathematics **61**, 751–775 (2000)
47. Hillen, T., Painter, K.: A user's guide to PDE models for chemotaxis. J. Math. Biol. **58**(1-2), 183–217 (2009)
48. Keller, E.F., Segel, L.A.: Initiation of slime mold aggregation viewed as an instability. J. Theor. Biol. **26**(3), 399–415 (1970). https://dx.doi.org/10.1016/0022-5193(70)90092-5
49. Keller, E.F., Segel, L.A.: Model for chemotaxis. Journal of Theoretical Biology **30**, 225–234 (1971)
50. Kershaw, D.S.: Flux Limiting Nature's Own Way: A New Method for Numerical Solution of the Transport Equation. Tech. rep., LLNL Report UCRL-78378 (1976)
51. Klar, A., Schneider, F., Tse, O.: Approximate models for stochastic dynamic systems with velocities on the sphere and associated Fokker–Planck equations. Kinetic and Related Models **7**(3), 509–529 (2014)
52. Le Bihan, D., Mangin, J.F., Poupon, C., Clark, C., Pappata, S., Molko, N., Chabriat, H.: Diffusion tensor imaging: concepts and applications. Journal of magnetic resonance imaging **13**(4), 534–546 (2001)
53. Levermore, C.D.: Relating Eddington factors to flux limiters. Journal of Quantitative Spectroscopy and Radiative Transfer **31**(2), 149–160 (1984)
54. Levermore, C.D.: Moment closure hierarchies for kinetic theories. Journal of Statistical Physics **83**, 1021–1065 (1996)
55. Mandal, B.B., Kundu, S.: Cell proliferation and migration in silk fibroin 3D scaffolds. Biomaterials **30**, 2956–2965 (2009)
56. Mark, J.C.: The spherical harmonics method, Part {I}. Tech. Rep. MT 92, National Research Council of Canada (1944)
57. Olbrant, E., Hauck, C.D., Frank, M.: A realizability-preserving discontinuous Galerkin method for the M1 model of radiative transfer. Journal of Computational Physics **231**(17), 5612–5639 (2012)
58. Painter, K., Hillen, T.: Mathematical modelling of glioma growth: the use of diffusion tensor imaging (DTI) data to predict the anisotropic pathways of cancer invasion. Journal of Theoretical Biology **323**, 25–39 (2013)
59. Pomraning, G.C.: The equations of radiation hydrodynamics. Pergamon Press (1973)

60. Ritter, J., Klar, A., Schneider, F.: Partial-moment minimum-entropy models for kinetic chemotaxis equations in one and two dimensions. J. Comp. Applied Math. **306**, 300–315 (2016)
61. Schneider, F., Alldredge, G., Frank, M., Klar, A.: Higher Order Mixed-Moment Approximations for the Fokker–Planck Equation in One Space Dimension. SIAM Journal on Applied Mathematics **74**(4), 1087–1114 (2014)
62. Schneider, F., Kall, J., Alldredge, G.: A realizability-preserving high-order kinetic scheme using WENO reconstruction for entropy-based moment closures of linear kinetic equations in slab geometry. Kinetic and Related Models **9**(1), 193–215 (2015)
63. Schneider, F., Kall, J., Roth, A.: First-order quarter- and mixed-moment realizability theory and Kershaw closures for a Fokker-Planck equation in two space dimensions. Kinetic and Related Models **10 (4)**, 1127–1161 (2017)

Kinetic Models for Pattern Formation in Animal Aggregations: A Symmetry and Bifurcation Approach

Pietro-Luciano Buono, Raluca Eftimie, Mitchell Kovacic, and Lennaert van Veen

Abstract In this study we start by reviewing a class of 1D hyperbolic/kinetic models (with two velocities) used to investigate the collective behaviour of cells, bacteria or animals. We then focus on a restricted class of nonlocal models that incorporate various inter-individual communication mechanisms, and discuss how the symmetries of these models impact the various types of spatially heterogeneous and spatially homogeneous equilibria exhibited by these nonlocal models. In particular, we characterise a new type of equilibria that was not discussed before for this class of models, namely a relative equilibria. Then we simulate numerically these models and show a variety of spatio-temporal patterns (including classic equilibria and relative equilibria) exhibited by these models. We conclude by introducing a continuation algorithm (which takes into account the models symmetries) that allows us to track the solutions bifurcating from these different equilibria. Finally, we apply this algorithm to identify a $\mathbf{D}_3$-symmetric steady-state solution.

P.-L. Buono
University of Ontario Institute of Technology, Oshawa, ON, Canada

Present address: Université du Québec à Rimouski, Rimouski, Qué, Canada
e-mail: luciano.buono@uoit.ca; Pietro-Luciano_Buono@uqar.ca

L. van Veen
University of Ontario Institute of Technology, Oshawa, ON, Canada
e-mail: lennaert.vanVeen@uoit.ca

R. Eftimie (✉)
University of Dundee, Dundee, UK
e-mail: r.a.eftimie@dundee.ac.uk

M. Kovacic
Simon Fraser University, Burnaby, BC, Canada
e-mail: mkovacic@sfu.ca

© Springer Nature Switzerland AG 2019
N. Bellomo et al. (eds.), *Active Particles, Volume 2*, Modeling
and Simulation in Science, Engineering and Technology,
https://doi.org/10.1007/978-3-030-20297-2_2

1 Introduction

Self-organised behaviours in animal communities have attracted the attention of researchers as well as general public for at least 2000 years. One of the earliest recordings of self-organised behaviours can be found in Pliny the Elder's book "Natural History" [54], which describes various aspects of collective dynamics in insects (e.g., bees that cluster "as they do, like a bunch of grapes, upon houses or temples"; book XI), birds (e.g., starlings that "fly in troops, as it were, and then to wheel round in a globular mass like a ball"; book X), fish (e.g., dolphins in [54], which "form among themselves a sort of general community"; book IX) and terrestrial animals (e.g., elephants that "always move in herds"; book VIII). Thus, people's attention has always been drawn to the aggregation patterns displayed by these animal aggregations, and the transitions between these patterns. However, these aggregation patterns are the result of how animals communicate with each other: how they perceive/sense their neighbours and how they respond to their neighbours' behaviours. Therefore, to understand the biological mechanisms behind the collective movement and aggregation of animals one needs to take into account also animal communication.

Mathematical models have been used for almost four decades to propose hypotheses (and then test them in silico) regarding the most important biological mechanisms that can explain the observed aggregation patterns [2, 3, 15, 19, 23, 28, 36, 45, 46]. While the initial models (and many of the current models) are described by individual based models (which track the position and velocity of every individual in the population) [15, 16, 19, 53, 58], the lack of analytical techniques that can be used to investigate the diverse patterns obtained with these models leads researchers to concentrate more and more on continuum models (which focus on the density of individuals at generic positions in space) [23–25, 47, 55, 56]. It should also be emphasised that current research pays also significant attention to multi-scale models that connect individual-based approaches to continuum approaches; see [9, 14, 20, 49] and the references therein. Moreover, the majority of these models for animal collective behaviours focus on the investigation of three basic type of social interactions among individuals (i.e., attraction, repulsion and orientation/alignment) and how these interactions lead to various aggregation patterns [25]. Since these social interactions generally act on different spatial ranges (e.g., repulsion acts on very short ranges, while attraction acts on longer ranges), some of these models are nonlocal. We need to emphasise that only few studies and mathematical models take into account how individuals perceive each other, and how these perception/sensing mechanisms influence the social interactions [10, 21–24].

In this study, we start by reviewing a class of 1D mathematical models of hyperbolic/kinetic type derived in [23] to describe self-organised behaviours in animal communities as a result of different types of animal communication. (Note that these models are also known as discrete-velocity kinetic models, since they can be recovered from Boltzmann-like kinetic models when we consider only two velocities, namely left and right; see [25].) Such models are known to exhibit a large variety of spatio-temporal patterns, ranging from stationary aggregations that can be time-variant or time-invariant to different types of moving aggregations (e.g., zigzags); see also the patterns in [11, 12, 22, 23]. Many of these patterns have complex dynamical features, which are still not fully understood in terms of invariant sets of phase space. In Sect. 3.2 we start discussing different spatially homogeneous and spatially heterogeneous equilibria exhibited by these nonlocal models (some of these equilibria not being studied before), and the symmetry properties of these equilibria. In Sect. 4 we show numerical simulations of different types of spatio-temporal patterns—including different equilibria—and how these patterns are influenced by the domain boundaries. We also present and discuss a new continuation algorithm (which takes into account the symmetries of the models) that can be used to trace the dynamics of the system from known equilibria to more dynamically exotic patterns in the parameter space.

Since generalisations of 1D hyperbolic models to 2D are not only more realistic but also more complex, their analytic investigation is more difficult. For this reason, we ignore them in this study.

2 Brief Review of 1D Hyperbolic/Kinetic Models for Collective Dynamics in Biology

The last 20 years have seen an increase in the use of hyperbolic/kinetic models that can describe various biological phenomena: from self-organised biological aggregations (i.e., aggregations in the absence of a leader or external stimuli; [23]) to chemotactic aggregations (i.e., aggregations in the presence of a chemotactic signal produced by the members of the group; see [27, 33, 35, 48]), migration of organisms based on food distribution [17], predator–prey dynamics [7, 8] or age-structured models [37, 38, 60].

One of the simplest classes of 1D hyperbolic/kinetic models with constant speeds (for individuals moving right and left inside a certain domain) and constant turning rates was introduced and discussed extensively in [32, 34, 35]. The general form of this model for right-moving (u^+) and left-moving (u^-) particles/animals, which includes a turning behaviour $(\lambda^\pm)$ as well as a birth/death processes $(h^\pm(u^+, u^-))$, is given by

$$\frac{\partial u^+}{\partial t} + \frac{\partial(\gamma u^+)}{\partial x} = f^+(u^+, u^-) = -\lambda^+ u^+ + \lambda^- u^- + \frac{1}{2}h^+(u^+, u^-), \qquad (1a)$$

$$\frac{\partial u^-}{\partial t} - \frac{\partial(\gamma u^-)}{\partial x} = f^-(u^+, u^-) = \lambda^+ u^+ - \lambda^- u^- + \frac{1}{2}h^-(u^+, u^-). \qquad (1b)$$

Here, the turning terms $\lambda^\pm$ describe the probability of right-moving individuals (u^+) to turn around and become left-moving (u^-), as well as the probability of left-

moving individuals to turn around and become right-moving. Since the change in
the movement direction of cells/bacteria/animals is not always constant, but might
depend on (local or nonlocal) interactions with other cells/bacteria/animals, Pfistner
[47] and Eftimie et al. [23, 24] considered models of type (1) with constant speeds
and nonlocal density-dependent turning rates

$$f^+[u^+, u^-] = -\lambda^+[u^+, u^-]u^+ + \lambda^-[u^+, u^-]u^-,$$

$f^- = -f^+$, and no birth/death dynamics ($h^\pm = 0$). Because of the complex spatial
and spatio-temporal dynamics exhibited by these nonlocal models, in Sect. 3 we will
review in more detail the nonlocal models introduced in [23, 24].

Models of type (1) for the movement of cells/bacteria/animals can be easily
coupled with reaction-diffusion models for the dynamics of a chemical ($c(x, t)$)
that controls the direction of movement of cells/bacteria/animals (chemical which
can be produced by the cells/animals themselves) [33, 48]:

$$\frac{\partial c(x, t)}{\partial t} = p(c, u^+, u^-) + D\frac{\partial^2 c(x, t)}{\partial x^2}, \tag{2}$$

where $p(c, u^+, u^-)$ describes the production/degradation of this chemical, and D
is its diffusion rate. The chemical can influence the speed of animals/cells (i.e.,
$\gamma[u^+, u^-, c]$ in (1)), their turning behaviour and even the birth–death dynamics of
the population (i.e., $f^\pm[u^+, u^-, c]$ in (1)).

Regarding the 1D hyperbolic predator–prey models, these do not usually consider
turning behaviour (i.e., $\lambda^\pm = 0$) [8], as the two populations are independent
(however, the 2D kinetic models can incorporate changes in the movement direction
of sub-populations in response to prey/predator behaviour; see [25]). In this case,
the functions $f^\pm[u^+, u^-]$ incorporate only the predator–prey dynamics between the
two populations. Usually, this dynamics is described by Lotka-Volterra-type terms
[18], but other terms such as Holling-type functional responses can also be used [8].
Moreover, the interactions between the prey and predator populations can affect the
speed of either prey or predator [18], as the animals speed up to avoid or to catch
up with the other population. Note here that not all 1D predator–prey models are of
the type (1). For example, Barbera et al. [8] derived a hyperbolic model where the
hyperbolic equations for the two populations are coupled with transport equations
for the dissipative fluxes.

A final type of hyperbolic model that we would like to mention briefly
describes age-structured populations. The hyperbolic age-structured models (of
the McKendrick-von Foerster type) have the general form [38]

$$\frac{\partial u(a, t)}{\partial t} + \frac{\partial u(a, t)}{\partial a} = -\lambda(a)u(a, t), \tag{3}$$

with $u(t, a)$ representing the density of the population of age a at time t, and
$\lambda(a)$ describing the mortality rate. The description of the model is completed with

conditions for the initial population $u(0, a) = Q(a)$, $a \geq 0$, and conditions for the newborn population:

$$u(0, t) = \int_{\alpha}^{\beta} u(x, t) m(x) dx, \tag{4}$$

with m the maternity function. Therefore, this class of age-structured models investigate the movement of a population through the age-space.

While all these different hyperbolic models can exhibit a large variety of spatial and spatio-temporal patterns ranging from stationary and moving aggregations of animals/cells (e.g., stationary pulses, travelling pulses, breathers, ripples, zigzags; see [23]) to networks of cells [27], thorough investigations of these patterns are still not the common approach in mathematical biology. For a more in-depth review of pattern formation in hyperbolic models in biology, and the analytical and numerical techniques available to investigate them, see [21, 59]. Existence of reduction methods (e.g., Centre Manifold reduction) for local bifurcations of various types of equations described in this section has been established for parabolic equations [31], for equations such as (1) in [13] and for hyperbolic age-structured models [44].

Next, we focus on a particular class of nonlocal mathematical models for self-organised biological aggregations, for which there are a few preliminary studies on the local bifurcation of patterns near codimension-1 and codimension-2 bifurcation points [11, 12].

3 Nonlocal 1D Hyperbolic Models for Self-organised Animal Aggregations via Communication

Here, we present in more detail a class of 1D nonlocal hyperbolic models derived to describe the formation and movement of various animal, cell and bacterial aggregations as a result of inter-individual communication [23, 24]. The evolution of densities of right-moving (u^+) and left-moving (u^-) individuals, which travel with constant velocity γ and change their movement direction from right to left (with rate λ^+) and from left to right (with rate λ^-) [23] is given by:

$$\partial_t u^+(x, t) + \partial_x(\gamma u^+(x, t)) = -\lambda^+[u^+, u^-]u^+(x, t) + \lambda^-[u^+, u^-]u^-(x, t), \tag{5a}$$

$$\partial_t u^-(x, t) - \partial_x(\gamma u^-(x, t)) = \lambda^+[u^+, u^-]u^+(x, t) - \lambda^-[u^+, u^-]u^-(x, t), \tag{5b}$$

$$u^\pm(x, 0) = u_0^\pm(x). \tag{5c}$$

The turning rates are defined as

$$\lambda^{\pm}[u^+, u^-] = \lambda_1 + \lambda_2 f(y_r^{\pm}[u^+, u^-] - y_a^{\pm}[u^+, u^-] + y_{al}^{\pm}[u^+, u^-]), \qquad (6)$$

$$= \left(\lambda_1 + \lambda_2 f(0)\right) + \lambda_2\left(f(y_r^{\pm} - y_a^{\pm} + y_{al}^{\pm}) - f(0)\right).$$

The terms $\lambda_1 + \lambda_2 f(0)$ and $\lambda_2(f(y^{\pm}) - f(0))$ describe the baseline random turning rate and the bias turning rate, respectively. The function f is a positive function saturating for large values of its argument (to describe the biologically realistic situation of bounded turning rates). An example of such function is $f(y) = 0.5 + 0.5 \tanh(y)$; see [11, 12, 23, 24]. These turning rates are influenced by the social interactions among individuals: attraction towards far-away neighbours ($y_a^{\pm}$), alignment with neighbours at intermediate distances ($y_{al}^{\pm}$) and repulsion from individuals at very close distances ($y_r^{\pm}$). Moreover, these social interactions depend on the perception of neighbours, which communicate via different mechanisms involving visual, sound, tactile or chemical signals. Table 1 shows the social interaction terms $y_{r,al,a}^{\pm}$ corresponding to four examples of communication mechanisms introduced in [23]. Note that in [23] the authors considered also a fifth mechanism (denoted M1), which combined attraction/repulsion forces as described by M2 and alignment forces as described by M4. Since this mechanism did not bring any new results in terms of pattern formation or model symmetry, it was ignored in more recent studies [11, 12] and thus we ignore it throughout this study too. The parameters $q_{r,a,al}$ are the magnitudes of the repulsive (r), attractive (a) and alignment (al) interactions. The kernels $K_{r,a,al}$ that model long-distance social interactions are given by Gaussian functions

$$K_j(s) = \frac{1}{2\pi m_j^2} e^{-(s-s_j)^2/(2m_j^2)}, \quad \text{with} \ \ j = r, a, al, \ \ \text{and} \ \ m_j = s_j/8, \qquad (7)$$

with s_j and m_j, $j = r, a, al$, describing the position and the width of the interaction ranges; see also Fig. 1.

The integrals in Table 1 can be re-written by defining the operator $\mathscr{I}_{i,\ell}^{\pm}(u^+(x), u^-(x), s)$, with $\ell = a, r, al$, to describe the integrand for model Mi, $i = 2, 3, 4, 5$. The superscript $\pm$ in $\mathscr{I}^{\pm}$ corresponds to the superscript in $y^{\pm}$. Thus, the social interaction terms become

$$y_{i,\ell}^{\pm}(u(x)) := \int_0^{\infty} K_{\ell}(s)\mathscr{I}_{i,\ell}^{\pm}(u^+(x), u^-(x), s)\, ds. \qquad (8)$$

Note that $\mathscr{I}^{\pm}$ satisfies the following relation:

$$\mathscr{I}_{i,\ell}^{\pm}(v_1^+(x) + v_2^+(x), v_1^-(x) + v_2^-(x), s) = \mathscr{I}_{i,\ell}^{\pm}(v_1^+(x), v_1^-(x), s)$$

$$+ \mathscr{I}_{i,\ell}^{\pm}(v_2^+(x), v_2^-(x), s),$$

for all $i = 2, 3, 4, 5$ and $\ell = a, r, al$.

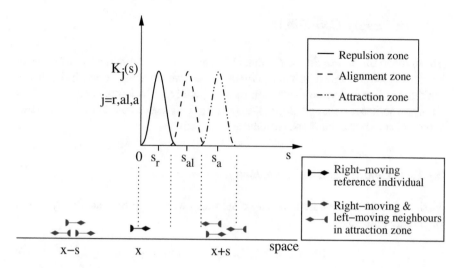

Fig. 1 Kernels K_j, $j = r, al, a$, that model the long-distance social interactions among individuals

Table 1 Nonlocal social interaction terms ($y_j^\pm$, $j \in \{a, al, r\}$) introduced in [23]

Mechanisms	Communic. models	Interaction terms: attraction ($y_a^\pm$), repulsion ($y_r^\pm$), alignment ($y_{al}^\pm$)
Omnidirectional perception	M2	$y_{a,r}^\pm = q_{r,a} \int_0^\infty K_{a,r}(s)\big(u(x \pm s) - u(x \mp s)\big)ds$
Omnidirectional emission		$y_{al}^\pm = q_{al} \int_0^\infty K_{al}(s)\big(u^\mp(x \mp s) + u^\mp(x \pm s) - u^\pm(x \mp s) - u^\pm(x \pm s)\big)ds$
Unidirectional perception	M3	$y_{r,a}^\pm = q_{r,a} \int_0^\infty K_{r,a}(s)u(x \pm s)ds$
Omnidirectional emission		$y_{al}^\pm = q_{al} \int_0^\infty K_{al}(s)\big(u^\mp(x \pm s) - u^\pm(x \pm s)\big)ds$
Omnidirectional perception	M4	$y_{r,a}^\pm = q_{r,a} \int_0^\infty K_{r,a}(s)\big(u^\mp(x \pm s) - u^\pm(x \mp s)\big)ds$
Unidirectional emission		$y_{al}^\pm = q_{al} \int_0^\infty K_{al}(s)\big(u^\mp(x \pm s) - u^\pm(x \mp s)\big)ds$
Unidirectional perception	M5	$y_{a,r}^\pm = q_{r,a} \int_0^\infty K_{a,r}(s)u^\mp(x \pm s)ds$
Unidirectional emission		$y_{al}^\pm = q_{al} \int_0^\infty K_{al}(s)u^\mp(x \pm s)ds$

Constants q_a, q_{al}, q_r describe the magnitudes of the attractive, alignment and repulsive interactions, respectively. Kernels $K_{a,al,r}(s)$ describe the spatial ranges for each of these social interactions. Note that $u = u^+ + u^-$ (total population density)

3.1 Boundary Conditions

To complete the description of the model (5), and because numerical simulations of system (5) are performed on a finite domain $[0, L]$, we need to describe the boundary conditions. For a detailed discussion of biologically realistic boundary conditions for hyperbolic systems, see [30, 34]. In the following we discuss in more detail two types of boundary conditions: periodic and reflective.

3.1.1 Periodic Boundary Conditions

First, we consider periodic boundary conditions, which approximate the dynamics on infinite domains:

$$u^+(0, t) = u^+(L, t), \quad u^-(L, t) = u^-(0, t). \tag{9}$$

Hillen [34] showed the existence of solutions for *local hyperbolic systems* that satisfy periodic, homogeneous Dirichlet and homogeneous Neumann boundary conditions. Since model (5) is nonlocal, next we confirm that the integrals (8) are well-defined for $u^\pm$ satisfying conditions (9). We reproduce the calculation found in [13]. First define the space

$$L_{per}^2 = \{u \in L^2(\mathbb{R}) \mid u(x) = u(x + L) \text{ for all } x \in [0, L)\}.$$

We now show that for the interaction kernels $K(s)$ as in (7) and for $v \in L_{per}^2$ and

$$\tilde{K}^\pm v(x) := \int_0^\infty K(s) v(x \pm s) \, ds, \tag{10}$$

we have $\tilde{K}^\pm v(x) \in L_{per}^2$. To this end, we write $v(x) = \sum_{n=-\infty}^\infty c_n e^{i k_n x}$, where $k_n = 2\pi n / L$. Then,

$$
\begin{aligned}
\tilde{K}^+ v(x) &= \int_0^\infty K(s) \sum_{n=-\infty}^\infty c_n e^{i k_n (x+s)} \, ds \\
&= \sum_{n=-\infty}^\infty c_n e^{i k_n x} \int_0^\infty K(s) e^{i k_n s} \, ds \\
&= \sum_{n=-\infty}^\infty c_n \hat{K}(n) e^{i k_n x}.
\end{aligned}
$$

Here, $\hat{K}(n)$ is the Fourier transform of $K(s)$, and $\hat{K}(\eta) \to 0$ as $|\eta| \to \infty$ exponentially fast (since $K(s)$ is Gaussian). Next we know that $|c_n|^2 < 1$ if $|n| > N$ for some $N \in \mathbb{N}$:

$$\sum_{n=0}^{\infty} |c_n \hat{K}(n)|^2 \leq \sum_{n=-N}^{N} |c_n|^2 |\hat{K}(n)|^2 + \sum_{n=-\infty}^{-(N+1)} |\hat{K}(n)|^2 + \sum_{n=N+1}^{\infty} |\hat{K}(n)|^2 < \infty.$$

Thus, $\tilde{K}^+ v(x) \in L^2_{per}$ and the same holds for $\tilde{K}^- v(x)$.

Remark 1 Note that if we choose to work with functions in C^0_{per} with the sup-norm $||v||_\infty = \sup\{|v(x)| \mid x \in [0, L]\}$, it is a straightforward exercise to show that $\tilde{K}^\pm$ is a bounded operator from C^0_{per} to itself.

3.1.2 Reflective Boundary Conditions

Another type of boundary condition that is commonly used for systems of hyperbolic models (both in biology and physics; see, for example, [39, 40]) is the homogeneous Neumann condition. On the domain $[0, L/2]$, this condition reads

$$u^+(0, t) = u^-(0, t), \quad u^+(L/2, t) = u^-(L/2, t), \quad t \geq 0. \tag{11}$$

These Neumann (reflective) conditions describe the case where cells/animals cannot leave the domain and turn around at the boundary [30, 34, 43]. In regard to the equivalence between periodic and reflective boundary conditions for local hyperbolic systems, Lutscher [43] and Hillen [34] showed that for solutions that satisfy the mirror symmetry condition

$$u^+(x) = u^-(L - x), \quad x \in [0, L], \tag{12}$$

if one considers $w_0^\pm$ the initial data on $[0, L/2]$ that satisfies the no-flux boundary conditions (11), then it can be shown that

$$u_0^\pm(x) = \begin{cases} w_0^\pm(x) & \text{for } x \in [0, L/2], \\ w_0^\mp(L - x) & \text{for } x \in [L/2, L], \end{cases}$$

defines initial data on $[0, L]$ that satisfies periodic boundary conditions. Moreover, considering solutions $u^\pm$ of a local version of (5) with periodic boundary conditions (9), one can construct restrictions $w^\pm(x, t) = u^\pm(x, t)$, for $x \in [0, L/2]$, which are solutions of the same system with no-flux boundary conditions (11).

The steady-state solutions of nonlocal system (5) described below do satisfy the mirror symmetry condition (12). Therefore, the results of the next sections obtained for periodic (or zero-flux) boundary conditions can be easily generalised to zero-flux (or periodic) conditions.

Fig. 2 An interval of length
L in between two observers
facing each other, where each
has its own coordinate
convention

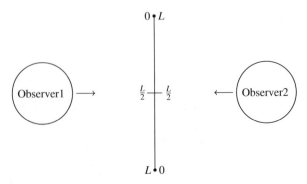

3.2 Symmetry of Hyperbolic Models for Self-organised Biological Aggregations

Because the model is built upon a transport term and the interactions between densities at various points, it is not surprising that the model has some symmetry properties. Indeed, as the densities are moving with constant speed, the location in space where the densities are interacting should not matter. This means the model should be translation invariant.

The other symmetry of the model can be understood by thinking of the domain $[0, L)$ as being in between two observers facing each other, but each respecting its own convention. That is, each observer looking at its own left boundary of the interval identifies it as 0. Thus, $[0, L/2)$ for one observer corresponds to $(L/2, L]$ for the facing observer. See Fig. 2. Thus, right-moving individuals for one observer at location x are left-moving individuals at location $L - x$ for the other observer. Thus, the spatial reflection $x \mapsto L - x$ is linked to this symmetry. We now formalise these ideas.

We look at functions $u(x, t) = (u^+(x, t), u^-(x, t))$ satisfying the boundary condition $u(0, t) = u(L, t)$. We can define a translation operator T_θ with $\theta \in [0, L)$. With the reflection symmetry on $[0, L)$ defined above, we can also define an involution κ acting on $u(x, t)$:

$$T_\theta \cdot u(x, t) := u(1x - \theta, t) \text{ and}; \kappa \cdot (u^+(x, t), u^-(x, t)) := (u^-(L - x, t), u^+(L-x, t)). \tag{13}$$

Because of the periodic boundary conditions, the domain is a circle and T_θ generates a group isomorphic to **SO**(2), the group of rotations on the plane. A direct calculation shows that

$$T_\theta \circ \kappa = \kappa \circ T_\theta^{-1}. \tag{14}$$

That is, apart from $T_{L/2}$, all rotations do not commute with the involution κ. It is well-known (see [29]) that T_θ and κ along with condition (14) generate a group isomorphic to $O(2)$, where $O(2)$ is the orthogonal group and can be thought geometrically as the group of all symmetries of a circle. The operators T_θ and κ are *symmetries* of the model if for any solution $u(x, t)$, then $\kappa \cdot u(x, t)$ and $T_\theta \cdot u(x, t)$ are also solutions for all $\theta \in [0, L)$.

To show formally the symmetry of the model as in [12], one can verify in a straightforward way that the transport portion of the equation is invariant with respect to the translation operator T_θ, where $\theta \in [0, L)$. The turning functions are also symmetric with respect to T_θ because of translation invariance of the integrals. The κ-invariance is derived by verifying that κ acts on the social interaction terms $y_j^\pm(x)$ and yields $\kappa \cdot y_j^\pm(x) = y_j^\mp(L - x)$, and this κ action carries over to the turning functions $\lambda^\pm(x)$. Thus, system (5) is $O(2)$-equivariant for any of the models M2, M3, M4, M5 described in Table 1.

We now define some useful quantities for identifying and classifying the symmetry of solutions. Consider now the action of a group Γ on a function space X. The *isotropy subgroup* of the point $u \in X$ is

$$\Gamma_u := \{\rho \in \Gamma \mid \rho \cdot u = u\}.$$

The symmetry of solutions of (5) is encoded in the isotropy subgroup. To each isotropy subgroup $\Sigma \subset \Gamma$ we define the *fixed-point subspace* of Σ as

$$\text{Fix}(\Sigma) = \{u \in X \mid \sigma \cdot u = u \quad \text{for all } \sigma \in \Sigma\}.$$

Fixed-point subspaces are invariant for the dynamics in Γ-equivariant systems, which can be easily verified by the following calculation. Let $u \in \text{Fix}(\Sigma)$ and let F be Γ-equivariant. Then

$$F(u) = F(\sigma \cdot u) = \sigma \cdot F(u).$$

Let $\Sigma \subset \Gamma$ be a subgroup, then the Σ *group orbit* of an element $u \in X$ is defined as

$$\Sigma u := \{\sigma \cdot u \mid \sigma \in \Sigma\}.$$

In particular, elements lying in the same group orbit have conjugate isotropy subgroups. That is,

$$\Sigma_{\rho \cdot u} = \rho \Sigma_u \rho^{-1}.$$

A *relative equilibrium* is a group orbit invariant for the solution operator of the differential equation; that is, a semiflow-invariant group orbit. In the next section, we present some results about equilibrium (or steady-state) solutions and relative equilibrium solutions.

3.2.1 Equilibria and Relative Equilibria

Equilibrium solutions of (5) are found by setting $\partial_t u^{\pm} = 0$ and solving the remaining integro-differential system. Suppose that $u^*(x) = (u_*^+(x), u_*^-(x))$ is an equilibrium solution with isotropy subgroup Σ. We consider the following two situations for group orbits $O(2)u^*$ of an equilibrium solution u^*:

1. If $\Sigma = O(2)$, then $O(2)u^*$ contains a unique element, and if $\Sigma = SO(2)$ the group orbit has two elements.
2. If Σ is a discrete subgroup of $O(2)$, then $O(2)u^*$ is diffeomorphic to a circle.

In the second situation, we have a circle of equilibrium solutions which is an example of a relative equilibrium, one for which the dynamics is given by the zero vector field on the circle. For a relative equilibrium with a nonzero vector field, one can show that it is a constant vector field which corresponds to the drift generated by the $SO(2)$ action. It is straightforward to see that adding the two equations in (5) and integrating with respect to x, a steady-state solution $(u_*^+(x), u_*^-(x))$ must then satisfy $u_*^+(x) = u_*^-(x) + C$, where C is a constant. We can now state our new result.

Theorem 1 *Suppose that* $O(2)u^*$ *is a relative equilibrium with finite isotropy subgroup* Σ.

1. *If there exists* θ *such that* $\theta\kappa \subset \Sigma$, *then* $O(2)u^*$ *is a circle of equilibrium solutions and* $u^+(x) = u^-(x)$.
2. *If* $\theta\kappa \not\subset \Sigma$ *for all* $\theta \in SO(2)$, *then generically, the dynamics on* $O(2)u^*$ *is given by a* $SO(2)$ *drift.*

Before we proceed with the proof, we recall the following facts. The normaliser subgroup of Σ is defined as the largest group in Γ where Σ is normal. In particular, $N(\Sigma) = \{g \in \Gamma \mid g\Sigma g^{-1} \subset \Sigma\}$. The normaliser subgroup is also the largest subgroup in Γ which leaves $\text{Fix}(\Sigma)$ invariant. That is, for any $g \in N(\Sigma)$ and $u \in \text{Fix}(\Sigma)$, then $g \cdot u \in \text{Fix}(\Sigma)$. The normaliser subgroup is also important for determining the flow on a relative equilibrium as we now explain (see, for instance, Ashwin and Melbourne [4] for all details).

Let x_0 be in the domain of the semiflow generated by solutions of the initial value problem (IVP) and Γx_0 be a relative equilibrium. We denote by $L\Gamma$ the Lie algebra associated with the Lie group Γ. Then $f(x_0) = \zeta x_0$ for some $\zeta \in L\Gamma$. One can refine this result as follows. If x_0 has isotropy subgroup Σ, then $f(x_0) = \zeta x_0$ with $\zeta \in L(N(\Sigma))$; see Proposition 6.1 in [4]. Moreover, generically, the relative equilibrium is foliated by maximal tori in Γ; see Theorem 6.3 in [4]. We can now proceed with the proof of Theorem 1.

Proof If u^* has isotropy subgroup Σ and $\theta\kappa \subset \Sigma$, then $\Sigma = D_n$ for some n and $N(\Sigma) = D_{2n}$. Therefore, $L(N(\Sigma)) = 0$ and so $f(u^*) = 0$. Hence we have a group orbit of equilibrium solutions. In the other case, $\Sigma = \mathbb{Z}_n$ for some n, or $\Sigma = 1$ and $N(\mathbb{Z}_n) = SO(2)$. The Lie algebra $L(N(\Sigma))$ is isomorphic to $\mathbb{R}$ and by the genericity result stated before this proof we have a maximal torus of dimension one and so $f(x_0) = \zeta x_0$ with ζ nonzero, thus leading to a nonzero drift. $\qquad\square$

For homogeneous steady-state solutions, let us first define the total conserved population density

$$A = \frac{1}{L} \int_0^L (u_*^+(x, t) + u_*^-(x, t))\, dx.$$

Then, the homogeneous steady-state solutions are of the form $(u_*^+, u_*^-) = (A/2, A/2)$ and $(u_*^+, u_*^-) = (A^*, A^{**})$, where $A^* \neq A^{**}$ and $A^* + A^{**} = A$. These two types of solutions have isotropy subgroups $\mathbf{O}(2)$ and $\mathbf{SO}(2)$, respectively. We graph these homogeneous steady state solutions in Fig. 3, for u_*^+ versus different model parameters. (Due to model symmetry, similar results can be obtained if we graph $u_*^- = A - u_*^+$ versus model parameters.) We note that all models (M2–M5) described in Table 1 display the steady state with $\mathbf{O}(2)$ symmetry (where $u_*^+ = u_*^-$). In regard to the states with $\mathbf{SO}(2)$ symmetry, the models can exhibit them depending on the parameter values. For example, all models M2–M5 can display states with $u_*^+ \neq u_*^-$ when we vary q_{al} and fix all other parameters (as shown in Fig. 3a). However, only some models can display such states when we vary A (see models M2 and M4 in panel (b)), λ_1 (see models M2, M3 and M4 in panel (c)) or λ_2 (see models M2 and M4 in panel (d)).

It is also possible to find non-homogeneous symmetric steady-state solutions with isotropy subgroup $\mathbf{D}_n$. Such solutions for $n = 1$ and $n = 3$ are observed in [12, 41].

We now discuss relative equilibria with a nonzero $\mathbf{SO}(2)$ drift. In this case, the isotropy subgroup of the solution is either $\mathbb{Z}_n$ for some $n \geq 2$ or 1. Such solutions can arise, for instance, from a symmetry-breaking bifurcation from a $\mathbf{D}_n$ or a $\mathbb{Z}_2$ symmetric group orbit of equilibria. The observed pattern is in fact similar to a rotating wave, but as we show below, the speed of rotation must be slow near bifurcation. Meanwhile, for a typical rotating wave arising from an $\mathbf{O}(2)$ symmetric Hopf bifurcation the speed of rotation is arbitrary as it depends on the purely imaginary part of the eigenvalue at bifurcation.

The bifurcations from relative equilibria can be studied by the method described in Krupa [42] for the case of ODEs with compact group symmetry and in Fiedler et al. [26] for the PDE case, including the case of noncompact groups. Consider a relative equilibrium Γu^* with u^* having isotropy subgroup Σ. The idea is that at each point u_0 of the relative equilibrium, the phase space can be decomposed into the tangent space to the relative equilibrium at u_0, written $T_{u_0} \Gamma u^*$, and a Σ-invariant normal component V contained in the orthogonal complement to $T_{u_0} \Gamma u^*$; see Fig. 4.

By this construction, a local centre bundle $\Gamma \times V$ parametrises a Γ-invariant *tube* in a neighbourhood of the relative equilibrium Γu^*. Note that the isotropy subgroup Σ of u^* acts linearly and orthogonally on V. Using these coordinates, the dynamics of the original PDE in the tube is given by

$$\dot{g} = g h_1(v), \qquad \dot{v} = h_2(v),$$

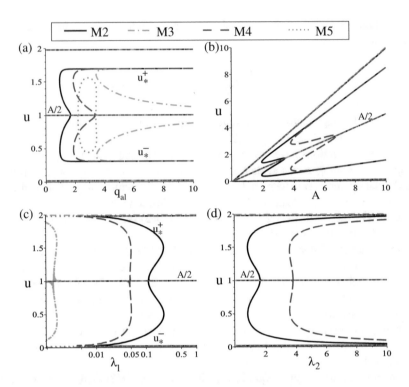

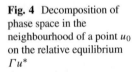

Fig. 3 Spatially homogeneous steady states $u = u_*^+$ displayed by the four models (M2, M3, M4, M5) described in Table 1, as we vary different model parameters: (**a**) q_{al}; (**b**) A; (**c**) λ_1; (**d**) λ_2. The fixed parameters have the following values: $q_a = 1.0$, $q_r = 0.1$, $q_{al} = 1.0$, $\lambda_1 = 0.2$, $\lambda_2 = 0.9$

Fig. 4 Decomposition of phase space in the neighbourhood of a point u_0 on the relative equilibrium Γu^*

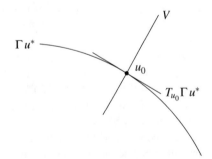

where $(g, v) \in \Gamma \times V$, $L\Gamma$ is again the Lie algebra of Γ, $h_1 : V \to L\Gamma$ and $h_2 : V \to V$. The smoothness of h_1 and h_2 is preserved from the original PDE. According to this representation of the dynamics, relative equilibria in a neighbourhood of Γu^* are obtained by solving $h_2(v) = 0$.

In the case of an **O(2)** group orbit of equilibria given by **O(2)**u^*, the normal component V is a $\mathbf{D}_n$-invariant codimension-one subspace of phase space. Thus, one obtains the system

$$\dot{g} = gh_1(v), \qquad \dot{v} = h_2(v),$$

where h_1 and h_2 are $\mathbf{D}_n$-symmetric. Because u^* is an equilibrium, we have $h_1(u^*) = h_2(u^*) = 0$.

Consider a one-parameter family of equations (5) parametrised by $\lambda \in \mathbb{R}$ and for which $h_2(u^*, \lambda_0) = 0$. Steady-state bifurcations at the $\mathbf{D}_n$-symmetric equilibrium (u^*, λ_0) follow the generic $\mathbf{D}_n$ symmetric bifurcation theory as described in Golubitsky et al. [29]. In particular, steady-state bifurcation problems depend on the parity of n and the kernel of the group action on the eigenspace of the zero eigenvalue of the linearisation at the bifurcation point. That is, a steady-state bifurcating from a $\mathbf{D}_n$ equilibrium generically has isotropy subgroup $\mathbf{D}_p$, where p is a divisor of n.

Suppose that for λ near λ_0, $h_2(v^*(\lambda), \lambda) = 0$, then generically $h_1(v^*(\lambda), \lambda) \neq 0$ and in fact must be very close to 0 by continuity. This means that the drift of the relative equilibrium is slow and this is what one should observe in a numerical simulation of a $\mathbf{SO(2)}$ rotating pattern near bifurcation.

4   Numerical Results

In this section, we present an overview of model dynamics using a finite-difference based time-stepping scheme, as well as some sample results of the continuation of invariant solutions obtained using a pseudo-spectral scheme. While the former approach is comparatively easy to implement and allows for quick scans of asymptotically stable solutions, the latter has superior convergence properties that are helpful for the accurate approximation of equilibrium and time-periodic solutions, both stable and unstable.

4.1   Model Simulations

For the numerical simulations of the dynamics displayed by model (5), we discretise the spatial domain $[0, L]$ into equally spaced intervals of length Δx, and the time is advanced in steps Δt. The nonlocal terms are discretised using Simpson's method [50]. At the boundaries, the integrals are either wrapped around the domain (for the periodic BCs) or they are reflected back into the domain (for the Neumann BCs). Finally, the advection terms are discretised with the help of the classical upwind/downwind method [50]. Note that we chose Δx and Δt such that the Courant-Friedrichs-Lewy condition is satisfied (i.e., $|\gamma \Delta x / \Delta t| \leq 1$).

Remark 2   We note that while the numerical implementation of the periodic BCs (9) preserves the total population density, the numerical implementation of the reflective BCs (11) for the upwind/downwind method does not preserve this total density. To this end, one needs to slightly change these conditions to:

$$u^+(0,t) = \alpha u^-(0,t), \quad u^-(0,t) = \beta u^+(L,t), \tag{15}$$

with α and β given as follows:

$$\alpha = \frac{\frac{\gamma \Delta x}{\Delta t} + \Delta x \lambda^-[u^+(0,t), u^-(0,t)]}{\frac{\gamma \Delta x}{\Delta t} + \Delta x \lambda^+[u^+(0,t), u^-(0,t)]}, \quad \beta = \frac{\frac{\gamma \Delta x}{\Delta t} + \Delta x \lambda^+[u^+(L,t), u^-(L,t)]}{\frac{\gamma \Delta x}{\Delta t} + \Delta x \lambda^-[u^+(L,t), u^-(L,t)]}, \tag{16}$$

where Δx and Δt are the space and time steps used in the numerical discretisation of the domain.

Finally, the initial conditions for the numerical simulations discussed in the next section are random perturbations of two types of spatially homogeneous steady states (u_*^+, u_*^-):

- states with **O(2)** symmetry: $u_*^+ = u_*^- = A/2 = 1$;
- states with **SO(2)** symmetry: $u_*^+ > u_*^-$ or $u_*^+ < u_*^-$ (the exact steady state values depend on model parameters: $q_{al}, \gamma, \lambda_1, \lambda_2$, as shown in Fig. 3).

The parameters used for the numerical simulations are summarised in Table 2. Any differences from these values are discussed in the figure captions.

4.1.1   Spatio-Temporal Patterns

We start the presentation of various types of spatio-temporal patterns by focusing on the impact of different boundary conditions. First, we focus on the case where there are a few number of aggregation patterns inside the domain. Figure 5 shows that the patterns localised in the middle of the domain, such as (i) feathers (where individuals leave periodically the aggregations) and (ii) the stationary pulses, are not influenced by the boundary conditions (either (a) periodic or (b) reflective). In contrast, the moving aggregation patterns, such as rotating waves (iii), are affected by the presence of the boundaries. When there are a small number of travelling aggregations inside the domain, these aggregations turn around when they reach the reflective boundary, and move in the opposite direction (thus giving rise to a zigzagging pattern). In contrast, when there are a large number of such travelling aggregations inside the domain, the social interactions between adjacent aggregations and the reflective domain boundaries give rise to stationary aggregations—as shown in Fig. 6d.

———————————————————————————————————▶

Fig. 5 (continued) $q_{al} = 0$, $q_a = 6.0$, $q_r = 6.4$ [23]; (ii) Stationary pulse patterns obtained with model M2, for $q_{al} = 0$, $q_a = 4$, $q_r = 0.5$; (iii) Travelling pulse (i.e., rotating wave) pattern (left panel), and zigzag pattern (right panel), obtained with model M2, for $q_{al} = 2$, $q_a = 2$, $q_r = 2$; (iv) Modulated travelling pulse (i.e., modulated rotating wave) pattern (left panel), and zigzag pattern (right panel), obtained with model M4 for $q_{al} = 2.0$, $q_a = 4.0$, $q_r = 4.0$

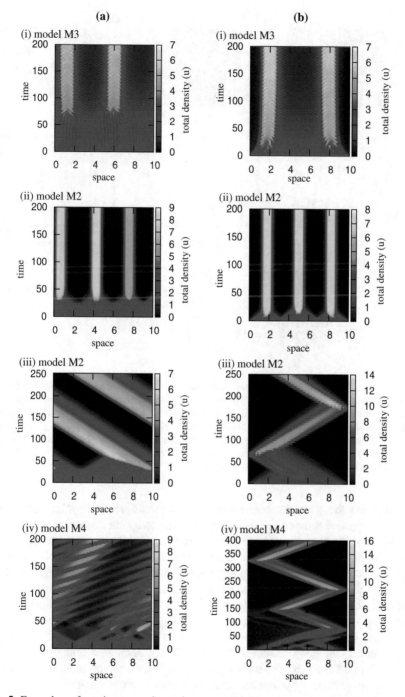

Fig. 5 Examples of stationary and moving aggregation patterns, for different boundary conditions. (**a**) Periodic boundary conditions (9) on domain $[0, L]$; (**b**) Neumann (reflective) boundary conditions (15) on domain $[0, L]$. (**i**) Feather-like patterns obtained with model M3, for

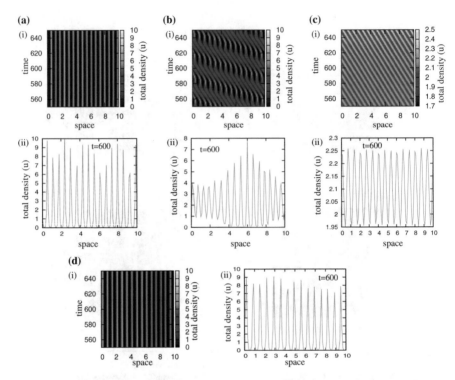

Fig. 6 Example of spatio-temporal patterns obtained with model M4 (see Table 1) for different q_{al} values, while keeping $q_r = q_a = 0.0$. (**a**) Stationary pulses, obtained for $q_{al} = 2.05$ and periodic BCs; (**b**) Semi-zigzags obtained for $q_{al} = 2.15$ and periodic BCs; (**c**) Rotating waves obtained for $q_{al} = 2.18$ and periodic BCs; (**d**) Stationary pulses, obtained for $q_{al} = 2.05$, $q_{al} = 2.15$ and $q_{al} = 2.18$ with Neumann BCs. Sub-panels (i) show the full spatio-temporal pattern, while sub-panels (ii) show a snapshot of the spatial pattern at $t = 600$. Here $\lambda_1 = 0.2/0.3$, $\lambda_2 = 0.9/0.3$. All other parameter values are as described in Table 2

Next, we focus on the case when there are multiple aggregations inside the domain. Figure 6 shows a few examples of spatio-temporal patterns obtained when we increase the strength of alignment interactions (while ignoring any repulsive-attractive interactions: $q_a = q_r = 0$). As a transient behaviour we can observe three types of patterns: (a) stationary aggregations, (b) semi-zigzagging aggregations (i.e., stop-and-go waves, where individuals at different positions in space move in one direction, then they stop, only to start moving again in the same direction), (c) travelling aggregations (i.e., rotating waves). Regarding the asymptotic behaviour (observed for $t > 1000$-not shown here) it is possible to have only stationary aggregations (for $q_{al} = 2.05$ and $q_{al} = 2.18$) and semi-zigzag aggregations (for $q_{al} = 2.15$). Therefore, the semi-zigzag aggregations, which represent a relative equilibrium pattern, are stable and seem to exist in a narrow parameter region.

Next, we investigate numerically the difference between the modulated rotating waves and the semi-zigzag waves. Figure 7 shows on the left the spatio-temporal evolution of the pattern obtained with model M4 (and periodic BCs) when $q_{al} = 2.15$ (and $q_a = 0$, $q_r = 0$, $\lambda_1 = 0.667$, $\lambda_2 = 3.0$ and all other parameter values as in Table 2). For $t < 250$ the model exhibits unstable modulated rotating waves, which then evolve for $t > 250$ into stable semi-zigzags (i.e., relative equilibria). The right panels show time snapshots of these two different patterns (for $t = 150$ and $t = 310$). We can see here that both patterns exhibit relatively similar spatial modulations. The difference is that for the semi-zigzags, the high-amplitude solutions (corresponding to the stationary pulses) have a minimum at $u^{\pm}(x, t) \approx 0$ (same as in Fig. 6a(ii)), while for the modulated travelling waves the high amplitude solutions have a minimum at $u(x, t) = (u^+ + u^-)(x, t) \gg 0$.

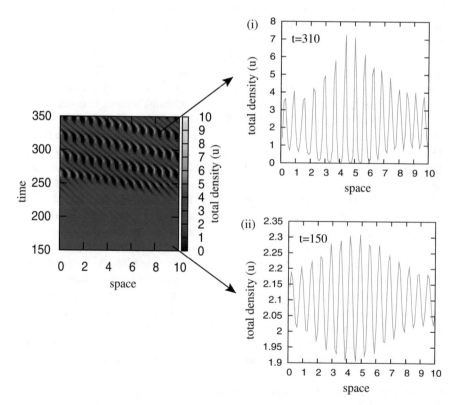

Fig. 7 Transitions between two spatio-temporal patterns obtained with model M4 with periodic BCs, when $q_{al} = 2.15$ (and $q_a = 0$, $q_r = 0$, $\lambda_1 = 0.667$, $\lambda_2 = 3.0$ and all other parameter values as in Table 2). For $t < 250$ we observe modulated rotating waves, while for $t > 250$ we observe semi-zigzags (or relative equilibria). Sub-panels (i) and (ii) show time snapshots of these two different patterns

4.2   Continuation Algorithm

From Figs. 5, 6 and 7 it seems clear that the various nonlocal models discussed before admit invariant solutions such as equilibria, time-periodic orbits and travelling waves. In order to accurately approximate such solutions, we use a Fourier spectral representation of Eqs. (5a)–(5c). The series expansion is formally given by the semi-discrete Fourier transform (SDFT)

$$u^{\pm}(x, t) = \sum_{k=\infty}^{\infty} \hat{u}_k^{\pm}(t) e^{2\pi ikx/L} \; ; \; \hat{u}_k^{\pm}(t) = \frac{1}{L} \int_{x=0}^{L} u^{\pm}(x, t) e^{-2\pi ikx/L} \, \mathrm{d}x. \tag{17}$$

In practice, we retain a finite number of Fourier coefficients, say $|k| \leq M < \infty$. In that case, the above transformation is approximated by the discrete Fourier transform (DFT) and a truncation error is introduced. However, it is known that if $u \in C^P[0, L]$ then its Fourier coefficients satisfy $|\hat{u}_k| = O(k^{-P})$, and if $u \in C^\infty[0, L]$ then $|\hat{u}_k| = O(\exp[-\alpha k])$ for some $\alpha > 0$ [57]. The truncated Fourier series represents a solution periodic on the domain $[0, L]$, and below we will only consider invariant solutions that satisfy periodic boundary conditions.

The discretised version of Eqs. (5a)–(5c) is obtained by taking the SDFT of the left-hand and right-hand sides. While this is a trivial computation for the linear terms, the nonlinear terms are more involved. We evaluate them in four steps. First, the integral terms are evaluated on the Fourier basis. Since all integrals have the form of convolutions or cross-covariances, they can be evaluated by taking element wise products of the Fourier coefficients of $u^{\pm}$ and the kernels, as done in Sect. 3.1.1. Secondly, the argument of the nonlinear function f, i.e. $y_r^{\pm} - y_a^{\pm} + y_{al}^{\pm}$, is computed on a regular spatial grid using the fast Fourier transform algorithm for the inverse DFT. Thirdly, the nonlinear function is evaluated on the grid and, finally, the DFT of the result is computed to find the contribution of the nonlinear terms to the dynamics. This trick of evaluating some terms on the Fourier basis and some on the spatial grid, and using the (inverse) FFT to move from one basis to the other is often called the pseudo-spectral approach. The most computationally costly step in the process is the FFT, which has an order of complexity of $O(n \ln[n])$, which makes it feasible to evaluate the discretised equations with up to $M = 4096$ modes in less than a second on an average laptop running Matlab, the environment in which our continuation code is written.

The result of this discretisation can be written compactly as

$$\dot{z} = F(z, \lambda), \qquad z(t) = \phi(z(0), t, \lambda), \tag{18}$$

where we have introduced the flow ϕ and the vector Λ of system parameters λ. The vector $z \in \mathbb{R}^{2M}$ holds the real and imaginary parts of the Fourier coefficients of u^+ and u^-. We discretise time by applying the trapezoidal method to the linear terms, and the explicit Euler method to the nonlinear terms. The resulting method is only first order accurate in time, but since only short integrations are necessary for the computation of invariant solutions this is not a limitation.

Equilibria and periodic solutions relative to the translation symmetry T_θ satisfy

$$\phi(z, P, \Lambda) - T_\theta z = 0. \tag{19}$$

For relative equilibria, P is arbitrary and θ/P is the propagation speed, while for relative periodic solutions P is the period and θ the shift per period. For fixed parameters, each solution to Eq. (19) that is not spatially homogeneous is embedded in a continuous family because of the translational symmetry in space and, for periodic orbits, in time. If we add phase conditions to remove these degeneracies and vary a single component λ of Λ, we can approximate a curve of solutions using the arclength continuation (see, e.g., [1]). Suppose that we know a solution $(z_p, P_p, \lambda_p, \theta_p)$ and a vector $\varXi$ approximately tangent to the family of solutions, we use Newton iteration so solve

$$\phi(z, P, \lambda) - T_\theta z = 0,$$

$$\psi(z, P, \lambda, \theta) = 0, \tag{20}$$

$$\left.\frac{\mathrm{d}}{\mathrm{d}\theta} T_\theta z_p\right|_{\theta=0} \cdot (z - z_p) = 0,$$

$$\varXi \cdot (z - z_p, P - P_p, \lambda - \lambda_p, \theta - \theta_p) = 0, \tag{21}$$

where, with a slight abuse of notation, we have listed only one system parameter as variable of the flow. The first phase condition is taken to be a Poincaré plane of intersection for periodic orbits, and simply set to $P - c = 0$ for constant c for relative equilibria. Equation (21) constitutes a well-posed system of $2M + 3$ equations and unknowns. In the implementation we follow [52] and use a Krylov subspace method to approximate the Newton update step. The convergence of Krylov subspace methods depends on the spectrum of the invariant solution under consideration. For equilibrium solutions, we know that this spectrum consists of isolated eigenvalues of finite multiplicity and that there are no accumulation points [13]. The solutions we compute have only a small number of unstable eigenvalues, and thus we expect the spectrum to accumulate in the left half of the complex plane. In practice, we find that the real part of the eigenvalues tends to minus infinity. The corresponding eigenvalues of the linearisation of Eq. (19), which involves the flow over time P, are thus clustered around minus unity and the Krylov iterations tend to converge rapidly.

Our application of pseudo-spectral discretisation and Newton-Krylov continuation to a dynamical system with spatial interactions described by convolutions is similar to the work of Avitabile et al. on a neural mass model [6, 51]. In their model, the convolution represents the integration of synaptic signals originating from a neighbourhood of a given neuron, weighed by the connection density. There is one important difference between the neural field model and the aggregation model. In the former, a sigmoidal functional is applied to the input signal before the convolution is computed. In the aggregation model, we first evaluate the convolution and then apply the sigmoidal functional. Since applying the functional tends to lead to steep interfaces, while taking the convolution is a smoothing operation, the nonlocal contribution tends to be less smooth in the current model.

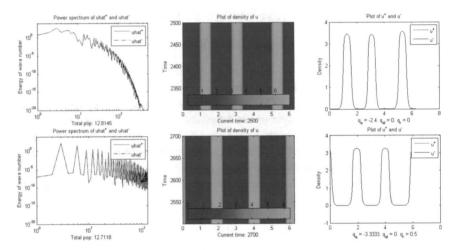

Fig. 8 Top: Power spectrum (left), density plot through time (middle), and final time plot of densities (right) of a solution close to a three bump equilibrium. Bottom: Power spectrum (left), density plot through time (middle), and final time (right) of the corrected solution now showing a three bump equilibrium with the $\mathbf{D}_3$-symmetry more apparent. Note that the q_a value is inserted with a negative sign in the code as one can see below the right pictures

It is important to note that, in addition to zero eigenvalues related to exact symmetries of the system, small eigenvalues can exist for certain solutions. For instance, the solution with three pulses shown in Fig. 5(ii) is close to a $\mathbf{D}_3$-invariant solution. Such a solution will generically have a single zero eigenvalue caused by the translational invariance. Numerically, however, we find three eigenvalues close to zero, with eigenvectors corresponding to translations of individual pulses. This is understandable if we consider the fact that the distance between two pulses is about $d = 3$ on a domain of size $L = 10$ (as considered throughout all these numerical simulations). The largest interaction radius is $s_a = 1 = d/3$. Thus, the change to the population density inside each pulse due to the other pulses is small, and the pulses are almost decoupled. Three more small eigenvalues exist that correspond to perturbations that change the height of each of the pulses independently. These small eigenvalues correspond to a near-degeneracy of the linearisation of Eq. (21), which hampers the convergence of the Newton-Krylov continuation. A practical approach is to restrict the sequence of approximate solutions to lie in the fixed point subspace for the $\mathbf{D}_n$ symmetry. For the three pulse solution, for instance, we can include in z only the Fourier coefficients of wave numbers $k = 3j$, $j \in [\lceil -M/3 \rceil, \lceil M/3 \rceil]$. Thus, we solve a system of equations smaller by a factor of three and exclude the symmetry-breaking, near-zero eigenvalues. This approach was used by Aston et al. to compute $\mathbf{D}_n$ symmetric rotating waves [5].

As an example for the application of this algorithm to the identification of new spatially heterogeneous equilibria, Fig. 8 (top) shows a solution (of model M2) approximated by time-stepping for $q_a = 2.4$, $q_{al} = 0$ and $q_r = 0$. Figure 8 (bottom) shows a corrected solution much closer to a $\mathbf{D}_3$ symmetric equilibrium at $q_a = 3.3333$, $q_{al} = 0$ and $q_r = 0.5$. One can see in the power spectrum for the corrected solution that the energy in the high wave numbers does not decay exponentially as they should according to the theory. This is due to the extra eigenvalues near zero described above.

5 Summary and Discussion of the Results

In this study we reviewed a class of 1D nonlocal hyperbolic models introduced in [23] to describe self-organised animal behaviours, as a result of different types of inter-individual communication. We focused on the symmetry of these models, and discussed some examples of spatio-temporal patterns that can be obtained with two different types of boundary conditions, namely periodic and reflective conditions. In the context of spatially heterogeneous equilibria, we focused on the existence of relative equilibria, and showed in Theorem 1 the conditions for the existence of these equilibria (which have not been previously investigated for this class of 1D nonlocal hyperbolic/kinetic models). This analytical result was complemented by various numerical simulations showing these types of patterns, and the similarity (in terms of spatial distributions of population density) between relative equilibria patterns and modulated rotating waves.

Finally, we introduced a new continuation algorithm (which considered the symmetry of the solutions) that allowed us to approximate more accurately the equilibria exhibited by this nonlocal class of kinetic models. In Fig. 8 we showed an application of this algorithm to one particular case, the three-pulse solution identified numerically in Fig. 5(ii). We started with the equilibrium solution shown in the top panels in Fig. 8, and we followed it as it developed into another similarly looking three-bump equilibrium solution, which showed the $\mathbf{D}_3$-symmetry more clearly (i.e., the three peaks in the bottom panel of Fig. 8 all have the same amplitude, while the right peak in the top panel is slightly higher than the other two peaks). While one could apply this algorithm to trace the evolution of known equilibria to more exotic patterns, such an attempt is beyond the scope of this study. Rather, our aim was to briefly describe this algorithm developed for the nonlocal class of kinetic models (5), and to exemplify its applicability to the identification of steady state solutions with $\mathbf{D}_3$-symmetry (while emphasising also the degeneracy of the eigenvalues for the $\mathbf{D}_n$ symmetric solutions).

Acknowledgements PLB and LvV acknowledge the financial support from NSERC in the form of a Discovery Grant.

Appendix

In Table 2 we summarise the parameters that appear in the nonlocal models (5).

Table 2 Summary of model parameters that appear in Eq. (5) and in Table 1, together with their description and the values used throughout the numerical simulations

Param.	Value	Description
γ	0.1	Average speed of individuals
λ_1	0.2	Approximation of random turning rate
λ_2	0.9	Approximation of density-directed turning rate
q_r	0–10	Strength of repulsive interactions
q_a	0–10	Strength of attractive interactions
q_{al}	0–10	Strength of alignment interactions
s_r	0.25	Parameter that gives the mid of the repulsion zone
s_{al}	0.5	Parameter that gives the mid of the alignment zone
s_a	1.0	Parameter that gives the mid of the attraction zone
m_r	$s_r/8$	Parameter that controls the width of the repulsion zone
m_{al}	$s_{al}/8$	Parameter that controls the width of the alignment zone
m_a	$s_a/8$	Parameter that controls the width of the attraction zone
L	10	Domain length

References

1. E. L. Allgower and K. Georg. *Introduction to Numerical Continuation Methods*. SIAM, 2003.
2. I. Aoki. A simulation study on the schooling mechanism in fish. *Bull. Jpn. Soc. Sci. Fish*, pages 1081–1088, 1982.
3. S. Arganda, A. Pérez-Escudero, and G.G. de Polavieja. A common rule for decision making in animal collectives across species. *Proc. Matl. Acad. Sci.*, 109:20508–20513, 2012.
4. P. Ashwin and I. Melbourne. Noncompact drift for relative equilibria and relative periodic orbits. *Nonlinearity*, 10:595, 1997.
5. P. J. Aston, A. Spence, and W. Wu. Bifurcation to rotating waves in equations with $O(2)$– symmetry. *SIAM J. Appl. Math.*, 52:792–809, 1992.
6. D. Avitable and K.C.A. Wedgwood. Macroscopic coherent structures in a stochastic neural network: from interface dynamics to coarse-grained bifurcation analysis. *J. Math. Biol.*, 75(4):885–928, 2017.
7. E. Barbera, G. Consolo, and G. Valenti. A two or three compartments hyperbolic reaction-diffusion model for the aquatic food chain. *Math. Biosci. Eng.*, 12(3):451–472, 2015.
8. E. Barbera, C. Currò, and G. Valenti. Wave features of a hyperbolic prey-predator model. *Math. Methods Appl. Sci.*, 33(12):1504–1515, 2010.
9. N. Bellomo, A. Bellouquid, and M. Delitala. From the mathematical kinetic theory of active particles to multiscale modelling of complex biological systems. *Math. Comput. Model.*, 47(7-8):687–698, 2008.
10. A. Berdhal, C.J. Torney, C.C. Ioannou, J.J. Faria, and I.D. Couzin. Emergent sensing of complex environments by mobile animal groups. *Science*, 339(6119):574–576, 2013.
11. P-L. Buono and R. Eftimie. Analysis of Hopf/Hopf bifurcations in nonlocal hyperbolic models for self-organised aggregations. *Math. Models Methods Appl. Sci.*, 24(2):327–357, 2014.
12. P-L. Buono and R. Eftimie. Codimension-two bifurcations in animal aggregation models with symmetry. *SIAM J. Appl. Dyn. Syst.*, 13(4):1542–1582, 2014.
13. P.-L. Buono and R. Eftimie. Lyapunov-Schmidt and centre-manifold reduction methods for nonlocal PDEs modelling animal aggregations. In B. Tony, editor, *Mathematical Sciences with Multidisciplinary Applications. Springer Proceedings in Mathematics & Statistics*, volume 157, pages 29–59. Springer, Cham, 2016.

14. D. Burini, L. Gibelli, and N. Outada. A kinetic theory approach to the modeling of complex living systems. In N. Bellomo, P. Degond, and E. Tadmor, editors, *Active Particles*, volume 1, pages 229–258. Birkhäuser, Basel, 2017.

15. D.S. Calovi, U. Lopez, S. Ngo, C. Sire, H. Chaté, and G. Theraulaz. Swarming, schooling, milling: phase diagram of data-driven fish school model. *New Journal of Physics*, 16:015026, 2014.

16. H. Chaté, F. Ginelli, G. Grégoire, F. Peruani, and F. Raynaud. Modeling collective motion: variations on the Vicsek model. *The European Physics Journal B*, 64(3-4):451–456, 2008.

17. S.-H. Choi and Y.-J. Kim. A discrete velocity kinetic model with food metric: chemotaxis travelling waves. *Bull. Math. Biol.*, 79(2):277–302, 2017.

18. R.M. Colombo and E. Rossi. Hyperbolic predators vs. parabolic prey. *Communications in Mathematical Sciences*, 13(2):369–400, 2015.

19. I.D. Couzin, J. Krause, R. James, G.D. Ruxton, and N.R. Franks. Collective memory and spatial sorting in animal groups. *J. Theor. Biol.*, 218:1–11, 2002.

20. P. Degond, A. Frouvelle, S. Merino-Aceituno, and A. Trescases. Quaternions in collective dynamics. *Multiscale Model. Simul.*, 16(1):28–77, 2018.

21. R. Eftimie. Hyperbolic and kinetic models for self-organised biological aggregations and movement: a brief review. *J. Math. Biol.*, 65(1):35–75, 2012.

22. R. Eftimie. Simultaneous use of different communication mechanisms leads to spatial sorting and unexpected collective behaviours in animal groups. *J. Theor. Biol.*, 337:42–53, 2013.

23. R. Eftimie, G. de Vries, and M.A. Lewis. Complex spatial group patterns result from different animal communication mechanisms. *Proc. Natl. Acad. Sci.*, 104(17):6974–6979, 2007.

24. R. Eftimie, G. de Vries, M.A. Lewis, and F. Lutscher. Modeling group formation and activity patterns in self-organising collectives of individuals. *Bull. Math. Biol.*, 69(5):1537–1566, 2007.

25. R. Fetecau. Collective behaviour of biological aggregations in two dimensions: a nonlocal kinetic model. *Math. Models Methods Appl. Sci.*, 21:1539–1569, 2011.

26. B. Fiedler, S. Björn, A. Scheel, and C. Wulff. Bifurcation form relative equilibria of nonimpact group actions: Skew products, meanders, and drifts. *Documenta Mathematica*, 1:479–505, 1996.

27. F. Filbet, P. Laurencot, and B. Perthame. Derivation of hyperbolic models for chemosensitive movement. *J. Math. Biol.*, 50(2):189–207, 2005.

28. A. Filella, F. Nadal, C. Sire, E. Kanso, and C. Eloy. Model of collective fish behavior with hydrodynamic interactions. *Phys. Rev. Lett.*, 120:198101, 2018.

29. M. Golubitsky, I. Stewart, and D.G. Schaeffer. *Singularities and Groups in Bifurcation Theory. Volume 2*. Springer-Verlag New York Inc., 1988.

30. K.P. Hadeler. Reaction transport equations in biological modeling. *Mathematical and Computer Modelling*, 31(4-5):75–81, 2000. Proceedings of the Conference on Dynamical Systems in Biology and Medicine.

31. M. Haragus and G. Iooss. *Local bifurcations, centre manifolds, and normal forms in infinite-dimensional systems*. Springer, 2010.

32. T. Hillen. Invariance principles for hyperbolic random walk systems. *J. Math. Anal. Appl.*, 210(1):360–374, 1997.

33. T. Hillen. Hyperbolic models for chemosensitive movement. *Mathematical Models and Methods in Applied Sciences*, 12(07):1007–1034, 2002.

34. T. Hillen. Existence theory for correlated random walks on bounded domains. *Canad. Appl. Math. Quart*, 18(1):1–40, 2010.

35. T. Hillen and K.P. Hadeler. Hyperbolic systems and transport equations in mathematical biology. In Gerald Warnecke, editor, *Analysis and Numerics for Conservation Laws*, pages 257–279. Springer Berlin Heidelberg, 2005.

36. A. Huth and C. Wissel. The simulation of fish schools in comparison with experimental data. *Ecol. Model*, 75/76:135–145, 1994.

37. H. Inaba. Threshold and stability results for an age-structured epidemic model. *J. Math. Biol.*, 28:411–434, 1990.

38. B.L. Keyfitz and N. Keyfitz. The McKendrick partial differential equation and its uses in epidemiology and population study. *Math. Comput. Modelling*, 26(6):1–9, 1997.
39. I. Kmit. Fredholm solvability of a periodic Neumann problem for a linear telegraph equation. *Ukrainian Mathematical Journal*, 65(3), 2013.
40. I. Kmit and L. Recke. Hopf bifurcation for semilinear dissipative hyperbolic systems. *J. Differential Equations*, 257:264–309, 2014.
41. M. Kovacic. On matrix-free pseudo-arclength continuation methods applied to a nonlocal PDE in 1+1d with pseudo-spectral time-stepping. Master's thesis, University of Ontario Institute of Technology, 2013.
42. M. Krupa. Bifurcations of relative equilibria. *SIAM J. Math. Anal.*, 21(6):1453–1486, 1990.
43. F. Lutscher. Modeling alignment and movement of animals and cells. *J. Math. Biol.*, 45:234–260, 2002.
44. P. Magal and S. Ruan. On integrated semigroups and age structured models in L^p spaces. *Differential Integral Equations*, 20(2):197–239, 2007.
45. J.K. Parrish and L. Edelstein-Keshet. Complexity, pattern, and evolutionary trade-offs in animal aggregations. *Science*, 284(2):99–101, 1999.
46. J.K. Parrish, S.V. Viscido, and D. Grünbaum. Self-organised fish schools: an examination of emergent properties. *Biol. Bull.*, 202:296–305, 2002.
47. B. Pfistner. A one dimensional model for the swarming behaviour of Myxobacteria. In G. Hoffmann W. Alt, editor, *Biological Motion. Lecture Notes on Biomathematics*, pages 556–563. Springer, Berlin, 1990.
48. M. Pineda, C.J. Weijer, and R. Eftimie. Modelling cell movement, cell differentiation, cell sorting and proportion regulation in Dictyostelium discoideum aggregations. *J. Theor. Biol.*, 370:135–150, 2015.
49. W. Pönisch, C.A. Weber, G. Juckeland, N. Biais, and V. Zaburdaev. Multiscale modeling of bacterial colonies: how pili mediate the dynamics of single cells and cellular aggregates. *New Journal of Physics*, 19:015003, 2017.
50. W.H. Press, S.A. Teukolsky, W.T. Vetterling, and B.P. Flannery. *Numerical Recipes: The Art of Scientific Computing*. Cambridge University Press, 2007.
51. J. Rankin, D. Avitabile, J. Baladron, G. Faye, and D. J. B. Lloyd. Continuation of localized coherent structures in nonlocal neural field equations. *SIAM J. Sci. Comp.*, 36:B70–B93, 2014.
52. J. Sánchez Umbría and M. Net. Numerical continuation methods for large–scale dissipative dynamical systems. *Eur. Phys. J. – Spec. Top.*, 225:2465–2486, 2016.
53. A.P. Solon, H. Chaté, and J. Tailleur. From phase to microphase separation in flocking models: the essential role of nonequilibrium fluctuations. *Phys. Rev. Lett.*, 114(6):068101, 2015.
54. Pliny the Elder. *The Natural History*. H.G. Bohn, London, 1855. (Translated by John Bostock M.D. and F.R.S. Henry T. Riley Esq.).
55. C.M. Topaz and A.L. Bertozzi. Swarming patterns in a two-dimensional kinematic model for biological groups. *SIAM J. Appl. Math*, 65(1):152–174, 2004.
56. C.M. Topaz, A.L. Bertozzi, and M.A. Lewis. A nonlocal continuum model for biological aggregation. *Bull. Math. Biol.*, 68:1601–1623, 2006.
57. L. N. Trefethen. *Spectral methods in MATLAB*. SIAM, 2000.
58. T. Vicsek, A. Czirók, E. Ben-Jacob, I. Cohen, and O. Shochet. Novel type of phase transition in a system of self-driven particles. *Phys. Rev. Lett.*, 75:1226, 1995.
59. M. Witten, editor. *Hyperbolic Partial Differential Equations. Populations, reactors, tides and waves: theory and applications*. Pergamon, 1983.
60. D.J. Wollkind. Applications of linear hyperbolic partial equations: predator-prey systems and gravitational instability of nebulae. *Mathematical Modelling*, 7:413–428, 1986.

Aggregation-Diffusion Equations: Dynamics, Asymptotics, and Singular Limits

José A. Carrillo, Katy Craig, and Yao Yao

Abstract Given a large ensemble of interacting particles, driven by nonlocal interactions and localized repulsion, the mean-field limit leads to a class of nonlocal, nonlinear partial differential equations known as *aggregation-diffusion equations*. Over the past 15 years, aggregation-diffusion equations have become widespread in biological applications and have also attracted significant mathematical interest, due to their competing forces at different length scales. These competing forces lead to rich dynamics, including symmetrization, stabilization, and metastability, as well as sharp dichotomies separating well-posedness from finite time blow-up. In the present work, we review known analytical results for aggregation-diffusion equations and consider singular limits of these equations, including the slow diffusion limit, which leads to the constrained aggregation equation, and localized aggregation and vanishing diffusion limits, which lead to metastability behavior. We also review the range of numerical methods available for simulating solutions, with special attention devoted to recent advances in deterministic particle methods. We close by applying such a method—the *blob method for diffusion*—to showcase key properties of the dynamics of aggregation-diffusion equations and related singular limits.

J. A. Carrillo
Imperial College London, London, UK
e-mail: carrillo@imperial.ac.uk

K. Craig
University of California, Santa Barbara, CA, USA
e-mail: kcraig@math.ucsb.edu

Y. Yao (✉)
Georgia Institute of Technology, Atlanta, GA, USA
e-mail: yaoyao@math.gatech.edu

© Springer Nature Switzerland AG 2019
N. Bellomo et al. (eds.), *Active Particles, Volume 2*, Modeling
and Simulation in Science, Engineering and Technology,
https://doi.org/10.1007/978-3-030-20297-2_3

1 Introduction

Many phenomena in the life sciences, ranging from the microscopic to macroscopic level, exhibit surprisingly similar structures. Behavior at the microscopic level, including ion channel transport, chemotaxis, and angiogenesis, and behavior at the macroscopic level, including herding of animal populations, motion of human crowds, and bacteria orientation, are both largely driven by long-range attractive forces, due to electrical, chemical, or social interactions, and short-range repulsion, due to dissipation or finite size effects.

Various modeling approaches at the agent-based level, from cellular automata to Brownian particles, have been used to describe these phenomena. To pass from microscopic models [60, 83, 87] to continuum descriptions requires analysis of the mean-field limit, as the number of agents becomes large [33, 52, 54, 101, 102]. This approach leads to a continuum kinematic equation for the evolution of the density of individuals $\rho(x, t)$ known as the *aggregation equation*,

$$\frac{\partial \rho}{\partial t} + \nabla \cdot (\rho u) = 0, \quad u = -\nabla W * \rho, \tag{1}$$

where $W : \mathbb{R}^d \longrightarrow \mathbb{R}$ is a symmetric interaction potential. Typical examples of interaction potentials W are given by repulsive-attractive power laws

$$W(x) = \frac{|x|^A}{A} - \frac{|x|^B}{B}, \quad 2 \geq A > B > -d, \tag{2}$$

or Morse potentials [87],

$$W(x) = -C_A e^{-|x|/\ell_A} + C_R e^{-|x|/\ell_R}, \quad C = C_R/C_A > 1, \quad \ell = \ell_R/\ell_A < 1, \quad C\ell^d < 1.$$

The continuum equation (1) may also be modified to include linear or nonlinear diffusion terms, which arise from two possible modeling assumptions: noise at the level of interacting particles (linear diffusion) or scaling hypotheses on a repulsion potential that depend on the inter-particle distance (nonlinear diffusion) [32, 125, 139]. The latter case can be seen via the following formal argument: take $W_\nu = W + 2\nu\delta_0$ with W an attractive potential. In this case, W_ν induces strongly localized repulsion of strength ν, along with a nonlocal attractive force via the potential W. Formally, the corresponding PDE is given by

$$\rho_t = \nu \Delta \rho^2 + \nabla \cdot (\rho \nabla (\rho * W)). \tag{3}$$

This equation can be obtained from particle approximations [32, 38, 125] in a limit in which the potential has a repulsive part that becomes concentrated at the origin as the number of particles increases, with an overall mean-field scaling for the forces. See [55] and the references therein for the rigorous relation between (1) and (3).

Incorporating either linear or nonlinear diffusion into the aggregation equation (1), we arrive at a partial differential equation of the form

$$\frac{\partial \rho}{\partial t} = -\nabla \cdot (\rho u) = \nabla \cdot \left[\rho \nabla \left(U'(\rho) + W * \rho \right) \right],$$

where the function $U(\rho)$ determines the type of diffusion. In particular, $U(s) = s \log s$ corresponds to linear diffusion and $U(s) = s^m/(m-1)$, $m > 0$ corresponds to nonlinear diffusion [141]. Nonlocal partial differential equations of this form are becoming widespread in both modeling and theory, and a key question from the biological perspective is how to identify the proper mechanisms for collective motion among the many potential equations describing the behavior [62, 112, 124].

In this chapter, we consider a class of partial differential equations with nonlocal interactions and nonlinear diffusion, known as *aggregation-diffusion equations*

$$\rho_t = \Delta \rho^m + \nabla \cdot (\rho \nabla (\rho * W)), m \geq 1. \tag{4}$$

Formally, equations of this form have a gradient flow structure with respect to the Wasserstein metric d_2 on the space of probability measures with finite second moment $\mathscr{P}_2(\mathbb{R}^d)$ [3, 66, 106, 126, 142]. In particular, defining the free energy

$$E[\rho] = \underbrace{\frac{1}{m-1} \int_{\mathbb{R}^d} \rho^m dx}_{\mathscr{S}_m[\rho]} + \underbrace{\frac{1}{2} \int_{\mathbb{R}^d} \rho(\rho * W) dx}_{\mathscr{W}[\rho]}, \tag{5}$$

one may rewrite the aggregation diffusion equation (4) as

$$\frac{d\rho}{dt} = -\nabla_{d_2} E(\rho), \qquad \nabla_{d_2} E(\rho) = -\nabla \cdot \left(\rho \nabla \frac{\delta E}{\delta \rho} \right), \tag{6}$$

where ∇_{d_2} denotes the gradient with respect to the Wasserstein metric and $\nabla \frac{\delta E}{\delta \rho}$ denotes the first variation of the free energy with respect to ρ. The first term in the free energy (5) is the entropy $\mathscr{S}_m[\rho]$, which gives rise to the diffusion term, where we use the convention $\frac{1}{m-1} \rho^{m-1} = \log \rho$ for $m = 1$. The second term in the energy (5) is the interaction energy $\mathscr{W}[\rho]$, which gives rise to the aggregation term.

An archetypical example of aggregation-diffusion equation (4) is the Keller–Segel model of chemotaxis in mathematical biology, which describes the collective motion of cells (usually bacteria or slime mold) that are attracted by a self-emitted chemical substance [108]. Let $\rho(x, t)$ denote the population density of a cell colony on a two-dimensional surface subject to Brownian motion, and assume these cells have an additional drift velocity due to their tendency to move towards higher concentrations of a chemical attractant $c(x, t)$. The cells themselves are continually emitting this chemical attractant, and it diffuses across the surface and decays with rate α. This leads to the following system of parabolic equations:

$$\begin{cases} \rho_t = \Delta\rho - \nabla \cdot (\rho\nabla c), \\ \varepsilon c_t = \Delta c - \alpha c + \rho. \end{cases} \tag{7}$$

In fact, the same equation was also proposed by Patlak, as a mathematical model for random walk with persistence and external bias [128].

Since the chemical attractant reaches its equilibrium much faster than the bacteria density, it is common to take $\varepsilon \to 0$, simplifying the second equation so that (7) becomes a parabolic-elliptic system. This simplification allows us to write $c = -W * \rho$, where W is the Newtonian potential (if $\alpha = 0$) or the Bessel potential (if $\alpha > 0$). Finally, substituting c into the first equation gives us a single equation for $\rho(x, t)$ in the form of (4) with $m = 1$. Variations of the Keller–Segel model, such as the consideration of volume effects, lead to a range of aggregation-diffusion equations of the form (4), see [40].

We begin in Sect. 2 by reviewing analytical results for aggregation-diffusion equations, focusing on conditions that ensure solutions are globally well-posed or blow-up in finite time. The majority of these analytical results consider interaction kernels of one of the following forms:

- Power-law kernel:

$$W_k(x) = \begin{cases} \frac{|x|^k}{k} & \text{if } k \neq 0 \\ \ln|x| & \text{if } k = 0. \end{cases} \tag{8}$$

- Integrable kernel: $W \in L^1(\mathbb{R}^d)$.

In both cases, the associated free energy functional plays an important role in the study of well-posedness of solutions and properties of steady states. Formally taking time derivative along a solution, we deduce

$$\frac{d}{dt}E[\rho] = -\int_{\mathbb{R}^d} \rho \left| \nabla\left(\frac{m}{m-1}\rho^{m-1} + \rho * W\right) \right|^2 dx \leq 0.$$

This energy dissipation inequality (in the integral form) can be made rigorous for weak solutions and can also be seen as a consequence of the equation's underlying Wasserstein gradient flow structure. We begin, in Sect. 2.1, by considering the classical Keller–Segel equation in two dimensions, providing heuristic arguments illustrating the dichotomy between well-posedness and finite time blow-up, as well as summarizing rigorous results. We then discuss the case for more general aggregation diffusion equations in Sect. 2.2 and consider long time behavior of solutions in Sect. 2.3.

In Sect. 3, we consider singular limits of aggregation diffusion equations in two limiting regimes. We begin in Sect. 3.1 by considering the *slow diffusion limit*, as the diffusion exponent $m \to +\infty$, which leads to the constrained aggregation equation. Then, in Sect. 3.2, we discuss singular limits that affect the balance between aggregation and diffusion, causing solutions to exhibit metastable behavior. In Sect. 3.2.1, we consider the localized attraction limit, and in Sect. 3.2.2, we consider the vanishing diffusion limit.

In Sect. 4, we review recent work on numerical methods for aggregation diffusion equations, including the recent blob method for diffusion, developed by Carrillo et al. [55]. We conclude in Sect. 5 by applying this numerical method to illustrate properties of the singular limits discussed in Sect. 3.

2 Well-Posedness, Steady States, and Dynamics

We now give a brief overview of analytical results for aggregation diffusion equations (4), describing conditions that ensure well-posedness or finite-time blow-up of solutions , existence or non-existence of steady states, and long time behavior of solutions. We begin in Sect. 2.1 by considering the classical two-dimensional *Keller–Segel equation*, where the interaction kernel in Eq. (4) is given by the Newtonian potential, $W(x) = \frac{1}{2\pi} \ln |x|$. The analysis in this particular case will serve as a guideline to understand the equation's behavior for more general interaction kernels. In Sect. 2.2, we review results classifying well-posedness and finite blow-up for general interaction kernels. Finally, in Sect. 2.3, we discuss the steady states and dynamics of solutions in the diffusion-dominated regime.

2.1 Keller–Segel Equation with Linear Diffusion in $\mathbb{R}^2$

We begin with the classical Keller–Segel equation,

$$\rho_t = \Delta \rho + \nabla \cdot \left(\rho \nabla \left(\frac{1}{2\pi} \ln |x| * \rho \right) \right) \quad \text{in } \mathbb{R}^2 \times [0, T). \tag{9}$$

As the primary interest of the present work is aggregation diffusion equations of the form (4), we will not discuss the vast literature on the parabolic-parabolic Keller–Segel equation; see instead [15, 98].

A fundamental property of the Keller–Segel equation is that solutions are subject to a remarkable dichotomy depending on their mass $M = \int \rho(x)dx$. In particular, solutions exist globally in time if $M > 8\pi$, whereas they blow up in finite time when $M < 8\pi$. The fact that the critical value of the mass is 8π can be seen from the following formal scaling argument, which we will generalize to a range of interaction kernels W in the following Sect. 2.2.

As described in the introduction, the Keller–Segel equation (9) is formally the Wasserstein gradient flow of the energy

$$E[\rho] = \mathscr{S}_m[\rho] + \mathscr{W}[\rho] = \int_{\mathbb{R}^2} \rho \log \rho dx + \frac{1}{4\pi} \int_{\mathbb{R}^2} \rho(\ln |x| * \rho) dx. \qquad (10)$$

In particular, solutions of the Keller–Segel equation are characterized by the fact that they move in the direction of steepest descent of the energy (10). Given $\rho \in L^1 \cap L^\infty(\mathbb{R}^d)$, consider the following dilation of ρ, which concentrates the density while preserving its mass:

$$\rho_\lambda(x) := \lambda^2 \rho(\lambda x), \quad \lambda \gg 1.$$

Then, $E[\rho_\lambda]$ and $E[\rho]$ are related in the following way:

$$E[\rho_\lambda] = \mathscr{S}_m[\rho_\lambda] + \mathscr{W}[\rho_\lambda] = (\mathscr{S}_m[\rho] + 2M \log \lambda) + \left(\mathscr{W}[\rho] - \frac{M^2 \log \lambda}{4\pi} \right)$$

$$= E[\rho] + M \log \lambda \left(2 - \frac{M}{4\pi} \right).$$

Consequently,

$$\lim_{\lambda \to \infty} E[\rho_\lambda] = \begin{cases} +\infty & \text{when } M < 8\pi, \\ -\infty & \text{when } M > 8\pi. \end{cases}$$

Thus, when $M < 8\pi$, it is not energy favorable for the solution to concentrate to a δ-function, so one expects global well-posedness. On the other hand, when $M > 8\pi$, it is possible for the solution to blow up in finite time.

Motivated by this formal argument, we now summarize the rigorous results.

Subcritical Mass, $M < 8\pi$
Global well-posedness for subcritical mass was obtained by Dolbeault and Perthame [86] and Blanchet et al. [31], improving an earlier result of Jäger and Luckhaus [104], which showed global existence for smaller mass. The key idea behind this approach is to use the logarithmic Hardy-Littlewood-Sobolev inequality [48]

$$\int_{\mathbb{R}^2} \rho \log \rho dx + \frac{2}{M} \int_{\mathbb{R}^2} \rho(\rho * \ln |x|) dx \geq -C(M). \qquad (11)$$

Combining (11) with the fact that $M < 8\pi$ and the energy decreases along solutions, $E[\rho(t)] \leq E[\rho_0]$, one can obtain an a priori upper bound of the entropy $\int \rho \log \rho dx$. This uniform in time equi-integrability allows one to use iterative arguments to upgrade the bound to a uniform L^∞ bound, ensuring global well-posedness.

In addition to global well-posedness, long-time behavior for solutions with subcritical mass has also been well studied. Blanchet et al. [31] showed that the solution will indeed follow the scaling of the heat equation as $t \to \infty$, converging to a self-similar profile in the rescaled variables. Furthermore, the rate of convergence is exponential [30, 44, 90].

Supercritical Mass, $M > 8\pi$
In the case of supercritical mass, any initial data $\rho_0 \in L^1_+((1 + |x|^2)dx)$ will experience finite-time blow-up [31, 86]. This was proved by Blanchet, Dolbeault, and Perthame by tracking the evolution of the second moment $M_2[\rho(t)] := \int \rho(x, t)|x|^2 dx$. During the existence of the solution, one has

$$\frac{d}{dt}M_2[\rho(t)] = 4M\left(1 - \frac{M}{8\pi}\right),$$

so that the second moment becomes negative in finite time, indicating that the solution must break down by this time. Note that the above virial argument does not give the nature of the blow-up. Subsequent studies showed that solutions must concentrate at least 8π of their mass into a single point at the blow-up time [12, 135], as well as considering finer properties of the blow-up profile [97, 131, 134].

Critical Mass, $M = 8\pi$
At the critical mass, the free energy functional $E[\rho]$ is invariant under dilations, and there exists a one-parameter family of steady states $\bar{\rho}_\lambda(x) := \frac{M}{\pi}\frac{\lambda}{(\lambda+|x|^2)^2}$, all with infinite second moment, which are the unique global minimizers of the free energy, up to a translation. While solutions of the Keller–Segel equation exist globally in time [29], solutions with finite initial second moment exhibit an infinite time aggregation that is stable under perturbations [92]. On the other hand, if the initial data has infinite second moment and is sufficiently close to a stationary solution $\bar{\rho}_\lambda$, then it converges to $\bar{\rho}_\lambda$, with quantitative rate [27, 47].

2.2 Well-Posedness vs. Blow-Up for General Interaction Kernels

We now describe how the same dichotomy that separates global well-posedness from finite time blow-up for solutions of the classical Keller–Segel equation can be adapted to general aggregation diffusion equations (4), in arbitrary dimensions. We begin by considering the case of attractive power-law kernels W_k of the form (8). Since blow-up occurs locally near a specific point, the regimes of global well-posedness vs. finite time blow-up are primarily determined by the singularity of interaction kernel near the origin. We conclude by discussing results for more general interaction kernels.

As in the Keller–Segel equation case, let us again consider the behavior of the free energy functional $E[\rho]$ under mass-preserving dilation $\rho_\lambda(x) = \lambda^d \rho(\lambda x)$. As before, both the entropy $\mathscr{S}_m[\rho]$ and the interaction energy $\mathscr{W}_k[\rho]$ possess simple homogeneity properties under this transformation:

$$\mathscr{S}_m[\rho_\lambda] = \begin{cases} \lambda^{(m-1)d}\mathscr{S}_m[\rho] & \text{if } m > 1, \\ \mathscr{S}_m[\rho] + dM \log \lambda & \text{if } m = 1, \end{cases} \tag{12}$$

and

$$\mathscr{W}_k[\rho_\lambda] = \begin{cases} \lambda^{-k}\mathscr{W}_k[\rho] & \text{if } k \neq 0, \\ \mathscr{W}_k[\rho] - M^2 \log \lambda & \text{if } k = 0. \end{cases} \tag{13}$$

Comparing the scaling properties of $\mathscr{S}_m$ and $\mathscr{W}_k$ for different values of diffusion exponent m and interaction range k motivates the consideration of three regimes, separated by the critical diffusion exponent $m_c = 1 - k/d$. We now summarize what is known in each regime, regarding well-posedness vs. blow-up.

Diffusion Dominated Regime, $m > m_c$

In the diffusion dominated regime, Calvez and Carrillo [40] and Sugiyama [137] considered the Newtonian potential in arbitrary dimensions $d \geq 3$, proving global well-posedness and a uniform-in-time L^∞ bound for any initial data $\rho_0 \in L^1 \cap L^\infty(\mathbb{R}^d)$. Bedrossian, Rodríguez, and Bertozzi generalized this result to general power-law kernels that are no more singular than Newtonian, $2 - d \leq k \leq 0$ [13, 14]. For kernels that are more singular than Newtonian, $-d < k < 2 - d$, Zhang recently proved that solutions remain globally bounded if either $k > 1 - d$ or $m < 2$ [146].

Aggregation Dominated Regime, $1 \leq m < m_c$

In the aggregation-dominated regime, most results consider the case when the interaction potential is Newtonian, $k = 2 - d$. Sugiyama proved that for arbitrarily small mass $M > 0$, there exist solutions with mass M that blow up in finite time [137] . On the other hand, Bedrossian showed that, if the initial data is sufficiently flat (even if the mass is large), solutions exist globally, and dissipate with porous medium equation scaling as $t \to \infty$ [10].

Subsequently, Bian and Liu showed that solutions with small $L^p(\mathbb{R}^d)$ norm, $p = \frac{d(2-m)}{2}$, exist globally in time, where p is chosen so that diffusion and aggregation are balanced under the scaling that keeps the L^p norm invariant [25]. The special case $m = \frac{2d}{2+d}$ was studied further by Chen, Liu, and Wang, who showed that solutions with small L^p norm exist globally, whereas solutions with large L^p norm must blow up in finite time [71]. Chen and Wang then generalized this result to all $\frac{2d}{2+d} < m < 2 - \frac{2}{d}$ [72]. Finally, for general interaction potentials that are no more singular than Newtonian, $2 - d \leq k \leq 0$, Bedrossian, Rodríguez, and Bertozzi showed that there exist solutions with arbitrarily small mass that blow up in finite time [14].

Fair Competition Regime, $m = m_c$

In the fair competition regime, aggregation diffusion equations exhibit a dichotomy in terms of the mass of solutions that is similar to the classical Keller–Segel equation in $\mathbb{R}^2$. In this case, the Hardy-Littlewood-Sobolev inequality with optimal constant $C(k, d)$,

$$\iint_{\mathbb{R}^d \times \mathbb{R}^d} \rho(x)\rho(y)W_k(x-y)dxdy \le C(k,d)M^{2-m_c}\|\rho\|_{m_c}^{m_c}$$

plays the same role as the logarithmic Hardy-Littlewood-Sobolev inequality (11) for the Keller–Segel equation. In particular, combining this inequality with the fact that the free energy functional decreases along solutions gives a critical value of the mass $M_c = M_c(k,d)$, for which one has an a priori bound on $\|\rho\|_{m_c}$ if $M < M_c$, which then leads to uniform in time L^∞ bound. The results on well-posedness vs. blow-up can be summarized as follows:

- **Subcritical Mass**, $M < M_c$. Blanchet, Carrillo, and Laurençot proved that, for Newtonian interaction $k = 2-d$, solutions remain bounded globally in time [28], and there exist self-similar solutions decaying in time with the porous medium equation. Bedrossian, Rodriguez, and Bertozzi generalized this result to power-law kernels no more singular than Newtonian, $2 - d \le k \le 0$. Furthermore, Bedrossian proved that, under certain conditions of the initial data, solutions decay to the self-similar spreading solutions of the porous medium equation with an explicit convergence rate [10].
- **Supercritical Mass**, $M > M_c$. For kernels that are no more singular than Newtonian, there exist solutions with mass M that blow up in finite time [14, 28]. Bedrossian and Kim showed that, in the Newtonian case $k = 2-d$, all radial solutions blow up in finite time [11]. However, it remains unknown whether all non-radial solutions must also blow up.
- **Critical Mass** $M = M_c$. In the Newtonian case, Blanchet, Carrillo, and Laurençot showed that solutions are globally well-posed, and their L^∞ norm is globally bounded in time [28]. Furthermore, there is a family of compactly supported stationary solutions that are dilations of each other and all of which are global minimizers of the free energy. Calvez, Carrillo, and Hoffmann extended the latter result to general power-law interaction potentials $-d < k < 0$ [41, 42]. Finally, in the Newtonian case, Yao showed that every radial solution with compact support will converge to some stationary solution in this family [144], though the asymptotic behavior of non-radial solutions remains unclear.

General Interaction Potentials

We close this section by discussing what is known about well-posedness of aggregation diffusion equations for general interaction potentials W, which are not necessarily of power-law form. If the interaction potential is purely attractive, Bedrossian, Rodríguez, and Bertozzi fully develop the well-posedness theory for kernels with singularity up to and including Newtonian [13, 14]. On the other hand, if the interaction potential is a repulsive-attractive power-law potential (2) for $A \ge 2 - d$, Carrillo and Wang prove a global-in-time L^∞ bound for all $m \ge 1$, under the assumption that solutions exist locally in time [70].

General well-posedness results were recently obtained by Craig and Topaloglu, without any assumptions on the specific structure of the attractive and repulsive parts of the interaction potential [82]. In particular, under weak regularity assumptions on the interaction potential, which roughly correspond to being no more singular than the Newtonian potential, they prove that solutions of aggregation diffusion equations exist and provide sufficient conditions for global in time uniqueness. In particular, this extends the existence theory for power-law interaction kernels $W_k(x) = |x|^k/k$ to include unbounded initial data, as long as it belongs to domain of the energy, $\rho_0 \in D(E)$, as well as obtaining existence for repulsive-attractive power-law potentials (2), $2 \geq A > B \geq 2 - d$, complementing previous work by Carrillo and Wang.

2.3 Steady States and Dynamics in the Diffusion-Dominated Regime

In this subsection, we restrict our attention to the diffusion-dominated regime, where global well-posedness is known. The following questions naturally arise: Are there stationary solutions for all masses? And if so, are they unique up to translations? Do they determine the long time asymptotics of solutions? In this section, we review the existing work on these questions. We begin, in Sect. 2.3.1, by showing how the same dilation scaling argument that separated global well-posedness from finite time blow-up in the diffusion vs. aggregation dominated regimes leads to a second dichotomy in terms of existence of global minimizers. For kernels that grow at least as fast as a power-law W_k at infinity, global minimizers exist for m in the diffusion dominated regime. On the other hand, for *globally integrable* interaction kernels, the value of m does not depend on the specific structure of the kernel at hand, and the critical diffusion exponent $m_c = 2$ separates existence from nonexistence of global minimizers. In Sect. 2.3.2 we summarize the precise statements of these results, and in Sect. 2.3.3, we discuss in which regimes it is known that solutions of aggregation diffusion equations asymptotically converge to these steady states.

2.3.1 A Formal Scaling Argument: Power-Law Kernels vs. Integrable Kernels

Due to the gradient flow structure of aggregation diffusion equations with respect to the energy $E[\rho(t)]$, a natural approach for studying long time behavior of solutions is to begin by finding global minimizers of the energy. As explained in Sect. 2.2, considering the energy's scaling under dilations $\rho_\lambda(x) = \lambda^d \rho(\lambda x)$ reveals that in the diffusion-dominated regime, it is not energy favorable for the density to become concentrated. Consequently, if the energy lacks a global minimizer for a given mass, it must be due to lack of compactness—that is, there is a minimizing sequence $\{\rho_n\}_{n \in \mathbb{N}}$ which is not tight, up to translation.

As in the case for well-posedness vs. blow-up, considering the scaling of the energy under dilations reveals a dichotomy separating integrable kernels and kernels that grow like a power-law at infinity. In the pure power-law case W_k, $-d < k \leq 0$, Eqs. (12) and (13) for the scaling of the energy show that, when $m > m_c := 1 -$

k/d, $E[\rho_\lambda]$ is increasing as $\lambda \to 0^+$. In other words, it is not energy favorable for the mass to spread to infinity. Thus, for power-law kernels W_k in the diffusion dominated regime, we expect a global minimizer for any given mass. Furthermore, if the kernel grows faster than a power law W_k at infinity, then the long range attractive interactions are even stronger, and a similar scaling argument immediately yields that, for any $m \geq 1$, we again expect a global minimizer for any mass.

On the other hand, if W is a globally integrable kernel, the interaction energy $\mathcal{W}[\rho]$ scales differently. In particular, we obtain

$$\mathcal{W}[\rho_\lambda] = \frac{\lambda^d}{2} \int_{\mathbb{R}^d} \rho(\rho * \tilde{W}_\lambda) dx, \quad \tilde{W}_\lambda(x) = \lambda^{-d} W(\lambda^{-1} x).$$

If W is integrable, then as $\lambda \to 0$, $\tilde{W}_\lambda$ approaches a Dirac mass in distribution, and for bounded and continuous densities ρ, $\lim_{\lambda \to 0^+} \lambda^{-d} \mathcal{W}[\rho_\lambda] = \frac{1}{2} \left(\int W dx \right) \int \rho^2 dx$. As a result, for $0 < \lambda \ll 1$, we obtain

$$\mathcal{W}[\rho_\lambda] = \frac{\lambda^d}{2} \left(\int_{\mathbb{R}^d} W dx \right) \int_{\mathbb{R}^d} \rho^2 dx + o(\lambda^d).$$

Comparing this with the scaling for $\mathcal{S}_m[\rho_\lambda]$ from Eq. (12), we formally obtain that $m = 2$ is the critical power separating the energies for which it is favorable ($m < 2$) vs. unfavorable ($m > 2$) for the mass to spread to infinity. Thus, we expect this critical exponent to also determine non-existence vs. existence of global minimizers.

At the critical power $m = 2$, the balance between diffusion and aggregation is more delicate. In this case,

$$E[\rho_\lambda] = \lambda^d \int_{\mathbb{R}^d} \rho^2 dx \left(1 + \frac{1}{2} \int_{\mathbb{R}^d} W dx \right) + o(\lambda^d).$$

Therefore, if $\int_{\mathbb{R}^d} W dx < -2$, we formally expect existence of global minimizers, whereas $\int_{\mathbb{R}^d} W dx > -2$ leads to non-existence.

2.3.2 Existence/Non-existence of Global Minimizers

We now turn to the precise statements of what is known concerning existence vs. non-existence of global minimizers. Note that once one obtains existence of a global minimizer, applying Riesz's rearrangement inequality directly yields that such global minimizer must be radially decreasing for all purely attractive kernels.

If the interaction potential is a power-law W_k, $-d < k \leq 0$, and m is in the diffusion-dominated regime, $m > m_c := 1 - k/d$, then for any mass M, there exists a global minimizer for the energy $E[\rho]$ in the class

$$\mathcal{Y} := \{\rho \in L^1_+(\mathbb{R}^d) \cap L^m(\mathbb{R}^d), \|\rho\|_1 = M, \int_{\mathbb{R}^d} x\rho(x) = 0\}.$$

Lions was the first to show this, in the case the interaction is potential is Newtonian and $d \geq 3$, via a concentration compactness argument [115, 116]. Subsequently, Bedrossian generalized the result to all purely attractive potentials no more singular than Newtonian potential [9]. Carrillo, Castorina, and Volzone extended the result to the Newtonian potential in two dimensions [50]. For kernels more singular than Newtonian, recent work by Carrillo, Hoffmann, Mainini, and Volzone showed existence of global minimizers for power law kernels W_k with $-d < k < 2 - d$. Furthermore, the same work proved that, for all power law kernels $-d < k \leq 0$, global minimizers are compactly supported [64].

If W is an integrable, bounded, and purely attractive kernel, Bedrossian showed that, for all $m > 2$ and any mass M, there exists a global minimizer of (5) [9]. At the critical power $m = 2$, existence vs. non-existence of global minimizers depends on the value of $\int W dx$. Bedrossian and Burger, Di Francesco, and Franek showed that, for any mass, a global minimizer of (5) exists if and only if $\int W dx < -2$ [9, 35]. Subsequently, Kaib showed that for $1 < m < 2$, there is a compactly supported global minimizer if $-\int W dx$ is sufficiently large [107]. More recently, Carrillo, Delgadino, and Patacchini showed that, for $m = 1$, there is no steady state (thus no global minimizer) for any bounded interaction kernel [58].

We conclude by observing that, in all the above results where W is no more singular than the Newtonian potential at the origin, whenever a global minimizer for a given mass is known to exist, standard arguments ensure that it must be a steady state in the weak sense. For power-law kernels more singular than Newtonian, in the diffusion-dominated regime, Carrillo, Hoffmann, Mainini, and Volzone showed that global minimizers are steady states for all $1 - d \leq k < 2 - d$; for more singular kernels, $-d < k < 1 - d$, they also show that, under additional constraints on m, global minimizers have sufficient regularity to be steady states [64].

2.3.3 Steady States and Dynamics

Once existence of a global minimizer is known, it is natural to ask whether it is a global attractor of the evolution equation. Note that a necessary condition for this to be true is that steady states are unique, up to a translation. Compared to the existence theory, much less is known about uniqueness of steady states. Using continuous Steiner symmetrization techniques, Carrillo, Hittmeir, Volzone, and Yao showed that, for purely attractive kernels, all stationary solutions in $L_+^1 \cap L^\infty(\mathbb{R}^d)$ are radially decreasing [63]. Therefore, if the kernel is attractive, it suffices to study the uniqueness question among the radial class of functions.

In the particular case of the Newtonian interaction potential, uniqueness of steady states among radial functions was first shown by Lieb and Yau [114], using the specific structure of the Newtonian kernel to obtain an explicit ordinary differential equation for the mass distribution function of solutions. For general power-law kernels in the diffusion-dominated regime, uniqueness of steady states is known when $d = 1$ [64] but remains unclear for higher dimensions, aside from the

Newtonian case. For integrable, purely attractive kernels, Kaib and Burger, Di Francesco, and Franek proved uniqueness of steady states at the critical power $m = 2$ when $\int W dx < -2$ using the Krein-Rutman theorem, under the additional assumptions that W is smooth and $W'(r)$ vanishes only at 0 [35, 107]. Aside from these special cases, the uniqueness of steady states remains open.

We close by reviewing what is known about the dynamics of solutions in the diffusion-dominated regime—in particular, when solutions converge to the previously described steady states and global minimizers. Formally, if uniqueness of steady states, up to translation, is known for a given mass, one would expect that any solution would converge to a translation of it. On the other hand, if there are no global minimizers, one would expect the solution to spread to infinity. However, in practice, long-time asymptotics are only known in special cases. Clarifying the precise conditions on the interaction kernel that lead to convergence towards equilibrium or spreading to infinity is a challenging open problem related to sharp conditions on the confinement of the mass.

For the Newtonian potential in two dimensions in the diffusion-dominated regime $m > 1$, Carrillo, Hittmeir, Volzone, and Yao showed that every solution with finite initial second moment must converge to a (unique) stationary solution, which has the same mass and center of mass as the initial data [63]. The proof relies on a global-in-time bound of the second moment, leveraging the structure of the two-dimensional Newtonian potential. When $d \geq 3$, although the uniqueness of steady states is also known [63], without any uniform-in-time mass confinement bounds, it is unclear how to show that the mass cannot escape to infinity as $t \to \infty$. However, in the case of radial solutions, Kim and Yao succeeded in showing that that all radial solutions converge to the global minimizer exponentially fast, by building sub/super solutions for the mass concentration functions [111].

For an integrable, purely attractive kernel, at the critical power $m = 2$, Di Francesco and Jaafra obtained the following results on the asymptotic behavior in dimension one [85]. If $\int_{\mathbb{R}} W dx < -2$, then the (unique) steady state is locally asymptotically stable in 2-Wasserstein distance, and if $\int_{\mathbb{R}} W dx > -2$ (where non-existence of global minimizer is known), all solutions with finite energy must decay to zero locally in L^2 and almost everywhere in $\mathbb{R}$ as $t \to \infty$. When $m \neq 2$, also in dimension one, numerical results by Burger, Fetecau, and Huang suggest that solutions are attracted to the global minimizer when $m > 2$ (with some coarsening and metastability, see next section), and when $1 < m < 2$, the steady states (if any) have a limited basin of attraction [36].

3 Singular Limits

We now turn from the well-posedness theory and long-time behavior discussed in the previous sections to consider dynamics of solutions of aggregation diffusion equations in two limiting regimes. In Sect. 3.1, we consider the behavior of solutions in the *slow diffusion limit*, sending the diffusion exponent $m \to +\infty$.

In this case, the effect of diffusion transforms into a hard height constraint on the density, leading to the *constrained aggregation equation*. In Sect. 3.2, we consider singular limits that affect the balance between aggregation and diffusion, leading to metastable behavior, as solutions spend long time scales converging towards a local equilibrium, before quickly transitioning to converge to a global equilibrium.

3.1 Constrained Aggregation and the Slow Diffusion Limit

We begin by considering the behavior of solutions of aggregation diffusion equations as the diffusion exponent $m \to +\infty$. Formally, this leads to an aggregation equation with a height constraint $\rho \leq 1$ on the density,

$$\begin{cases} \rho_t = \nabla \cdot (\rho \nabla(\rho * W)) & \text{if } \rho < 1, \\ \text{``}\rho \leq 1 \text{ always.''} \end{cases} \tag{14}$$

This relation between the above constrained aggregation equation and aggregation diffusion equations can be formally understood by noting that the diffusion exponent $m \geq 1$ controls the strength of diffusion at different heights of the density. Since

$$\lim_{m \to \infty} \rho^m = \begin{cases} 0 & \text{for } \rho < 1, \\ +\infty & \text{for } \rho > 1, \end{cases}$$

in the $m \to \infty$ limit, the diffusion term $\Delta \rho^m$ vanishes in the set $\{\rho < 1\}$, whereas it becomes strong enough in the set $\{\rho > 1\}$ to prevent the density from growing above height 1. In what follows, we will clarify the sense in which we enforce the height constraint. (See Fig. 1 for a numerical illustration of how degenerate diffusion with m large approximates a height constraint.)

For various partial differential equations with a degenerate diffusion term $\Delta \rho^m$, it is well known that the asymptotic limit $m \to \infty$ imposes a height constraint $\rho \leq 1$, and the evolution of the free boundary $\partial\{\rho = 1\}$ is often determined by a Hele–Shaw type free boundary problem. In particular, for the porous medium equation $\rho_t = \Delta \rho^m$, this is known as the Mesa Problem, and it has been proved that the solution pair (ρ_m, p_m) (where $p_m := \frac{m}{m-1} \rho^{m-1}$ is the pressure) converges as $m \to \infty$ to a solution of the Hele–Shaw problem [39, 93, 94]. More recently, the analogous limit has been considered for the porous medium equation with growth $\rho_t = \Delta \rho^m + \rho \Phi(p_m)$, where the limiting equation describes the growth of a tumor with a restriction on the maximal cell density [109, 123, 129, 130].

Another viewpoint for understanding the $m \to \infty$ limit in the case of aggregation diffusion equations is through the gradient flow formulation. Recall that the aggregation diffusion equation (4) is formally the Wasserstein gradient flow of the free energy functional

$$E_m[\rho] = \frac{1}{m-1} \int_{\mathbb{R}^d} \rho^m dx + \frac{1}{2} \int_{\mathbb{R}^d} \rho(\rho * W) dx =: \mathscr{S}_m[\rho] + \mathscr{W}[\rho].$$

For any fixed ρ, it is easy to check that $\lim_{m \to \infty} E_m[\rho] = E_\infty[\rho]$, which is given by the constrained interaction energy

$$E_\infty[\rho] = \begin{cases} \frac{1}{2} \int_{\mathbb{R}^d} \rho(\rho * W) dx & \text{if } \|\rho\|_{L^\infty} \leq 1 \\ +\infty & \text{otherwise.} \end{cases}$$

Thus we formally expect the slow diffusion limit of aggregation diffusion equations should correspond to the Wasserstein gradient flow of $E_\infty[\rho]$, which is indeed the constrained aggregation equation (14). In fact, this is the precise sense in which we impose the height constraint "$\rho \leq 1$ always."

In the case when the nonlocal interaction energy $\frac{1}{2} \int_{\mathbb{R}^d} \rho(\rho * W) dx$ in E_∞ is replaced by a local potential energy $\int \rho(x) V(x) dx$, with a λ-convex potential $V(x)$, the gradient flow for E_∞ was introduced by Maury, Roudneff-Chupin, Santambrogio, and Venel as a model for pedestrian crowd motion [121, 122]. They showed that this gradient flow satisfies the transport equation $\rho_t + \nabla \cdot (\rho \mathbf{v}) = 0$ in the weak sense, where the velocity field $\mathbf{v}(\cdot, t)$ is given by the L^2 projection of ∇V onto the set of admissible velocities that do not increase the density in the saturated zone $\{\rho(\cdot, t) = 1\}$. Building upon this work, Alexander, Kim, and Yao showed that the gradient flow of E_∞ (with local potential V) can be approximated by the gradient flows of E_m, i.e. nonlinear Fokker-Planck equations, as $m \to \infty$, proving convergence in the Wasserstein distance with an explicit rate depending on m [1].

Building upon this work in the local case, Craig, Kim, and Yao studied the gradient flow of the constrained interaction energy E_∞, when W is an attractive Newtonian interaction potential [80]. Craig showed that the energy E_∞ is ω-convex, ensuring that its gradient flow is well-posed and can be quantitatively approximated by the discrete time Jordan-Kinderlehrer-Otto scheme [78, 106]. Craig, Kim, and Yao then showed that a patch solution of the constrained aggregation equation (where the initial data is given by a characteristic function) remains a patch for all time, and the evolution of the patch boundary satisfies a Hele-Shaw type free boundary problem. In addition, in two dimensions, these patch solutions converge to a characteristic function of a disk in the long-time limit. A key element in the proof was the approximation of the constrained aggregation equation by a sequence of nonlinear Fokker-Planck equations in the slow diffusion limit, where the local potential V depended on the choice of initial data. However, for nonconvex interaction potentials, including the Newtonian interaction potential W under consideration, convergence of aggregation diffusion equations to the constrained aggregation equation as $m \to +\infty$ remained open.

More recently, Craig and Topaloglu [82] proved that minimizers and gradient flows of E_m do converge to minimizers and gradient flows of E_∞ in the $m \to \infty$ limit for a general class of interaction potentials W, including both attractive and repulsive-attractive power-law potentials with singularity up to and including the Newtonian potential. This work was largely inspired by the following related question in geometric shape optimization: can we characterize *set-valued* or *patch* minimizers E_∞, i.e. minimizers which are of the form $\rho = 1_\Omega$ for $\Omega \subseteq \mathbb{R}^d$ of volume M? When the interaction potential is a repulsive-attractive power law (2), competition between the attraction parameter A and the repulsion parameter B determines existence, nonexistence, and qualitative properties of minimizers, providing a counterpoint to the well-studied nonlocal isoperimetric problem [74]. Due to the fact that solutions of aggregation diffusion equations approximate the constrained aggregation equation in the slow diffusion limit, numerical simulations of aggregation diffusion equations for large m can be used to explore qualitative properties of set-values minimizers of E_∞. (See Figs. 2 and 4 for numerical simulations illustrating critical mass behavior.)

Several open questions remain for the height constrained aggregation equation, stemming from the fact that posing the equation as a Wasserstein gradient flow offers only a very weak notion of solution. In the particular case of the attractive Newtonian interaction potential with patch initial data, Craig, Kim, and Yao's result provides a true characterization in terms of a partial differential equation [80]. It remains open whether this result can be extended to more general initial data, building on recent work in the local case [109, 110, 123]. An obstacle is the very low regularity of solutions and the variety of possible merging phenomena that may occur as the solution interacts with the height constraint. (See Fig. 3 for a numerical illustration of possible merging behavior.) Likewise, it is unknown how to extend this result to more general interaction potentials.

3.2 Singular Limits and Metastability

As a counterpoint to the previous section, which considered singular limits under which the effect of diffusion transforms into a height constraint, in the section, we consider singular limits affecting the balance between aggregation and diffusion, leading to various types of metastability behavior. In the absence of diffusion, much is known about the long-time behavior of solutions of the aggregation equation,

$$\rho_t = \nabla \cdot (\rho \nabla (\rho * W)), \tag{15}$$

for both purely attractive interaction potentials W and interaction potentials that are repulsive at short length scales and attractive at longer length scales. Solutions approach compactly supported stationary states, and the regularity of these equilibria is determined by the repulsive strength of the interaction potential at the origin [6, 37, 57, 61]. Furthermore, solutions may experience either finite or infinite time blow-up, concentrating at Dirac masses or on other lower dimensional sets [19–23, 59].

A natural question is how robust this behavior is under different perturbations, and in particular, how asymptotics are affected by the addition of linear or nonlinear diffusion. When are the above described stationary states preserved? And if steady states exist, how stable are they?

The appearance of metastable behavior in nonlinear aggregation-diffusion equations is apparent from numerical simulations when the interactions have nearly finite radius of perception, i.e. when the interaction kernel decays quickly at $+\infty$ [36, 51, 56, 89]. Moreover, for compactly supported attractive or repulsive-attractive interaction potentials with nonlinear diffusions, clustered solutions can form and the support of steady states can contain several disconnected components [2, 124]. Can the local effect of the nonlinear diffusion as repulsion lead to clusterization for fully attractive potentials? If the fully attractive potential satisfies $W'(r) > 0$ everywhere, then we know that all steady states must be radially decreasing [63], so this effect cannot happen for large times. So, can this clusterization happen as a metastable behavior in the dynamics?

While these questions remain mostly open, we now describe two different directions of recent progress. First, we will consider metastability that appears for porous medium diffusion ($m = 2$) when the attraction potential is localized via dilation. Next, we will discuss metastability of aggregation diffusion equations for linear diffusion ($m = 1$) as the amount of diffusion vanishes, analogous to the well-known vanishing viscosity limit in classical fluids equations.

3.2.1 Metastability in the Localized Attraction Limit

We begin by considering the behavior of aggregation diffusion equations when the interaction potential localizes as $\delta \to 0$,

$$\rho_t = \nu \Delta \rho^m + \nabla \cdot (\rho \nabla(\rho * W_\delta)), \ m > 1, \ W_\delta(x) = \delta^{-d} W(\delta^{-1}x), \ 0 < \delta \ll 1. \tag{16}$$

For simplicity, we consider the case when W is a bounded, strictly attractive interaction kernel. The attraction of W_δ localizes as $\delta \to 0$, since the dilation decreases attractive forces in the long range and increases them in a δ-neighborhood.

For this class of equations, Carrilllo, Hittmeir, Volzone, and Yao showed that, if a stationary state exists, it must be radially decreasing [63]. However, numerical results illustrate that the dynamical solution develops into non-radially decreasing clusters which are approximate steady states, where the length scale of each cluster goes to zero as $\delta \to 0$. Eventually, these clusters coalesce into a single cluster, which ultimately converges to a radially decreasing steady state; see Figs. 5 and 6. Similar metastability phenomena have been observed for fixed $\delta > 0$ in several related numerical studies [36, 51, 56]. The structure, duration, and existence of metastable states is strongly dependent on the smoothness of the interaction kernel W and the size of the diffusion coefficient ν. In particular, Fig. 7 illustrates how increasing the diffusion exponent can prevent the formation of steady states, and

Figs. 9 and 10 illustrate that a singularity in the interaction kernel W can also inhibit their formation.

Little is known about how the time scales of this metastability behavior relate to the localization parameter δ, the diffusion exponent $m > 1$, the diffusion coefficient ν, and the choice of interaction potential W. However, in the special case of aggregation-diffusion equations on a bounded interval, with no-flux boundary conditions, one can show that, for initial data given by a constant function, the length scale of the bumps in the resulting metastable state can be computed by analyzing the instability of the constant solution.

3.2.2 Metastability and the Vanishing Diffusion Limit

A natural context in which to examine robustness of solutions of the aggregation equation (15) under perturbations is to consider behavior of aggregation diffusion equations, with a small amount of linear diffusion,

$$\rho_t = \nu \Delta \rho + \nabla \cdot (\rho \nabla (\rho * W)) \quad \text{in } \mathbb{R}^d. \tag{17}$$

On the one hand, for λ-convex interaction potentials W, classical Γ-convergence techniques yield that solutions of (17) converge to solutions of the inviscid aggregation equation (15) as $\nu \to 0$, on any bounded time interval $[0, T]$ [3, 136]. Furthermore, Cozzi, Gie, and Kelliher extended this result to the case when W is the attractive Newtonian potential, on any time interval $[0, T]$, as long as the corresponding inviscid solution remains well-defined [76].

On the other hand, the analogous convergence result fails in the long time limit for all smooth, bounded attractive kernel W. For such kernels, solutions of the inviscid aggregation equation (15) always concentrate to one or more Dirac masses as $t \to +\infty$: the solution converges to a single Dirac mass if ρ_0 is initially within the perception range of W, otherwise it may cluster into multiple delta functions [124]. However, Carrillo, Delgadino, and Patacchini recently showed that, for all such interaction kernels, Eq. (17) has no steady states for any $\nu > 0$ [58]; see Fig. 8. Consequently, there is no hope of proving convergence to any nonzero equilibria as $t \to \infty$ for any $\nu > 0$. (Note that this result on nonexistence of steady states strongly uses the boundedness of W and non-compactness of $\mathbb{R}^d$.)

Therefore, when the interaction potential W is bounded, Eq. (17) illustrates a typical example of non-commutative double limits: $\nu \to 0$ and $t \to \infty$. For small $\nu > 0$, we expect solutions to remain close to the equilibrium of the inviscid aggregation equation (15) for a large time interval $[0, T]$, but to eventually dissipate away. Consequently, numerical simulations of (17) for small $\nu > 0$ may be quite misleading: it may appear that solutions are converging towards a stationary state before eventually becoming evident that mass is spreading at an extremely slow rate. It is still unknown how this time scale depends on $\nu > 0$ and how solutions dynamically transition from near the inviscid steady states to then glue together and spread to $+\infty$ [38].

Related metastability behavior has also been observed for interaction kernels that are unbounded away from the origin. For example, Evers and Kolkolnikov discovered that, in the case of the repulsive-attractive interaction potential $W(x) = \frac{x^4}{4} - \frac{x^2}{2}$ in $d = 1$, the effect of the diffusive regularization in Eq. (17) is to produce an extremely slow equilibration of the weights between the two stable points of the dynamics. To see this, recall that for the inviscid aggregation equation (15), any convex combination of Dirac masses at distance one apart is a stationary solution (since $W'(x) = x^3 - x$ has a zero at $x = 1$), and the mass need not be evenly distributed. However, for all $\nu > 0$, solutions of the aggregation diffusion equation (17) seek to equilibrate mass. Numerical simulations indicate that there is a unique equilibrium for all $\nu > 0$, which converges as $\nu \to 0$ to two Dirac masses with equal weight; see Fig. 11. Still, proving uniqueness of equilibria for the $\nu > 0$ case and understanding the timescales on which these effects take place remains entirely open.

4 Numerical Methods

As described in the previous sections, two key themes in the analysis of aggregation diffusion equations are the gradient flow structure and the competition between local and nonlocal effects. The gradient flow structure provides Lyapunov functionals, stability estimates, and a natural framework for considering singular limits, as in the case of the vanishing diffusion limit or the slow diffusion to height constraint limit. The delicate balance between aggregation and diffusion equations leads to a range of asymptotic behavior, from global existence to finite time blow-up, and rich structure of equilibrium configurations.

The same key themes arise in the development of numerical methods for aggregation diffusion and height constrained aggregation equations. Due to the competition between aggregation and diffusion or aggregation and a height constraint, numerical methods must be able to cope with solutions with very low regularity and to perform well enough globally in time to accurately approximate equilibrium profiles. With regard to the Wasserstein gradient flow structure, numerical methods must at a minimum respect the space to which solutions belong—preserving positivity and conserving mass—as well as ideally reproducing the energy decreasing property and the rate of energy decrease.

One of the first numerical methods developed for aggregation diffusion equations directly leveraged the Wasserstein gradient flow structure, and in particular, a time discretization of the gradient flow problem due to Jordan et al. [106]. In analogy with the implicit Euler method for approximating gradient flows in Euclidean space, a Wasserstein gradient flow (6) with initial data $\rho_0 \in \mathscr{P}_2(\mathbb{R}^d)$ may be approximated by solving the sequence of infinite dimensional minimization problems,

$$\rho^n = \operatorname*{argmin}_{v \in \mathscr{P}_2(\mathbb{R}^d)} \left\{ \frac{1}{2(t/N)} d_2^2(v, \rho^{n-1}) + E(v) \right\}, \quad n = 1, \ldots, N. \tag{18}$$

Under weak regularity and coercivity assumptions on E, we have

$$\lim_{N \to +\infty} d_2(\rho^N, \rho(t)) = 0,$$

where $\rho(t)$ is the solution of the Wasserstein gradient flow at time t [3]. Furthermore, provided that the energy satisfies generalized convexity requirements, such as λ-convexity or ω-convexity, one may obtain quantitative estimates on the rate of convergence [3, 46, 77, 78].

A first benefit of using the JKO scheme numerically is that, by definition of the sequence, the energy is decreasing in ρ^n, which provides automatic stability estimates along the sequence. Furthermore, since the minimization problem is constrained over $\mathscr{P}_2(\mathbb{R}^d)$, this approach conserves mass and preserves positivity of solutions. More generally, the JKO scheme can be easily extended to any initial data with mass $\int \rho_0 = M > 0$ by replacing $\mathscr{P}_2(\mathbb{R}^d)$ with the set of finite measures with mass M and finite second moment. Finally, since this approach to numerical methods most closely respects the equation's underlying gradient flow structure, such methods are a natural choice for simulating singular limits of aggregation diffusion equations, including height constrained aggregation equations [17, 49].

As the JKO scheme (18) is already discrete in time, to apply it numerically, one simply needs to determine a method for discretizing the infinite dimensional minimization problem (18) in space. The key difficulty is computing the Wasserstein distance term accurately and efficiently. One of the first techniques took advantage of the Monge formulation of the Wasserstein term, discretizing the space of Lagrangian maps via finite difference [67, 88] or finite element approximations [69], or directly using a discretization of the Monge-Ampére operator [18]. A more recent approach is to leverage the Benamou-Brenier dynamical reformulation of the Wasserstein distance, which can then either be discretized via Benamou and Brenier's original ALG-2 method [16, 17] or via modern operator splitting techniques [56, 127]. Aside from these two main approaches, several other methods for discretizing the Wasserstein term have also been applied to simulate aggregation diffusion equations [26, 43, 96, 143], including, most recently, using the Kantorovich formulation of the Wasserstein distance, with entropic regularization to improve computational efficiency [49].

A second common approach to developing numerical methods for aggregation diffusion equations is to discretize the equations directly via classical Eulerian methods, based on similar schemes for related hyperbolic, kinetic, and degenerate parabolic equations. Leveraging the analogy with degenerate diffusion equations, Carrillo, Chertock, and Huang developed a finite volume method for aggregation diffusion equations [24, 51]. Building on this, recent work by Bailo, Carrillo, and Hu proposes an implicit in time fully discrete entropy decreasing finite volume schemes for general aggregation-diffusion equations [5]. Alternatively, using the analogy with hyperbolic conservation laws, Sun, Carrillo, and Shu proposed a finite element method, using a discontinuous Galerkin approach which gives higher order convergence to smooth solutions [138]. For the particular case of attractive Newtonian aggregation, Liu, Wang, and Zhou introduced a change of variables by which the equation shares structural similarities with Fokker-Planck equations and used this to develop an implicit in time finite difference scheme [118]. Each of the previous methods preserves positivity of solutions, and numerical experiments indicate they also succeed in dissipating energy, providing stability for numerical solutions and allowing the methods to accurately capture asymptotic behavior.

A third approach to numerical methods for aggregation diffusion equations, and the approach most closely related to the microscopic interacting particle system underlying the partial differential equation, is to consider particle approximations for aggregation diffusion equations. Much of the existing work in this direction has leveraged the structural similarity between aggregation diffusion equations and equations from classical fluids, particularly the Navier-Stokes equation in vorticity form [22, 79]. In the absence of diffusion, the simplest type of particle method for the aggregation equation

$$\rho_t = \nabla \cdot (\rho \nabla (\rho * W)) \quad \text{in } \mathbb{R}^d, \tag{19}$$

proceeds by discretizing the initial datum ρ_0 as a finite sum of N Dirac masses,

$$\rho_0 \approx \rho_0^N = \sum_{i=1}^{N} \delta_{x_i} m_i, \qquad x_i \in \mathbb{R}^d, \quad m_i \geq 0, \tag{20}$$

where δ_{x_i} is a Dirac mass centered at $x_i \in \mathbb{R}^d$, and then evolving the locations of the Dirac masses according to the velocity field from Eq. (19),

$$\rho^N(t) = \sum_{i=1}^{N} \delta_{x_i(t)} m_i, \quad \dot{x}_i = -\sum_{j \neq i} \nabla W(x_i - x_j) m_j. \tag{21}$$

For $W \in C^1(\mathbb{R}^d)$, the particle solution $\rho^N(t)$ is then a weak solution of the aggregation equation (15) with initial data ρ_0^N, and in particular, it is a Wasserstein gradient flow of the interaction energy $\mathscr{W}[\rho]$, so that the energy $\mathscr{W}[\rho]$ automatically decreases along $\rho_N(t)$. Furthermore, results on the mean-field limit for the aggregation equation (15) ensure that, for a range of λ-convex or power law interaction kernels W, as the approximation of the initial data improves, $\rho_0^N \xrightarrow{d_2} \rho_0$, the particle solution converges to the exact solution of the aggregation equation $\rho^N(t) \xrightarrow{d_2} \rho(t)$ on bounded time intervals [53, 59, 101]. The particle method (21) provides a semi-discrete numerical method, and in order to obtain a fully discrete scheme, one may use a variety of fast solvers to integrate the system of ODEs.

In order to develop methods with higher-order accuracy and capture competing effects in repulsive-attractive systems, recent work has considered enhancements of standard particle methods inspired by techniques from classical fluid dynamics, including *vortex blob methods* and *linearly transformed particle methods* [45, 79, 95]. Bertozzi and the second author's blob method for the aggregation equation obtained a higher order accurate method for singular interaction potentials W by convolving W with a mollifier $\varphi_\varepsilon(x) = \varphi(x/\varepsilon)/\varepsilon^d$, $\varepsilon > 0$. In terms of the Wasserstein gradient flow perspective this translates into regularizing the interaction energy $(1/2) \int \rho(W * \rho)\, dx$ as $(1/2) \int \rho(W * \varphi_\varepsilon * \rho)\, dx$.

To extend particle methods to aggregation *diffusion* equations, one has to confront a fundamental obstacle: unlike in the pure aggregation case, solutions of aggregation diffusion equations with particle initial data (20) do not remain particles. One way to address this issue is to simulate the aggregation and diffusion terms separately, via a splitting scheme, using a classical finite volume method for the diffusion term [145]. Another natural approach in the case of linear diffusion ($m = 1$) is to consider a stochastic particle method, in which Brownian motion is added to the differential equation for the motion of the particles (21) [99, 101, 103, 119]. The main practical disadvantages of such stochastic methods is that the simulations must be averaged over a large number of runs to compensate for the randomness of the approximation and such methods have not been extended to the case of degenerate diffusion $m > 1$.

More recently, two *deterministic* particle methods for aggregation diffusion equations have been introduced, inspired by classical particle-in-cell methods in fluid, kinetic, and plasma physics equations [73, 75, 84, 113, 117, 120, 132, 133]. The first, due to Carrillo, Huang, Patacchini, Sternberg, and Wolansky, approximates one-dimensional aggregation diffusion equations by discretizing the energy using non-overlapping balls centered at the particles [65, 68]. The second, due to Carrillo, Craig, and Patacchini, extends naturally to all dimensions $d \geq 1$ by considering a regularization of the energy similar to Craig and Bertozzi's blob method for the aggregation equation, leading to a *blob method for diffusion* [55].

We now turn to a precise description of this blob method for diffusion, which we apply in the next section to generate our numerical examples. Given a Gaussian mollifier $\varphi_\varepsilon(x)$, the regularized energy is defined by

$$E_\varepsilon(\rho) = \int \frac{(\varphi_\varepsilon * \rho)^{m-1}}{m - 1}\, d\rho + \frac{1}{2} \int (W * \rho)\rho.$$

Unlike gradient flows of the original aggregation diffusion energy $E(\rho)$ defined in Eq. (5), gradient flows of $E_\varepsilon(\rho)$ for $\varepsilon > 0$ with particle initial data remain particles for all time, and the evolution of the particles is determined by the following system of ordinary differential equations:

$$\dot{x}_i = -\sum_{j=1}^{N} \nabla W(x_i - x_j)m_j + \sum_{j=1}^{N} \nabla \varphi_\varepsilon(x_i - x_j)m_j \left(\sum_{k=1}^{N} \varphi_\varepsilon(x_j - x_k)m_k\right)^{m-2}$$

$$\tag{22}$$

$$+ \left(\sum_{j=1}^{N} \varphi_\varepsilon(x_i - x_j)m_j\right)^{m-2} \left(\sum_{j=1}^{N} \nabla \varphi_\varepsilon(x_i - x_j)m_j\right). \tag{23}$$

As $\varepsilon \to 0$, Carrillo, Craig, and Patacchini show that, for any lower-semicontinuous interaction potential W, the regularized energies E_ε Γ-converge to the unregularized energy E for all $m \geq 1$. Furthermore, for W λ-convex and $m \geq 2$, they show that gradient flows of the regularized energies E_ε are well-posed. Finally, provided that sufficient a priori estimates hold along the flow, gradient flows of the regularized energies converge to the solution of the aggregation diffusion equation with initial data ρ_0 as $\varepsilon \to 0$ and $d_2(\rho_0^N, \rho_0) \to 0$. More recently, Craig and Topaloglu have demonstrated numerically that this method also provides a robust approach for simulating aggregation diffusion equations, by sending the diffusion exponent $m \to +\infty$ as the regularization and the discretization of the initial data are refined [82].

5   Simulations via the Blob Method for Diffusion

In this section, we apply the blob method for diffusion to illustrate several properties of the singular limits discussed in Sect. 3. We begin in Sect. 5.1 by describing the details of our numerical implementation, which include various refinements over previous works, such as regridding to reduce the number of particles required for convergence [55, 82]. In Sect. 5.2, we provide several numerical examples of the slow diffusion limit and properties of the constrained aggregation equation, particularly critical mass behavior relating to open problems in geometric shape optimization [34, 82, 91]. In Sect. 5.3, we give numerical examples illustrating the relationship between singular limits and metastability behavior, both as aggregation becomes localized and as diffusion vanishes.

5.1   Numerical Implementation

We now describe our numerical implementation of the blob method for diffusion, which we perform in Python, using the Numpy, SciPy, and Matplotlib libraries [100, 105, 140]. As explained in the previous section, there are two key steps in the blob method for diffusion: first, one must approximate the initial data ρ_0 by a sum of

Dirac masses (20); then, one must evolve the locations of those Dirac masses by solving a system of ordinary differential equations (22)–(23).

We begin by describing the approximation of the initial data (20). In the following, we typically consider examples in which the solution is entirely supported on the spatial domain $[-1, 1]$. Unless otherwise specified, we approximate the initial data on the computational domain $[-1.1, 1.1]$ by partitioning the domain into N intervals and placing a Dirac mass at the center of each interval, weighted according to the integral of the initial data over the interval. We let h denote the width of these initial intervals.

In order to solve the system of ordinary differential equations (22)–(23), we take the mollifier φ_ε to be a Gaussian

$$\varphi_\varepsilon(x) = \frac{1}{(4\pi\varepsilon^2)^{d/2}} e^{-|x|^2/4\varepsilon^2}.$$

We then solve the system of ordinary differential equations using the SciPy `solve_ivp` implementation of the backward differentiation formula (BDF) method. Similarly to analogous work on blob methods in the fluids case [4], we observe that the numerical error due to the choice of time discretization is of lower order than the error due to the regularization and spatial discretization. When the interaction kernel W has a Newtonian or stronger singularity (e.g. $W(x) = |x|$ or $W(x) = \log|x|$ in one dimension), we also mollify the interaction potential by convolution with the mollifier, as Bertozzi and the second author demonstrated this provides higher order rates of convergence [79].

In order to pass from the particle approximation

$$\rho_N(t) = \sum_{i=1}^{N} \delta_{x_i(t)} m_i$$

to a density that can be compared with exact solutions and visualized, we convolve our particle approximation with the mollifier φ_ε, leading to the following approximate density, that is defined on all of Euclidean space:

$$\rho_{\varepsilon,N}(x) = \sum_{i=1}^{N} \varphi_\varepsilon(x - x_i(t)) m_i. \tag{24}$$

As in the previous work on the Euler equations [7, 8] and the aggregation equation [79], we observe the fastest rate of convergence if the regularization parameter ε scales according to the initial grid spacing h according to

$$h^{1-p} \le \varepsilon, \text{ for } 0 < p \ll 1. \tag{25}$$

In the fluids community, it is well known that this relationship (25) between the interparticle distance and the regularization needs to hold globally in time for the numerical simulation to agree well with exact solutions. In the previous work on aggregation and aggregation diffusion equations [55, 79, 82], this was accomplished by taking the spatial discretization h very small. In the present work, we accomplish this by using the formula for the approximate density (24) to re-initialize our particle approximation (20) whenever the maximum interparticle distance exceeds $1.5h$. Numerical examples illustrate that such remeshing is much more computationally efficient than taking h very small.

In the following numerical examples, we choose our initial data to be given by characteristic functions or Barenblatt profiles, which we define as follows:

$$1_\Omega (x) = \begin{cases} 1 & \text{if } x \in \Omega \\ 0 & \text{otherwise.} \end{cases}$$

$$\rho_\alpha (x, \tau) = \tau^{-d\beta} \left(\kappa - \frac{\beta}{2} \left(\frac{\alpha - 1}{\alpha} \right) \tau^{-2\beta} |x|^2 \right)_+^{(\alpha-1)^{-1}}, \quad \beta = \frac{1}{2 + d(\alpha - 1)},$$

with $\alpha > 1$, $\tau > 0$, and $\kappa = \kappa(\alpha, d) > 0$ chosen so that $\int \rho_\alpha \, dx = 1$.

5.2 Numerical Examples: Height Constrained Aggregation

We begin with several numerical examples illustrating the slow diffusion limit and properties of the height constrained aggregation equation. See Sect. 3.1 for a discussion of this singular limit and the constrained aggregation equation.

In Fig. 1, we consider the repulsive attractive interaction kernel $W(x) = |x|^4/4 - |x|$ with degenerate diffusion, illustrating how the asymptotic behavior of the solution depends on the diffusion exponent m for three different choices of mass of the initial data. For all three choices of the mass, we observe that, once the diffusion exponent reaches $m = 800$, the degenerate diffusion acts effectively as a height constraint. In each case, the mass of the initial data determines whether the height constraint is active. Motivated by this result, in all future simulations, we approximate a height constraint via degenerate diffusion with diffusion exponent $m = 800$. We compute the equilibrium profile in Fig. 1 by solving the aggregation diffusion equation with initial data given by a multiple of a Barenblatt profile $M\rho_2(x, 0.15)$, where M determines the mass. We compute the evolution up to time $T = 6.0$, with maximum time step size $k = 10^{-3}$. We discretize the domain with $N = 1000$ particles for $m = 2$ and $N = 500$ particles for $m > 2$, and we regularize the diffusion and singular interaction terms with $\varepsilon = 0.9$.

In Fig. 2, we illustrate the long time behavior of solutions to the height constrained aggregation equation for the same repulsive-attractive interaction kernel and masses considered in Fig. 1. To simulate the height constraint, we take $m = 100$ for mass 0.6 and $m = 800$ for masses 1.0 and 1.4. In the left column, we plot the trajectories of particles computed via the blob method. In the right column,

Equilibria for Varying Diffusion Exponent, $W(x) = |x|^4/4 - |x|$

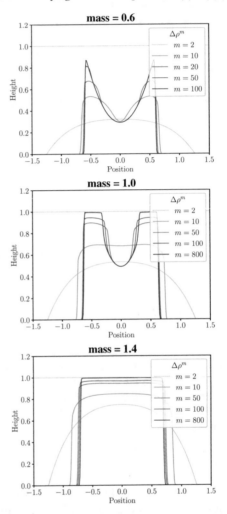

Fig. 1 We simulate asymptotic behavior of aggregation diffusion equations with repulsive-attractive interaction kernel and for varying diffusion exponents. In all three examples, diffusion exponent $m = 800$ is sufficient to impose the height constraint. The mass of the solution determines whether the height constraint is active

we plot the reconstructed density. For small mass 0.6, the equilibrium is in a liquid phase ($|\{\rho_\infty = 1\}| = 0$); for large mass 1.4, the equilibrium is in a solid phase ($|\{\rho_\infty = 1\}| = \int \rho_\infty$); and an intermediate phase exists for mass 1.0 ($0 < |\{\rho_\infty = 1\}| < \int \rho_\infty$). This agrees with previous numerical simulations of height constrained equilibria [81], as well as analytical results on the existence of set valued minimizers [34, 91]. See Fig. 4 for further analysis of how equilibria of the

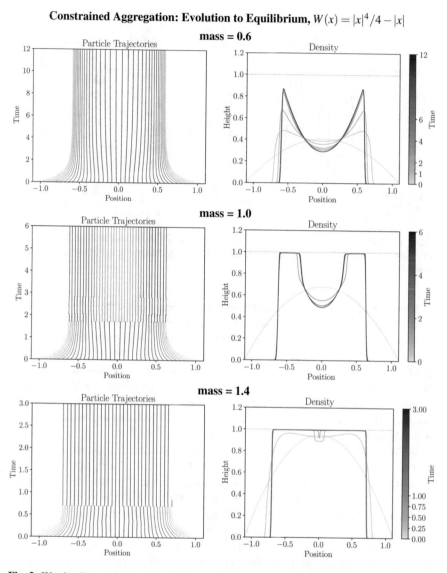

Fig. 2 We simulate evolution to equilibrium of height constrained aggregation equations with a repulsive-attractive interaction kernel. The small mass equilibrium is in liquid phase, the large mass equilibrium is in solid phase (a characteristic function on a set), and in between the equilibrium is in an intermediate phase

height constrained problem depend on the mass, and in particular how the strength of the short range repulsion affects this relationship.

In Fig. 3, we simulate the evolution to equilibrium of the height constrained aggregation equation with attractive Newtonian interaction. We observe different

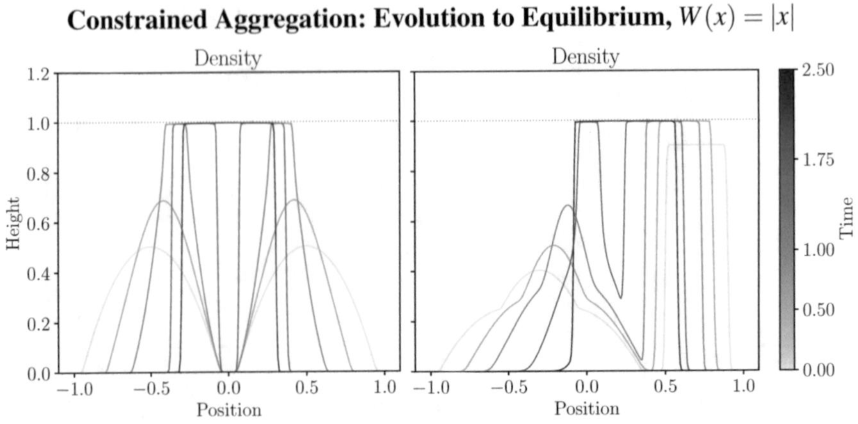

Constrained Aggregation: Evolution to Equilibrium, $W(x) = |x|$

Fig. 3 Evolution to equilibrium of height constrained aggregation with Newtonian interaction for two choices of initial data, illustrating different possible merging behavior as density reaches height constraint

possible merging behaviors as the density reaches the height constraint. On the left, we consider initial data given by the sum of two Barenblatt profiles $(0.3)\rho_2(x, 0.01)$ centered at ± 0.5. In this case, the bumps first hit the height constraint and then merge. On the right, we consider initial data given by the sum of two Barenblatts $(0.15)\rho_2(x, 0.01)$ centered at $x = 0.1$ and $x = 0.5$ and a characteristic function $(0.9)1_{[-0.2, 0.2]}(x)$ centered at $x = 0.7$. In this case, merging begins before the height constraint becomes active. We compute the evolution with a maximum time step size $k = 10^{-4}$, discretizing the domain with $N = 400$ particles and regularizing the diffusion and singular interaction terms with $\varepsilon = 0.99$.

In Fig. 4, we illustrate equilibria of the height constrained aggregation equation for repulsive-attractive interaction kernel $W(x) = |x|^4/4 - |x|^p/p$, showing how these equilibria depend on the repulsion exponent p and the mass of the solution. For strong short range repulsion, as in the cases $p = 0$ and $p = 1$, we observe the following behavior as the mass increases: the equilibrium is initially in liquid phase $(|\{\rho_\infty = 1\}| = 0)$, transitions to an intermediate phase $(0 < |\{\rho_\infty = 1\}| < \int \rho_\infty)$, and ultimately reaches a solid phase $(|\{\rho_\infty = 1\}| = \int \rho_\infty)$, where it is a characteristic function on an interval. However, for weaker short range repulsion, as in the case $p = 2$, we observe that the equilibrium is in the solid phase for all sufficiently small masses, then transitions to an intermediate phase, and ultimately returns to a solid phase. This provides numerical evidence that Lieb and Frank's result on the existence of liquid phase equilibria for interaction kernels that are at least as singular as the Newtonian potential is sharp, in that it requires this strength of short range repulsion in order to hold [91].

To compute the equilibrium profiles illustrated in Fig. 4, we compute solutions of aggregation diffusion equations for $m = 800$ with patch initial data $\left(\frac{M}{2.1}\right)1_{[-1.05, 1.05]}$, where the scalar M determines the mass of the initial data. The

Constrained Aggregation: Equilibria for Varying Masses, $W(x) = |x|^4/4 - |x|^p/p$

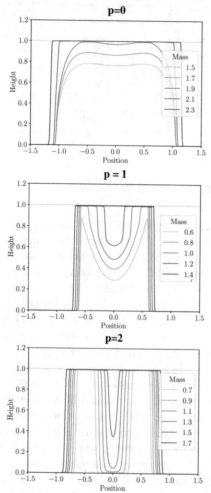

Fig. 4 We illustrate how the mass of the initial data affects the equilibrium configuration. For $p = 0$ and $p = 1$, as the mass increases, the equilibrium transitions from a liquid phase to an intermediate phase to a solid phase. However, for $p = 2$, the equilibrium begins in solid phase, transitions to an intermediate phase, and then returns to solid phase. This provides numerical evidence that Lieb and Frank's result on the existence of liquid phase equilibria for singular interaction potentials is sharp, in the sense that it requires this strength of short range repulsion in order to hold [91]

maximum time for which we compute the evolution depends on the value of p and the mass of the initial data, as described in Table 1. For $p = 0$, we take spatial discretization and regularization $N = 300$ and $\varepsilon = 0.99$. For $p = 1$, we take $N = 500$, $\varepsilon = 0.9$. For $p = 2$, we take $N = 600$, $\varepsilon = 0.85$.

Table 1 Values of maximum time T and minimum time step k for each simulation in Fig. 4

Mass	T	k	Mass	T	k	Mass	T	k
$p=0$			$p=1$			$p=2$		
1.5	1	10^{-3}	0.6	12	10^{-3}	0.7	28	10^{-4}
1.7	1	10^{-3}	0.8	6	$10^{-3.5}$	0.9	20	10^{-4}
1.9	1	10^{-3}	1.0	6	$10^{-3.5}$	1.1	12	10^{-4}
2.1	0.25	10^{-4}	1.2	3	$10^{-3.5}$	1.3	12	10^{-4}
2.3	0.25	10^{-4}	1.4	3	$10^{-3.5}$	1.5	2.5	10^{-5}
						1.7	0.75	10^{-5}

5.3 Numerical Examples: Metastability

We now turn to numerical examples illustrating metastability behavior in singular limits, considering both localized aggregation and vanishing diffusion. See Sect. 3.2 for a discussion of the interplay between singular limits and metastability.

In Fig. 5, we consider the initial dynamics of solutions for a localized, attractive Gaussian interaction potential. The diffusion is of porous medium type $m = 2$, weighted with diffusion coefficient $\nu = 0.25$, as in Eq. (16). In each case, we take the initial data to be $1_{[-.6,.6]}$, with $N = 600$ particles, regularization $\varepsilon = 0.9$, and maximal time step $k = 10^{-4}$. For $\delta = 0.02$, we briefly observe the formation of six bumps, before two quickly merge and form a metastable five bump configuration. Next, for $\delta = 0.03$, we observe the formation of a metastable state with four bumps, which at time 0.1 quickly transitions into a second metastable state with three bumps. For $\delta = 0.04$, a three bump metastable state emerges immediately, and for $\delta = 0.05$, the solution forms a two bump metastable state. Note that none of these numerical examples illustrates convergence to equilibrium, since in each case, the equilibrium must be radially decreasing, as discussed in Sect. 3.2.1 [63].

In Fig. 6, we compute the convergence to equilibrium for two examples from the previous figure: $\delta = 0.04$ and $\delta = 0.05$. In the former case, the intermediate two bump metastable state is preserved on a timescale of order $\Delta t \sim 20$, before a near instantaneous transition to a single bump steady state. In the latter case, the two bump steady state, which Fig. 5 showed to form by time $t = 0.15$, is preserved past time $t = 120$ before rapidly transitioning to a single bump equilibrium. The long lived metastable state in the $\delta = 0.05$ case can be easily confused numerically with a steady state. Indeed, it was only the theoretical result ensuring that the equilibrium configuration must be radially decreasing that prompted us to run the simulations three orders of magnitude after the dynamics seem to stabilize ($t \in [0.15, 125.00]$).

We again consider metastable behavior in Fig. 7, this time considering an attractive Gaussian interaction kernel with fixed localization $\delta = 0.05$ and varying the diffusion coefficient ν for the porous medium type diffusion ($m = 2$). As in the previous examples, the structure and duration of the metastable steady states depend strongly on the competition between the aggregation and diffusion terms. In each

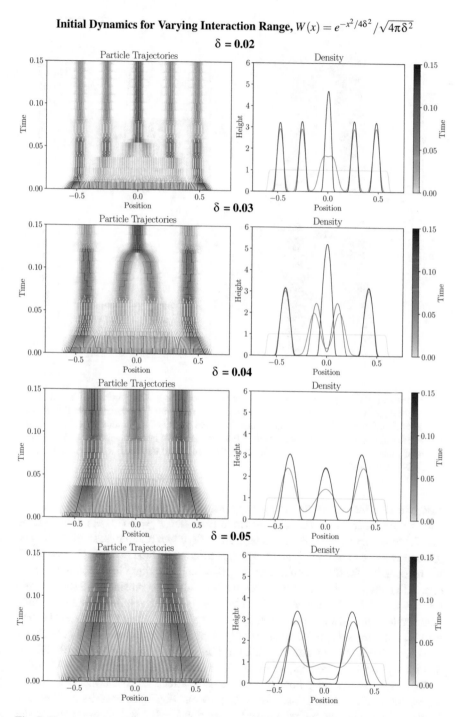

Fig. 5 For aggregation diffusion equations with porous medium diffusion ($m = 2$), the localization of the interaction kernel δ affects the number of clusters in each metastable state

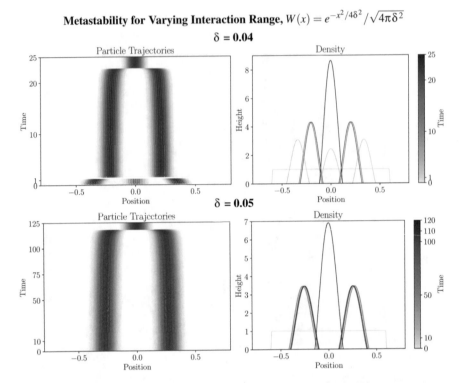

Fig. 6 When a purely attractive interaction kernel is localized, with porous medium diffusion ($m = 2$), solutions of aggregation diffusion equations can form long-lived metastable states. In the case $\delta = 0.04$, the two bump metastable state lasts two orders of magnitude after the dynamics numerically appear to stabilize, and in the $\delta = 0.05$ case, the two bump metastable state lasts three orders of magnitude after apparent stabilization. Both solutions ultimately converge to a radially decreasing stationary state

case, we take the initial data to be $1_{[-0.6,0.6]}$, and we choose $N = 500$ particles, with regularization $\varepsilon = 0.9$, and maximal time step $k = 10^{-4}$. For $\nu = 0.15$ in Fig. 7, we observe a three bump metastable state, a two bump metastable state, and ultimately a one bump steady state, which is reached by time $t = 2.5$. For $\nu = 0.25$, we recover the fourth example from Fig. 5, which is unique in that it is not until approximately $t \sim 120$ until the metastable steady state subsides and the solution reaches a radially decreasing profile. For $\nu = 0.35$, the solution has a two bump metastable state and a one bump steady state, which is reached by time $t = 0.6$, and for $\nu = 0.45$, the solution remains radially decreasing for all time and approaches a one bump steady state by time $t = 2.5$. For $\nu > 0.5$, DiFrancesco and Jaafra showed that steady states do not exist, as all solutions decay to zero locally in $L^2(\mathbb{R}^d)$ [85].

In Fig. 8, we consider how linear diffusion affects metastability behavior in aggregation diffusion equations. As in the previous Fig. 7, we consider an attractive Gaussian interaction kernel with fixed localization $\delta = 0.1$ and diffusion coefficient

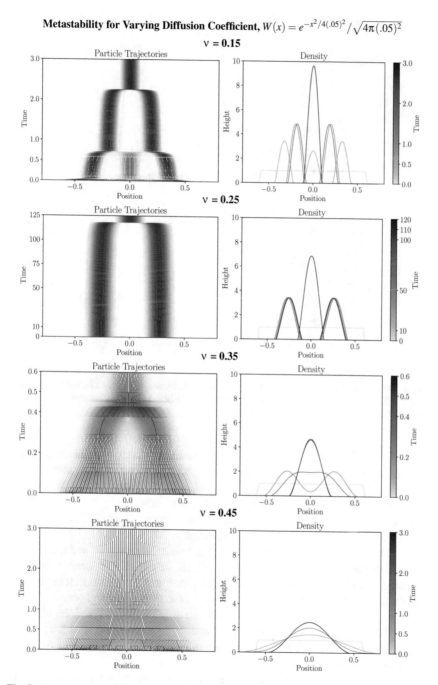

Fig. 7 For aggregation diffusion equations with porous medium diffusion ($m = 2$), varying the diffusion coefficient v leads to a variety of metastable profiles, ranging from one to three bumps, with duration varying more than two orders of magnitude, from $t = 0.6$ to $t = 120$

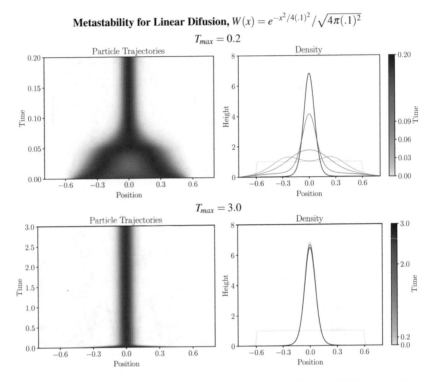

Fig. 8 For aggregation diffusion equations with a small amount of linear diffusion ($m = 1$, $\nu = 0.2$), we observe rapid convergence to a single bump metastable state (top), followed by very slow spreading as the bump diffuses away (bottom)

$\nu = 0.2$. Similarly to the porous medium case, we observe rapid convergence to a metastable state, in this case forming a single bump. However, unlike in the porous medium case, a steady state does not exist [58]. Instead, we observe that the metastable state spreads slowly as the bump diffuses away. In our simulation, we take the initial data to be $1_{[-0.6,0.6]}$, and we choose $N = 500$ particles, with regularization $\varepsilon = 0.9$, and maximal time step $k = 10^{-3}$.

In Fig. 9, we consider the behavior of localized aggregation with a singular interaction potential. Unlike in Fig. 5, in which we considered smooth Gaussian interactions, the strong short range aggregation prevents the solution from breaking apart into distinct clusters and causes the solution to quickly approach a radially decreasing equilibrium. In each simulation, we consider porous medium diffusion $m = 2$ with diffusion coefficient $\nu = 0.4$, and we take $N = 500$, $\varepsilon = 0.9$, and the maximum time step $k = 10^{-5}$. We consider the initial data $1_{[-0.5,0.5]}$ and restrict to the computational domain $[-0.55, 0.55]$.

In Fig. 10, we investigate the competition between aggregation and porous medium diffusion $m = 2$ as the interaction localizes for a third choice of interaction kernel: the attractive Newtonian potential. In terms of its singularity at the origin, this kernel lies between the very singular logarithmic kernel from

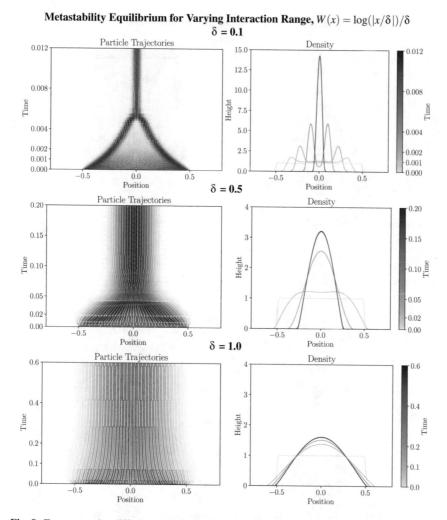

Fig. 9 For aggregation diffusion equations with a singular, attractive interaction kernel and porous medium diffusion ($m = 2$), the strong short range aggregation prevents the solution from forming distinct, metastable clusters, as observed in Fig. 5

Fig. 9 and the smooth Gaussian kernel from Fig. 5. However, unlike in both of these cases, we observe that the solution remains radially decreasing for all times, quickly approaches a single bump steady state. We do not observe any metastability behavior. For each simulation, we consider diffusion coefficient $\nu = 0.1$, $N = 600$, $\varepsilon = 0.9$, and the maximum time step $k = 10^{-5}$. We consider the initial data $1_{[-0.9,0.9]}$.

In our final example, Fig. 11, we illustrate metastability behavior for aggregation diffusion exponents with the repulsive-attractive interaction kernel $W(x) = |x|^4 -$

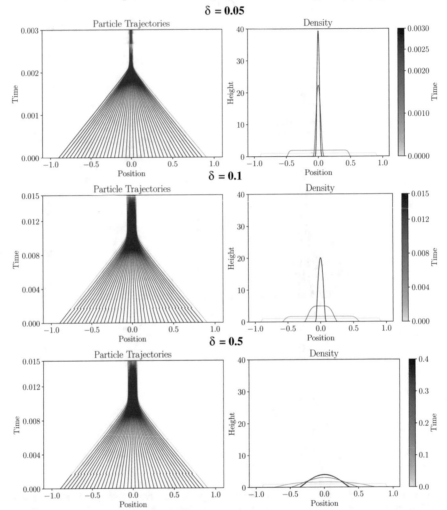

Fig. 10 Unlike in the cases of aggregation diffusion equations with either a Gaussian or logarithmic localized interaction kernel (see Figs. 5 and 9), for a Newtonian interaction kernel, the qualitative dynamics remain the same as the scale of the interaction varies: solutions with radially decreasing initial data remain radially decreasing for all time

Fig. 11 (continued) solution quickly forms two unequal bumps, which do not symmetrize for small diffusion coefficient ($\nu = 0.075^2/2$), but do symmetrize for larger diffusion coefficient ($\nu = 0.01$). For the height constrained problem, the solution does not equilibrate, as there is no diffusion to mediate this process

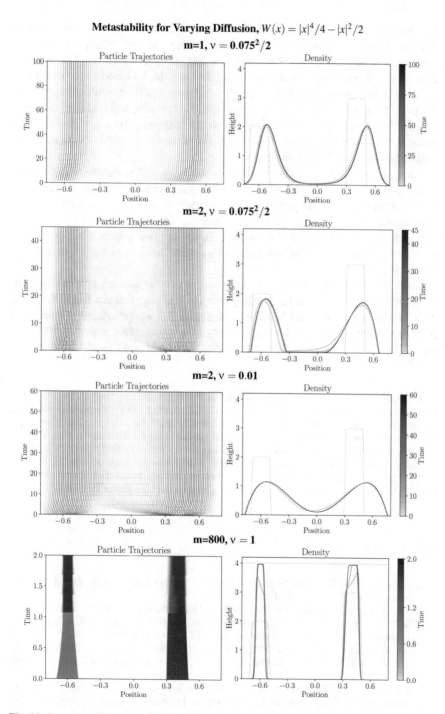

Fig. 11 For a repulsive-attractive interaction kernel and linear diffusion ($m = 1$), solutions quickly evolve to a metastable state of two bumps with unequal mass and then the mass between the two bumps slowly equilibrates over time; see [89]. For quadratic diffusion ($m = 2$), the

$|x|^2$ and various types of diffusion. For diffusion exponent $m = 1$, weighted with diffusion coefficient $\nu = 0.075^2/2$, we recover the metastability behavior observed by Evers and Kolokolnikov [89]: a solution that is initially given by characteristic functions with weight 0.2 centered at -0.6 and weight 0.3 centered at 0.4 quickly smooths to two bumps of unequal weights and then equilibrates slowly over time. We take $N = 600$ particles, with regularization $\varepsilon = 0.99$, and maximal time step $k = 10^{-3}$. Interestingly, the equilibration behavior appears to be driven entirely by particles with extremely small mass on the tails of the bumps and is therefore nearly indistinguishable at the particle level.

Next, in Fig. 11, we consider diffusion exponent $m = 2$, weighted with diffusion coefficient $\nu = 0.075^2/2$ and observe that the solution starting from characteristic functions with unequal weights does not symmetrize: the larger mass on the right-hand side is preserved asymptotically. However, when we increase the diffusion exponent to $\nu = 0.01$ in the third simulation, the solution symmetrizes quickly. In both cases, we take $N = 800$ particles, with regularization $\varepsilon = 0.99$, and maximal time step 10^{-3}.

In the fourth simulation of Fig. 11, we contrast the previous examples of degenerate diffusion $m = 1, 2$ with the height constrained aggregation equation. Naturally, without the mechanism of diffusion to spread mass, the unequal distribution of mass in the initial data is preserved asymptotically. We discretize the interval $[-0.9, 0.7]$ with $N = 600$ particles, with regularization $\varepsilon = 0.85$, and maximal times step $k = 10^{-4}$.

Acknowledgements We thank Matias Delgadino, Franca Hoffmann, Jingwei Hu, Francesco Patacchini, Ihsan Topaloglu, and Li Wang for useful discussions. JAC was partially supported by the EPSRC grant number EP/P031587/1. JAC is grateful to the Mittag-Leffler Institute for providing a fruitful working environment during the special semester *Mathematical Biology*. KC was supported by NSF DMS-1811012. YY was supported by NSF DMS-1715418. The authors acknowledge the American Institute of Mathematics (AIM) for supporting a visit during the early stages of this work. This work used the Extreme Science and Engineering Discovery Environment (XSEDE) Comet at the San Diego Supercomputer Center through allocation DMS180023.

References

1. D. Alexander, I. Kim, and Y. Yao. Quasi-static evolution and congested crowd transport. *Nonlinearity*, 27(4):823–858, 2014.
2. L. N. Almeida, F. Bubba, B. Perthame, and C. Pouchol. Energy and implicit discretization of the Fokker-Planck and Keller-Segel type equations. *preprint arXiv:1803.10629*, 2018.
3. L. Ambrosio, N. Gigli, and G. Savaré. *Gradient Flows in Metric Spaces and in the Space of Probability Measures*. Lectures in Mathematics ETH Zürich. Birkhäuser Verlag, Basel, 2008.
4. C. Anderson and C. Greengard. On vortex methods. *SIAM J. Numer. Anal.*, 22(3):413–440, 1985.
5. R. Bailo, J. A. Carrillo, and J. Hu. Fully discrete positivity-preserving and energy-dissipative schemes for nonlinear nonlocal equations with a gradient flow structure. *preprint arXiv:*, 2018.

6. D. Balagué, J. A. Carrillo, T. Laurent, and G. Raoul. Dimensionality of local minimizers of the interaction energy. *Arch. Ration. Mech. Anal.*, 209(3):1055–1088, 2013.

7. J. T. Beale and A. Majda. Vortex methods. I. Convergence in three dimensions. *Math. Comp.*, 39(159):1–27, 1982.

8. J. T. Beale and A. Majda. Vortex methods. II. Higher order accuracy in two and three dimensions. *Math. Comp.*, 39(159):29–52, 1982.

9. J. Bedrossian. Global minimizers for free energies of subcritical aggregation equations with degenerate diffusion. *Appl. Math. Lett.*, 24(11):1927–1932, 2011.

10. J. Bedrossian. Intermediate asymptotics for critical and supercritical aggregation equations and Patlak–Keller–Segel models. *Commun. Math. Sci.*, 9(4):1143–1161, 2011.

11. J. Bedrossian and I. C. Kim. Global existence and finite time blow-up for critical Patlak–Keller–Segel models with inhomogeneous diffusion. *SIAM J. Math. Anal.*, 45(3):934–964, 2013.

12. J. Bedrossian and N. Masmoudi. Existence, uniqueness and Lipschitz dependence for Patlak–Keller–Segel and Navier–Stokes in $\mathbb{R}^2$ with measure-valued initial data. *Arch. Rat. Mech. Anal.*, 214(3):717–801, 2014.

13. J. Bedrossian and N. Rodríguez. Inhomogeneous Patlak–Keller–Segel models and aggregation equations with nonlinear diffusion in $\mathbb{R}^d$. *Discrete Contin. Dyn. Syst. Ser. B*, 19(5):1279–1309, 2014.

14. J. Bedrossian, N. Rodríguez, and A. L. Bertozzi. Local and global well-posedness for aggregation equations and Patlak–Keller–Segel models with degenerate diffusion. *Nonlinearity*, 24(6):1683–1714, 2011.

15. N. Bellomo, A. Bellouquid, Y. Tao, and M. Winkler. Toward a mathematical theory of Keller-Segel models of pattern formation in biological tissues. *Math. Models Methods Appl. Sci.*, 25(9):1663–1763, 2015.

16. J. Benamou and Y. Brenier. A computational fluid mechanics solution to the Monge-Kantorovich mass transfer problem. *Numer. Math.*, 84:375–393, 2000.

17. J. Benamou, G. Carlier, and M. Laborde. An augmented Lagrangian approach to Wasserstein gradient flows and applications. *ESAIM: PROCEEDINGS AND SURVEYS*, 54:1–17, 2016.

18. J.-D. Benamou, G. Carlier, Q. Mérigot, and E. Oudet. Discretization of functionals involving the Monge-Ampère operator. *Numer. Math.*, 134(3):611–636, 2016.

19. D. Benedetto, E. Caglioti, and M. Pulvirenti. A kinetic equation for granular media. *RAIRO Modél. Math. Anal. Numér.*, 31(5):615–641, 1997.

20. A. L. Bertozzi, J. A. Carrillo, and T. Laurent. Blow-up in multidimensional aggregation equations with mildly singular interaction kernels. *Nonlinearity*, 22(3):683–710, 2009.

21. A. L. Bertozzi, J. B. Garnett, and T. Laurent. Characterization of radially symmetric finite time blowup in multidimensional aggregation equations. *SIAM J. Math. Anal.*, 44(2):651–681, 2012.

22. A. L. Bertozzi, T. Laurent, and F. Léger. Aggregation and spreading via the Newtonian potential: the dynamics of patch solutions. *Math. Models Methods Appl. Sci.*, 22(suppl. 1):1140005, 39, 2012.

23. A. L. Bertozzi, T. Laurent, and J. Rosado. L^p theory for the multidimensional aggregation equation. *Comm. Pure Appl. Math.*, 64(1):45–83, 2011.

24. M. Bessemoulin-Chatard and F. Filbet. A finite volume scheme for nonlinear degenerate parabolic equations. *SIAM J. Sci. Comput.*, 34(5):B559–B583, 2012.

25. S. Bian and J.-G. Liu. Dynamic and steady states for multi-dimensional Keller–Segel model with diffusion exponent $m > 0$. *Comm. Math. Phy.*, 323(3):1017–1070, 2013.

26. A. Blanchet, V. Calvez, and J. A. Carrillo. Convergence of the mass-transport steepest descent scheme for the subcritical Patlak-Keller–Segel model. *SIAM J. Numer. Anal.*, 46(2):691–721, 2008.

27. A. Blanchet, E. A. Carlen, and J. A. Carrillo. Functional inequalities, thick tails and asymptotics for the critical mass Patlak-Keller–Segel model. *J. Funct. Anal.*, 262(5):2142–2230, 2012.

28. A. Blanchet, J. A. Carrillo, and P. Laurençot. Critical mass for a Patlak–Keller–Segel model with degenerate diffusion in higher dimensions. *Calc. Var. Partial Differ. Equ.*, 35(2):133–168, 2009.
29. A. Blanchet, J. A. Carrillo, and N. Masmoudi. Infinite time aggregation for the critical Patlak–Keller–Segel model in $\mathbb{R}^2$. *Comm. Pure Appl. Math.*, 61(10):1449–1481, 2008.
30. A. Blanchet, J. Dolbeault, M. Escobedo, and J. Fernández. Asymptotic behaviour for small mass in the two-dimensional parabolic-elliptic Keller–Segel model. *J. Math. Anal. Appl.*, 361:533–542, 2008.
31. A. Blanchet, J. Dolbeault, and B. Perthame. Two-dimensional Keller–Segel model: optimal critical mass and qualitative properties of the solutions. *Electron. J. Diff. Eq.*, 2006:1–33, 2006.
32. M. Bodnar and J. J. L. Velazquez. An integro-differential equation arising as a limit of individual cell-based models. *J. Differential Equations*, 222(2):341–380, 2006.
33. F. Bolley, J. A. Cañizo, and J. A. Carrillo. Stochastic mean-field limit: non-Lipschitz forces and swarming. *Math. Models Methods Appl. Sci.*, 21(11):2179–2210, 2011.
34. A. Burchard, R. Choksi, and I. Topaloglu. Nonlocal shape optimization via interactions of attractive and repulsive potentials. *Indiana Univ. Math. J.*, 67(1):375–395, 2018.
35. M. Burger, M. Di Francesco, and M. Franek. Stationary states of quadratic diffusion equations with long-range attraction. *Comm. Math. Sci.*, 11(3):709–738, 2013.
36. M. Burger, R. Fetecau, and Y. Huang. Stationary states and asymptotic behavior of aggregation models with nonlinear local repulsion. *SIAM J. Appl. Dyn. Syst.*, 13(1):397–424, 2014.
37. J. A. Cañizo, J. A. Carrillo, and F. S. Patacchini. Existence of compactly supported global minimisers for the interaction energy. *Arch. Ration. Mech. Anal.*, 217(3):1197–1217, 2015.
38. J. A. Cañizo, J. A. Carrillo, and M. E. Schonbek. Decay rates for a class of diffusive-dominated interaction equations. *J. Math. Anal. Appl.*, 389(1):541–557, 2012.
39. L. A. Caffarelli and A. Friedman. Asymptotic behavior of solutions of $u_t = \Delta u^m$ as $m \to \infty$. *Indiana U. Math. J.*, 36(4):711–728, 1987.
40. V. Calvez and J. A. Carrillo. Volume effects in the Keller–Segel model: energy estimates preventing blow-up. *J. Math. Pures Appl.*, 86(2):155–175, 2006.
41. V. Calvez, J. A. Carrillo, and F. Hoffmann. Equilibria of homogeneous functionals in the fair-competition regime. *Nonlinear Anal.*, 159:85–128, 2017.
42. V. Calvez, J. A. Carrillo, and F. Hoffmann. The geometry of diffusing and self-attracting particles in a one-dimensional fair-competition regime. 2186:1–71, 2017.
43. V. Calvez and T. O. Gallouët. Particle approximation of the one dimensional Keller–Segel equation, stability and rigidity of the blow-up. *Discrete Contin. Dyn. Syst. Ser. A*, 36(3):1175–1208, 2015.
44. J. F. Campos and J. Dolbeault. Asymptotic estimates for the parabolic-elliptic Keller–Segel model in the plane. *J. Math. Anal. Appl.*, 39(5):806–841, 2014.
45. M. Campos-Pinto, J. A. Carrillo, F. Charles, and Y.-P. Choi. Convergence of a linearly transformed particle method for aggregation equations. *Numerische Mathematik*, 139:743–793, 2018.
46. E. A. Carlen and K. Craig. Contraction of the proximal map and generalized convexity of the Moreau-Yosida regularization in the 2-Wasserstein metric. *Math. and Mech. of Complex Systems*, 1(1):33–65, 2013.
47. E. A. Carlen and A. Figalli. Stability for a GNS inequality and the Log-HLS inequality, with application to the critical mass Keller–Segel equation. *Duke Math. J.*, 162(3):579–625, 2013.
48. E. A. Carlen and M. Loss. Competing symmetries, the logarithmic HLS inequality and Onofri's inequality on S^n. *Geom. Funct. Anal.*, 2(1):90–104, 1992.
49. G. Carlier, V. Duval, G. Peyré, and B. Schmitzer. Convergence of entropic schemes for optimal transport and gradient flows. *SIAM J. Math. Anal.*, 49(2):1385–1418, 2017.
50. J. A. Carrillo, D. Castorina, and B. Volzone. Ground States for Diffusion Dominated Free Energies with Logarithmic Interaction. *SIAM J. Math. Anal.*, 47(1):1–25, Jan. 2015.
51. J. A. Carrillo, A. Chertock, and Y. Huang. A finite-volume method for nonlinear nonlocal equations with a gradient flow structure. *Commun. Comput. Phys.*, 17(1):233–258, 2015.

52. J. A. Carrillo, Y.-P. Choi, and M. Hauray. The derivation of swarming models: mean-field limit and Wasserstein distances. In *Collective dynamics from bacteria to crowds*, volume 553 of *CISM Courses and Lect.*, pages 1–46. Springer, Vienna, 2014.

53. J. A. Carrillo, Y.-P. Choi, and M. Hauray. The derivation of swarming models: mean-field limit and Wasserstein distances. In *Collective Dynamics from Bacteria to Crowds: An Excursion Through Modeling, Analysis and Simulation*, volume 553 of *CISM Courses and Lect.*, pages 1–46. Springer Vienna, 2014.

54. J. A. Carrillo, Y.-P. Choi, M. Hauray, and S. Salem. Mean-field limit for collective behavior models with sharp sensitivity regions. *J. Eur. Math. Soc.*, 21:121–161, 2019.

55. J. A. Carrillo, K. Craig, and F. S. Patacchini. A blob method for diffusion. *arXiv preprint arXiv:1709.09195*, 2017.

56. J. A. Carrillo, K. Craig, L. Wang, and C. Wei. Primal dual methods for Wasserstein gradient flows. *arXiv preprint arXiv:1901.08081*, 2019.

57. J. A. Carrillo, M. G. Delgadino, and A. Mellet. Regularity of local minimizers of the interaction energy via obstacle problems. *Comm. Math. Phys.*, 343(3):747–781, 2016.

58. J. A. Carrillo, M. Delgadino, and F. S. Patacchini. Existence of ground states for aggregation-diffusion equations. *Analysis and applications*, 17:393–423, 2019.

59. J. A. Carrillo, M. Di Francesco, A. Figalli, T. Laurent, and D. Slepčev. Global-in-time weak measure solutions and finite-time aggregation for nonlocal interaction equations. *Duke Math. J.*, 156(2):229–271, 2011.

60. J. A. Carrillo, M. DiFrancesco, A. Figalli, T. Laurent, and D. Slepčev. Global-in-time weak measure solutions and finite-time aggregation for nonlocal interaction equations. *Duke Math. J.*, 156(2):229–271, 2011.

61. J. A. Carrillo, A. Figalli, and F. S. Patacchini. Geometry of minimizers for the interaction energy with mildly repulsive potentials. *Ann. IHP*, 34:1299–1308, 2017.

62. J. A. Carrillo, M. Fornasier, G. Toscani, and F. Vecil. Particle, kinetic, and hydrodynamic models of swarming. *Mathematical Modelling of Collective Behavior in Socio-Economic and Life Sciences*, pages 297–336, 2010.

63. J. A. Carrillo, S. Hittmeir, B. Volzone, and Y. Yao. Nonlinear aggregation-diffusion equations: radial symmetry and long time asymptotics. *arXiv preprint arXiv:1603.07767*, 2016, to appear in *Inventiones Mathematicae*.

64. J. A. Carrillo, F. Hoffmann, E. Mainini, and B. Volzone. Ground states in the diffusion-dominated regime. *Calc. Var. Partial Differ. Equ.*, 57(5):127, 2018.

65. J. A. Carrillo, Y. Huang, F. S. Patacchini, and G. Wolansky. Numerical study of a particle method for gradient flows. *Kinet. Relat. Models*, 10(3):613–641, 2017.

66. J. A. Carrillo, R. J. McCann, and C. Villani. Kinetic equilibration rates for granular media and related equations: entropy dissipation and mass transportation estimates. *Rev. Mat. Iberoam.*, 19(3):971–1018, 2003.

67. J. A. Carrillo and J. S. Moll. Numerical simulation of diffusive and aggregation phenomena in nonlinear continuity equations by evolving diffeomorphisms. *SIAM J. Sci. Comput.*, 31(6):4305–4329, 2009/10.

68. J. A. Carrillo, F. S. Patacchini, P. Sternberg, and G. Wolansky. Convergence of a particle method for diffusive gradient flows in one dimension. *SIAM J. Math. Anal.*, 48(6):3708–3741, 2016.

69. J. A. Carrillo, H. Ranetbauer, and M.-T. Wolfram. Numerical simulation of nonlinear continuity equations by evolving diffeomorphisms. *J. Comput. Phys.*, 327:186–202, 2016.

70. J. A. Carrillo and J. Wang. Uniform in time L^∞-estimates for nonlinear aggregation-diffusion equations. *arXiv preprint arXiv:1712.09541*, 2017.

71. L. Chen, J.-G. Liu, and J. Wang. Multidimensional degenerate Keller–Segel system with critical diffusion exponent 2n/(n+2). *SIAM J. Math. Anal.*, 44(2):1077–1102, 2012.

72. L. Chen and J. Wang. Exact criterion for global existence and blow up to a degenerate Keller-Segel system. *Doc. Math.*, 19:103–120, 2014.

73. A. Chertock. *A Practical Guide to Deterministic Particle Methods*. Available at http://www4.ncsu.edu/~acherto/papers/Chertock_particles.pdf.

74. R. Choksi, C. B. Muratov, and I. Topaloglu. An old problem resurfaces nonlocally: Gamow's liquid drops inspire today's research and applications. *Notices Amer. Math. Soc.*, 64(11):1275–1283, 2017.

75. G.-H. Cottet and P.-A. Raviart. Particle methods for the one-dimensional Vlasov–Poisson equations. *SIAM J. Numer. Anal.*, 21(1):52–76, 1984.

76. E. Cozzi, G.-M. Gie, and J. P. Kelliher. The aggregation equation with Newtonian potential: the vanishing viscosity limit. *J. Math. Anal. Appl.*, 453(2):841–893, 2017.

77. K. Craig. The exponential formula for the Wasserstein metric. *accepted to ESAIM COCV, preprint at http://arxiv.org/abs/1310.2912*.

78. K. Craig. Nonconvex gradient flow in the Wasserstein metric and applications to constrained nonlocal interactions. *Proc. Lond. Math. Soc.*, 114(1):60–102, 2017.

79. K. Craig and A. L. Bertozzi. A blob method for the aggregation equation. *Math. Comp.*, 85(300):1681–1717, 2016.

80. K. Craig, I. Kim, and Y. Yao. Congested aggregation via Newtonian interaction. *Arch. Rational Mech. Anal.*, 227(1):1–67, 2018.

81. K. Craig and I. Topaloglu. Convergence of regularized nonlocal interaction energies. *SIAM J. Math. Anal.*, 48(1):34–60, 2016.

82. K. Craig and I. Topaloglu. Aggregation-diffusion to constrained interaction: Minimizers & gradient flows in the slow diffusion limit. *arXiv preprint arXiv:1806.07415*, 2018.

83. F. Cucker and S. Smale. Emergent behavior in flocks. *IEEE Trans. Automat. Control*, 52(5):852–862, 2007.

84. P. Degond and F.-J. Mustieles. A deterministic approximation of diffusion equations using particles. *SIAM J. Sci. Statist. Comput.*, 11(2):293–310, 1990.

85. M. Di Francesco and Y. Jaafra. Multiple large-time behavior of nonlocal interaction equations with quadratic diffusion. *arXiv preprint arXiv:1710.08213*, 2017.

86. J. Dolbeault and B. Perthame. Optimal critical mass in the two-dimensional Keller–Segel model in $\mathbb{R}^2$. *C. R. Math. Acad. Sci. Paris*, 339(9):611–616, 2004.

87. M. R. D'Orsogna, Y.-L. Chuang, A. L. Bertozzi, and L. S. Chayes. Self-propelled particles with soft-core interactions: patterns, stability, and collapse. *Phys. Rev. Lett.*, 96(10):104302, 2006.

88. L. Evans, O. Savin, and W. Gangbo. Diffeomorphisms and nonlinear heat flows. *SIAM J. Math. Anal.*, 37(3):737–751, 2005.

89. J. H. Evers and T. Kolokolnikov. Metastable states for an aggregation model with noise. *SIAM J. Appl. Dyn. Syst.*, 15(4):2213–2226, 2016.

90. G. E. Fernández and S. Mischler. Uniqueness and long time asymptotic for the Keller-Segel equation: the parabolic-elliptic case. *Arch. Ration. Mech. Anal.*, 220(3):1159–1194, 2016.

91. R. L. Frank and E. H. Lieb. A 'liquid-solid' phase transition in a simple model for swarming, based on the 'no flat-spots' theorem for subharmonic functions. *Indiana Univ. Math. J.*, 2018. to appear.

92. T.-E. Ghoul and N. Masmoudi. Stability of infinite time blow up for the Patlak Keller Segel system. *preprint arXiv:1610.00456*, 2016.

93. O. Gil and F. Quirós. Convergence of the porous media equation to Hele–Shaw. *Nonlinear Anal Theory Methods Appl.*, 44(8):1111–1131, 2001.

94. O. Gil and F. Quirós. Boundary layer formation in the transition from the porous media equation to a Hele–Shaw flow. *Ann. Inst. H. Poincaré Anal. Non Linéaire*, 20(1):13–36, 2003.

95. J. Goodman, T. Y. Hou, and J. Lowengrub. Convergence of the point vortex method for the 2-D Euler equations. *Comm. Pure Appl. Math.*, 43(3):415–430, 1990.

96. L. Gosse and G. Toscani. Lagrangian numerical approximations to one-dimensional convolution-diffusion equations. *SIAM J. Sci. Comput.*, 28(4):1203–1227, 2006.

97. M. A. Herrero and J. J. Velázquez. Singularity patterns in a chemotaxis model. *Mathematische Annalen*, 306(1):583–623, 1996.

98. D. Horstmann. From 1970 until present: the Keller–Segel model in chemotaxis and its consequences. *Jahresber. Deutsch. Math.-Verein.*, 105(3):103–165, 2003.

99. H. Huang and J.-G. Liu. Error estimate of a random particle blob method for the Keller–Segel equation. *Math. Comp.*, 86(308):2719–2744.

100. J. D. Hunter. Matplotlib: a 2d graphics environment. *Comput. Sci. Eng.*, 9(3):90–95, 2007.

101. P.-E. Jabin. A review of the mean field limits for Vlasov equations. *Kinet. Relat. Models*, 7(4):661–711, 2014.

102. P.-E. Jabin and Z. Wang. Mean field limit and propagation of chaos for Vlasov systems with bounded forces. *J. Funct. Anal.*, 271(12):3588–3627, 2016.

103. P.-E. Jabin and Z. Wang. Mean field limit for stochastic particle systems. In *Active Particles. Vol. 1. Advances in Theory, Models, and Applications*, Model. Simul. Sci. Eng. Technol., pages 379–402. Birkhäuser/Springer, Cham, 2017.

104. W. Jäger and S. Luckhaus. On explosions of solutions to a system of partial differential equations modelling chemotaxis. *Trans. Amer. Math. Soc.*, 329(2):819–824, 1992.

105. E. Jones, T. Oliphant, P. Peterson, et al. *SciPy: Open source scientific tools for Python*, 2001–. Available at http://www.scipy.org/.

106. R. Jordan, D. Kinderlehrer, and F. Otto. The variational formulation of the Fokker–Planck equation. *SIAM J. Math. Anal.*, 29(1):1–17, 1998.

107. G. Kaib. Stationary states of an aggregation equation with degenerate diffusion and bounded attractive potential. *SIAM J. Math. Anal.*, 49(1):272–296, 2017.

108. E. F. Keller and L. A. Segel. Model for chemotaxis. *J. Theor. Biol.*, 30(2):225–234, 1971.

109. I. Kim and N. Požár. Porous medium equation to Hele–Shaw flow with general initial density. *Trans. Amer. Math. Soc.*, 370(2):873–909, 2018.

110. I. Kim, N. Požár, and B. Woodhouse. Singular limit of the porous medium equation with a drift. *arXiv preprint arXiv:1708.05842*, 2017.

111. I. Kim and Y. Yao. The Patlak-Keller–Segel model and its variations: properties of solutions via maximum principle. *SIAM J. Math. Anal.*, 44(2):568–602, 2012.

112. T. Kolokolnikov, J. A. Carrillo, A. Bertozzi, R. Fetecau, and M. Lewis. Emergent behaviour in multi-particle systems with non-local interactions [Editorial]. *Phys. D*, 260:1–4, 2013.

113. G. Lacombe and S. Mas-Gallic. Presentation and analysis of a diffusion-velocity method. In *Flows and Related Numerical Methods (Toulouse, 1998)*, volume 7 of *ESAIM Proc.*, pages 225–233. Soc. Math. Appl. Indust., Paris, 1999.

114. E. H. Lieb and H.-T. Yau. The Chandrasekhar theory of stellar collapse as the limit of quantum mechanics. *Comm. Math. Phy.*, 112(1):147–174, 1987.

115. P.-L. Lions. The concentration-compactness principle in the calculus of variations. the locally compact case, part 1. *Ann. Inst. H. Poincaré Anal. Non Linéaire*, 1(2):109–145, 1984.

116. P.-L. Lions. The concentration-compactness principle in the calculus of variations. the locally compact case, part 2. *Ann. Inst. H. Poincaré Anal. Non Linéaire*, 1(4):223–283, 1984.

117. P.-L. Lions and S. Mas-Gallic. Une méthode particulaire déterministe pour des équations diffusives non linéaires. *C. R. Acad. Sci. Paris Sér. I Math.*, 332(4):369–376, 2001.

118. J.-G. Liu, L. Wang, and Z. Zhou. Positivity-preserving and asymptotic preserving method for 2D Keller–Segel equations. *Mathematics of Computation*, 87:1165–1189, 2018.

119. J.-G. Liu and R. Yang. A random particle blob method for the Keller–Segel equation and convergence analysis. *Math. Comp.*, 86(304):725–745, 2017.

120. S. Mas-Gallic. The diffusion velocity method: a deterministic way of moving the nodes for solving diffusion equations. *Transp. Theory and Stat. Phys.*, 31(4-6):595–605, 2002.

121. B. Maury, A. Roudneff-Chupin, and F. Santambrogio. A macroscopic crowd motion model of gradient flow type. *Math. Models Methods Appl. Sci.*, 20(10):1787–1821, 2010.

122. B. Maury, A. Roudneff-Chupin, F. Santambrogio, and J. Venel. Handling congestion in crowd motion modeling. *Netw. Heterog. Media*, 6(3):485–519, 2011.

123. A. Mellet, B. Perthame, and F. Quiros. A Hele–Shaw problem for tumor growth. *J. Func. Anal.*, 273(10):3061–3093, 2017.

124. S. Motsch and E. Tadmor. Heterophilious dynamics enhances consensus. *SIAM Rev.*, 56(4):577–621, 2014.

125. K. Oelschläger. Large systems of interacting particles and the porous medium equation. *J. Diff. Eq.*, 88(2):294–346, 1990.

126. F. Otto. Doubly degenerate diffusion equations as steepest descent, manuscript. 1996.
127. N. Papadakis, G. Peyre, and E. Oudet. Optimal transport with proximal splitting. *SIAM. J. Image. Sci.*, 7(1):212–238, 2014.
128. C. S. Patlak. Random walk with persistence and external bias. *Bull. Math. Biophys.*, 15(3):311–338, 1953.
129. B. Perthame, F. Quirós, M. Tang, and N. Vauchelet. Derivation of a Hele–Shaw type system from a cell model with active motion. *Interfaces Free Bound.*, 16(4):489–508, 2014.
130. B. Perthame, F. Quirós, and J. L. Vázquez. The Hele–Shaw asymptotics for mechanical models of tumor growth. *Arch. Ration. Mech. Anal.*, 212(1):93–127, 2014.
131. P. Raphaël and R. Schweyer. On the stability of critical chemotactic aggregation. *Math. Ann.*, 359(1-2):267–377, 2014.
132. G. Russo. Deterministic diffusion of particles. *Comm. Pure Appl. Math.*, 43(6):697–733, 1990.
133. G. Russo. A particle method for collisional kinetic equations. I. basic theory and one-dimensional results. *J. Comput. Phys.*, 87(2):270–300, 1990.
134. T. Senba. Type II blowup of solutions to a simplified Keller–Segel system in two dimensional domains. *Nonlinear Anal. Theory Methods Appl.*, 66(8):1817–1839, 2007.
135. T. Senba and T. Suzuki. Weak solutions to a parabolic-elliptic system of chemotaxis. *J. Funct. Anal.*, 191(1):17–51, 2002.
136. S. Serfaty. Gamma-convergence of gradient flows on Hilbert and metric spaces and applications. *Discrete Contin. Dyn. Syst.*, 31(4):1427–1451, 2011.
137. Y. Sugiyama. The global existence and asymptotic behavior of solutions to degenerate quasilinear parabolic systems of chemotaxis. *Differential and Integral Equations*, 20(2):133–180, 2007.
138. Z. Sun, J. A. Carrillo, and C.-W. Shu. A discontinuous Galerkin method for nonlinear parabolic equations and gradient flow problems with interaction potentials, preprint. *Preprint*, 2017.
139. C. M. Topaz, A. L. Bertozzi, and M. A. Lewis. A nonlocal continuum model for biological aggregation. *Bull. Math. Biol.*, 68(7):1601–1623, 2006.
140. S. van der Walt, C. Colbert, and G. Varoquaux. The NumPy array: a structure for efficient numerical computation. *Comput. Sci. Eng.*, 13(2):22–30, 2011.
141. J. L. Vázquez. *The porous medium equation.* Oxford Mathematical Monographs. The Clarendon Press, Oxford University Press, Oxford, 2007. Mathematical theory.
142. C. Villani. *Topics in optimal transportation*, volume 58 of *Graduate Studies in Mathematics.* American Mathematical Society, Providence, RI, 2003.
143. M. Westdickenberg and J. Wilkening. Variational particle schemes for the porous medium equation and for the system of isentropic Euler equations. *M2AN Math. Model. Numer. Anal.*, 44(1):133–166, 2010.
144. Y. Yao. Asymptotic behavior of radial solutions for critical Patlak-Keller-Segel model and an repulsive-attractive aggregation equation. *Ann. Inst. H. Poincaré Anal. Non Linéaire*, 31:81–101, 2014.
145. Y. Yao and A. L. Bertozzi. Blow-up dynamics for the aggregation equation with degenerate diffusion. *Phys. D*, 260:77–89, 2013.
146. Y. Zhang. On a class of diffusion-aggregation equations. *arXiv preprint arXiv:1801.05543*, 2018.

High-Resolution Positivity and Asymptotic Preserving Numerical Methods for Chemotaxis and Related Models

Alina Chertock and Alexander Kurganov

Abstract Many microorganisms exhibit a special pattern formation at the presence of a chemoattractant, food, light, or areas with high oxygen concentration. Collective cell movement can be described by a system of nonlinear PDEs on both macroscopic and microscopic levels. The classical PDE chemotaxis model is the Patlak-Keller-Segel system, which consists of a convection-diffusion equation for the cell density and a reaction-diffusion equation for the chemoattractant concentration. At the cellular (microscopic) level, a multiscale chemotaxis models can be used. These models are based on a combination of the macroscopic evolution equation for chemoattractant and microscopic models for cell evolution. The latter is governed by a Boltzmann-type kinetic equation with a local turning kernel operator that describes the velocity change of the cells.

A common property of the chemotaxis systems is their ability to model a concentration phenomenon that mathematically results in solutions rapidly growing in small neighborhoods of concentration points/curves. The solutions may blow up or may exhibit a very singular, spiky behavior. In either case, capturing such singular solutions numerically is a challenging problem and the use of higher-order methods and/or adaptive strategies is often necessary. In addition, positivity preserving is an absolutely crucial property a good numerical method used to simulate chemotaxis should satisfy: this is the only way to guarantee a nonlinear stability of the method. For kinetic chemotaxis systems, it is also essential that numerical methods provide a consistent and stable discretization in certain asymptotic regimes.

A. Chertock (✉)
Department of Mathematics and Center for Research in Scientific Computation, North Carolina State University, Raleigh, NC, USA
e-mail: chertock@math.ncsu.edu

A. Kurganov
Department of Mathematics, Southern University of Science and Technology, Shenzhen, China

Mathematics Department, Tulane University, New Orleans, LA, USA
e-mail: kurganov@math.tulane.edu

© Springer Nature Switzerland AG 2019
N. Bellomo et al. (eds.), *Active Particles, Volume 2*, Modeling and Simulation in Science, Engineering and Technology,
https://doi.org/10.1007/978-3-030-20297-2_4

In this paper, we review some of the recent advances in developing of high-resolution finite-volume and finite-difference numerical methods that possess the aforementioned properties of the chemotaxis-type systems.

1 Introduction

Many microorganisms exhibit a special pattern formation in the presence of a chemoattractant, food, light, or areas with high oxygen concentration; see, e.g., [51–53, 59, 73, 75, 81, 88–90]. Collective movement of cells and organisms in response to chemical gradients, *chemotaxis*, has attracted a lot of attention due to its critical role in a wide range of biological phenomena; see, e.g., [25], where detailed comparison between different chemotactic mechanisms is provided.

The classical PDE chemotaxis model was introduced by Patlak in [73] and Keller and Segel [51, 52] and it is often referred to as the Patlak–Keller–Segel (PKS) model. The PKS model is derived at the macroscopic level in terms of the cell density and chemoattractant concentration and in the two-dimensional (2-D) case reads as

$$
\begin{cases}
\rho_t + \nabla \cdot (\chi \rho \nabla c) = \mu \Delta \rho, \\
\tau c_t = \alpha \Delta c - \beta c + \gamma \rho,
\end{cases}
\tag{1}
$$

where the cell density ρ and chemoattractant concentration c are functions of the spatial variables $x = (x, y) \in \Omega \subset \mathbb{R}^2$ and time t, $\mu > 0$ and $\alpha > 0$ are diffusion coefficients, $\chi > 0$ is the chemotactic sensitivity constant, and the constants $\gamma > 0$ and $\beta > 0$ stand for the production and degradation rate of the chemoattractant, respectively. The constant τ determines the type of the system: It is parabolic–parabolic if $\tau = 1$ and parabolic–elliptic for $\tau = 0$.

The PKS model (1) can be generalized to better describe the reality by taking into account several additional factors. For instance, one may consider a more realistic chemotactic sensitivity function $\chi = \chi(\rho, c)$ in the first equation of (1) as, e.g., in [42, 51, 62, 89, 90]. Some other factors, such as growth and death of cells, production and uptake of the chemoattractant by cells, presence of food and other chemicals in the system, may also be incorporated into the chemotaxis model, see, e.g., [89, 90]. We also refer the reader to [13, 21, 43–45, 55, 75], where several other modifications of the PKS system have been studied.

The most important phenomenon in chemotaxis is self-aggregation of cells (dramatic increase of ρ in a number of "centers"; see, e.g., [1, 8, 10, 11, 23, 78, 97]), which may occur even when the cells are initially distributed almost evenly over Ω. We note that the solution behavior depends on the value of the total mass, which under the assumption that no-flux boundary conditions are imposed is conserved:

$$M := \int_{\Omega} \rho(x, t)\, dx \equiv \int_{\Omega} \rho(x, 0)\, dx.$$

The behavior of the solutions of (1) also depends on the number of space dimensions. In the one-dimensional (1-D) case, global solutions exist for all initial conditions. In the 2-D case, the solution of (1) exists globally in time as long as the total mass M is initially below a critical threshold M_c. Otherwise, the solution may blow up in finite time; see, e.g., [12, 22, 36–39, 42, 44, 45, 49, 65, 75]. This blowup represents a mathematical description of the cell aggregation phenomenon that occurs in real biological systems; see, e.g., [1, 8, 10, 11, 22, 23, 67, 78]. In the parabolic–elliptic case ($\tau = 0$), the critical mass values M_c are explicitly available, while this is not the case for the parabolic–parabolic system ($\tau = 1$); see, e.g., [75]. The density ρ of the blowing up solutions of (1) becomes a linear combination of several Dirac δ-functions plus a regular part; see, e.g., [22, 40, 66].

While the blowup and the formation of the δ-function are not an unreasonable modeling of the cell aggregation phenomenon, they create enormous and sometimes unnecessary challenges to both numerics and analysis. As a result, a number of regularizations of the PKS system (1) have been introduced in the literature. Many of the regularized models admit bounded, global-in-time solutions that approach *spiky* steady states as time increases. Most of the regularized chemotaxis models can be put into the following form:

$$\begin{cases} \rho_t + \nabla \cdot (g(\rho)\, \mathbf{Q}\, (\chi \nabla c)) = \mu \Delta \rho, \\ \tau c_t = \alpha \Delta c - \beta c + \gamma \rho, \end{cases} \tag{2}$$

where $g > 0$ and $\mathbf{Q} = (Q_1, Q_2)$ are smooth functions of their arguments; see, e.g., [21, 43–45, 55, 62, 69, 76, 80, 95] and the references therein. Some of the typical examples include a chemotaxis model with a saturated chemotactic flux, logistic, signal- and density-dependent sensitivity models, and others.

A model with saturated chemotactic flux was proposed in [21], where the following functions g and $\mathbf{Q}$ were taken:

$$g(\rho) = \rho, \quad \mathbf{Q}(\chi \nabla c) = \begin{cases} \chi \nabla c, & \text{if } |\nabla c| \leq s^*, \\ \left(\dfrac{\chi|\nabla c| - s^*}{\sqrt{1 + (\chi|\nabla c| - s^*)^2}} + s^* \right) \dfrac{\nabla c}{|\nabla c|}, & \text{otherwise.} \end{cases} \tag{3}$$

Here, s^* is a switching parameter, which defines small gradient values, for which the system (2), (3) reduces to the original PKS system (1) so that the effect of saturated chemotactic flux function is felt at large gradient regimes only. This is expected to result in solutions which are spiky but yet bounded for all times; see, e.g., [21, 55].

The signal- and density-dependent sensitivity models can be obtained, for instance, by taking

$$g(\rho) = \rho \quad \text{and} \quad \boldsymbol{Q}(\chi \nabla c) = \chi \frac{\nabla c}{(1 + \kappa c)^2}$$

or

$$g(\rho) = \frac{\rho}{1 + \kappa \rho} \quad \text{and} \quad \boldsymbol{Q}(\chi \nabla c) = \chi \nabla c,$$

respectively, where κ is a (small) regularization parameter and $\kappa \to 0$ leads to the original PKS system (1). A global classical solution exists for these models and the regularization parameter κ allows one to conduct a detailed bifurcation analysis and study pattern formation and properties of nonuniform solutions; see, e.g., [43, 93, 94] and the references therein.

Generalizations of the PKS system (1) to multi-component chemotaxis models are also widely discussed in the literature; see, e.g., [29–31, 96] and the references therein. In this case, a mathematical model for, say, two noncompetitive biological species is governed by the following system of PDEs:

$$\begin{cases} (\rho_1)_t + \nabla \cdot (g_1(\rho_1) \boldsymbol{Q}(\chi_1 \nabla c)) = \mu_1 \Delta \rho_1, \\ (\rho_2)_t + \nabla \cdot (g_2(\rho_2) \boldsymbol{S}(\chi_2 \nabla c)) = \mu_2 \Delta \rho_2, \\ \tau c_t = \alpha \Delta c - \beta c + \gamma_1 \rho_1 + \gamma_2 \rho_2 = 0. \end{cases} \tag{4}$$

This model was proposed in [96] and then further studied both analytically [24, 29–33, 57] and numerically [16, 20, 55]. In (4), $\rho_1(\boldsymbol{x}, t)$ and $\rho_2(\boldsymbol{x}, t)$ denote the cell densities of the first and second species, $g_1 > 0$, $g_2 > 0$, $\boldsymbol{Q} = (Q_1, Q_2)$, and $\boldsymbol{S} = (S_1, S_2)$ are smooth functions of their arguments, $\chi_2 > \chi_1 > 0$ are the chemotactic sensitivity constants, $\mu_1 > 0$ and $\mu_2 > 0$ are diffusion coefficients, and $\gamma_1 > 0$ and $\gamma_2 > 0$ are the production rates for the first and second species, respectively.

Similarly to the one-species PKS model, solutions of (4) may either remain smooth (with decaying maxima of both ρ_1 and ρ_2) or blow up in a finite time as it was proven in [29, 31]. Moreover, in some cases only simultaneous blowup is possible, while in others the theory fails to predict the behavior of the solution. In addition, in the blow-up regime ρ_1 and ρ_2 may develop different types of singularities depending on the choice of functions g_1, g_2, $\boldsymbol{Q}$, and $\boldsymbol{S}$, values of χ_1 and χ_2, and on the total mass of each species:

$$M_1 := \int_\Omega \rho_1(\boldsymbol{x}, t) \, dx \, dy \quad \text{and} \quad M_2 := \int_\Omega \rho_2(\boldsymbol{x}, t) \, dx \, dy.$$

The classical PKS system (1) and the aforementioned related systems model the chemotaxis phenomenon on a macroscopic level. In many cases, this allows one to obtain a qualitatively accurate description of the chemotactic cell movement in an efficient way by solving the studied PDE models numerically. However, in certain practically relevant situations, a more accurate description may be required.

Typically microscopic models are used, such as the Fokker–Planck equations, the Langevin equations, or even some discrete particles models. We refer the reader to an overview paper [98] on collective behavior of active matter that includes swimming bacteria, chemotaxis effects, and many others; see also [6, 61, 74, 77].

In order to describe the chemotaxis at the cellular (microscopic) level, a class of Boltzmann-type kinetic equations has also been developed. A stochastic approach based on the velocity-jump process was introduced in [85] and was later used in the framework of kinetic chemotaxis models in [2, 70, 82]. The velocity-jump process characterizes the movement in two phases, namely run and tumble. During the run phase, the cells move (almost) linearly with constant speed and in the tumble phase, they reorient their motion with a new velocity and direction. The resulting nondimensionalized Boltzmann-type kinetic equation reads as (see, e.g., [15, 41, 71]):

$$
\begin{cases}
\varepsilon f_t + \boldsymbol{v} \cdot \nabla_{\boldsymbol{x}} f = \dfrac{1}{\varepsilon} \int\limits_V \left(T[c] f' - T^*[c] f \right) \mathrm{d}\boldsymbol{v}', \\[2ex]
\tau c_t = \alpha \Delta c - \beta c + \gamma \rho,
\end{cases}
\tag{5}
$$

where $f := f_\varepsilon(\boldsymbol{x}, t, \boldsymbol{v})$ is the probability density function (pdf) of cells at the position $\boldsymbol{x}$ with the velocity $\boldsymbol{v} = (u, v) \in \mathcal{V} \subset \mathbb{R}^2$ at a given time t, and $f' := f_\varepsilon(\boldsymbol{x}, t, \boldsymbol{v}')$. In (5), $T[c]$ and $T^*[c]$ are the turning kernel operators, which describe the velocity change at $(\boldsymbol{x}, t)$ from $\boldsymbol{v}'$ to $\boldsymbol{v}$ and from $\boldsymbol{v}$ to $\boldsymbol{v}'$, respectively, that is, $T[c] := T_\varepsilon[c](\boldsymbol{v}, \boldsymbol{v}')$ and $T^*[c] := T_\varepsilon[c](\boldsymbol{v}', \boldsymbol{v})$, and ε is a nondimensional scaling parameter (mean-free path), which provides the ratio of the mean running length between jumps to the typical observation length scale. In this model, it is assumed that the tumble (the reorientation) is a Poisson process with rate $\int_{\mathcal{V}} T^*[c] \mathrm{d}\boldsymbol{v}'$ and that $T^*[c] / \int_{\mathcal{V}} T^*[c] \mathrm{d}\boldsymbol{v}'$ is the probability density for a change in velocity from $\boldsymbol{v}$ to $\boldsymbol{v}'$, given that a reorientation occurs for a cell at position $\boldsymbol{x}$, velocity $\boldsymbol{v}$, and time t; see, e.g., [15]. Notice that the microscopic pdf f is related to the macroscopic cell density ρ in the following way:

$$
\rho(\boldsymbol{x}, t) := \int\limits_{\mathcal{V}} f(\boldsymbol{x}, t, \boldsymbol{v}) \, \mathrm{d}\boldsymbol{v}.
\tag{6}
$$

The question of convergence (in the singular limit as $\varepsilon \to 0$) of the kinetic model (5) to the PKS system (1) has been extensively studied. More precisely, the global-in-time convergence was proven in the parabolic–elliptic case in [15]. In the parabolic–parabolic case, only local convergence results were established; see [48]. We also refer the reader to [71] for more results on the limiting process.

Solutions of the kinetic chemotaxis system (5) exhibit a behavior similar to those of the PKS system (1). They depend, however, not only on the value of the initial mass M, but also on the specific kernel T, whose choice is crucial in the kinetic chemotaxis modeling; see, e.g., [9, 15]. At the same time, kinetic chemotaxis system may provide a more detailed description of the underlying cell dynamics and thus may be advantageous in a variety of applications.

While a large amount of effort has been expended on theoretical analysis of both macroscopic PDE and kinetic chemotaxis models over the past decades, the choice of numerical methods for these models is still rather limited. The main difficulty in numerical simulations of chemotaxis and related phenomena is associated with capturing blowing up or rapidly growing spiky solutions with high resolution and in an efficient manner.

Finite-volume [34] and finite-element [64, 79] methods were proposed for the PKS system (1) with the parabolic–elliptic ($\tau = 0$) coupling. A fractional step numerical method for fully time-dependent chemotaxis system from [90, 97] was proposed in [91]. Such splitting approach may, however, not be applicable for the system (1) since its convective part may loose hyperbolicity. As it has been demonstrated in [18], the latter is a generic situation for the PKS model with parabolic–parabolic ($\tau = 1$) coupling. Several methods for the parabolic–parabolic PKS system have been recently proposed: a family of high-order discontinuous Galerkin methods has been designed in [27, 28]; an implicit flux-corrected finite-element method has been developed in [84]. These methods achieve high-order of accuracy, but their high memory usage and computational costs are among their obvious drawbacks. Simpler and more efficient finite-volume and finite-volume-finite-difference methods were derived for the PKS system [16, 18], two-species chemotaxis [16, 55], and coupled chemotaxis-fluid models [17]. In [26], a modified version of the scheme from [18] is extended to the PKS system in irregular geometry using the upwind-difference potentials method. In [20], an adaptive moving mesh (AMM) finite-volume semi-discrete upwind method was developed and applied to two-species chemotaxis systems (4). Finally, several finite-volume methods have been recently introduced in [14, 19, 35] for 1-D and 2-D kinetic chemotaxis systems (5) with different turning kernels.

It should be observed that due to the lack of hyperbolicity of the convective part in many chemotaxis systems, enforcing nonlinear stability of the designed numerical methods may be a very challenging task. Indeed, unlike many other models, in which appearance of unphysical (small) negative values is numerically tolerable, in the chemotaxis models negative, even small negative values of the cell density ρ, produced by a numerical method, will trigger the development of negative cell density spikes, which in turn will make the computed solution completely irrelevant. Therefore, preserving positivity of the computed solutions is an absolutely crucial property of a good numerical method as this is the only way to guarantee its nonlinear stability. It is quite easy to design first-order positivity preserving schemes, but first-order schemes are typically impractical due to low resolution and low efficiency. Deriving high-order (high-resolution) schemes that possess this property is a significantly more complicated task, which was successfully achieved in a series of works presented in [16–18, 20, 21, 55].

In the context of kinetic chemotaxis system (5), additional difficulty in designing an efficient and accurate numerical method is related to the fact that the underlying system is stiff when $0 < \varepsilon \ll 1$. If an explicit numerical discretization is used, one may need to take both spatial and temporal discretization parameters to be proportional to $\mathscr{O}(\varepsilon)$ or even $\mathscr{O}(\varepsilon^2)$ due to severe stability restrictions, which may become unaffordable for small ε. Implicit discretizations, which are often uniformly stable for $0 < \varepsilon < 1$, may, on the other hand, be inconsistent with the limit problem and provide a wrong solution in the $\varepsilon \to 0$ limit. In order to overcome this difficulty, the so-called asymptotic preserving (AP) schemes, which yield a consistent approximation of the limiting macroscopic PKS system as $\varepsilon \to 0$ and are stable on a coarse spatio-temporal grid with the mesh parameters being independent of ε, have been recently introduced in [14, 19].

The goal of this review paper is to survey some of the recent advances in developing of high-resolution finite-volume and finite-difference numerical methods for chemotaxis-type systems that preserve positivity of the computed solutions and provide a consistent and stable discretization in certain asymptotic regimes.

The paper is organized as follows: In Sect. 2, we review positivity preserving hybrid finite-volume-finite-difference (FVFD) schemes. We begin in Sect. 2.1 with the second-order scheme, whose advantages and limitations are discussed and numerically illustrated in Sect. 2.1.1. The fourth-order FVFD scheme is presented in Sect. 2.2. Its advantages and disadvantages as well as additional moving-mesh resolution enhancement techniques are discussed in Sect. 2.2.1. In Sect. 3, we review the AP methods for kinetic chemotaxis systems. In order to construct the AP methods, we use the odd–even formulation (Sect. 3.1), Strang operator splitting (Sect. 3.2), the exact ODE solver for the stiff subsystem (Sect. 3.3.1), and second-order upwinding for the nonstiff subsystem (Sect. 3.3.2). Finally, the AP property is proven in Sect. 3.4.

2 Positivity Preserving Finite-Volume-Finite-Difference Methods

In this section, we describe second- and fourth-order hybrid FVFD schemes, which were originally proposed in [16] for the PKS model (1), but we adapt the description to the more general system (2) (see also [17, 21, 55]). To this end, we introduce the chemotactic velocities $U := c_x$ and $V := c_y$ and rewrite (2) in the equivalent form:

$$\begin{cases} \rho_t + (g(\rho)Q_1(\chi U) - \mu\rho_x)_x + (g(\rho)Q_2(\chi V) - \mu\rho_y)_y = 0, \\ \tau c_t = \alpha\Delta c - \beta c + \gamma\rho. \end{cases} \tag{7}$$

Note that in the particular case of linear functions g and $Q = (Q_1, Q_2)$, that is, if

$$g(\rho) = \rho, \quad Q_1(\chi U) = \chi U, \quad \text{and} \quad Q_2(\chi V) = \chi V, \tag{8}$$

the system (7) reduces to the PKS system (1).

Before proceeding with the presentation of the numerical schemes for (7), it should be pointed out that preserving the positivity of the computed cell density ρ is very important since appearance of negative values may trigger numerical instabilities as it was demonstrated in [18]. Indeed, in near blow-up regimes, the system (7) becomes convection-dominated and its convective part may lose hypervelocity. The latter can be illustrated by considering the parabolic–parabolic case ($\tau = 1$) and differentiating the second equation in (7) with respect to x and y and rewriting the system in the equivalent vector form as follows:

$$\begin{pmatrix} \rho \\ U \\ V \end{pmatrix}_t + \begin{pmatrix} g(\rho)Q_1(\chi U) \\ -\gamma\rho \\ 0 \end{pmatrix}_x + \begin{pmatrix} g(\rho)Q_2(\chi U) \\ 0 \\ -\gamma\rho \end{pmatrix}_y = \Delta \begin{pmatrix} \mu\rho \\ \alpha U \\ \alpha V \end{pmatrix} - \begin{pmatrix} 0 \\ \beta U \\ \beta V \end{pmatrix}. \quad (9)$$

One may now compute the x- and y-Jacobians of the convective fluxes, whose eigenvalues are given by

$$\left\{ 0, \quad \frac{g'(\rho)Q_1(\chi U) \pm \sqrt{(g'(\rho)Q_1(\chi U))^2 - 4\gamma g(\rho)Q_1'(\chi U)}}{2} \right\},$$

and

$$\left\{ 0, \quad \frac{g'(\rho)Q_2(\chi V) \pm \sqrt{(g'(\rho)Q_2(\chi V))^2 - 4\gamma g(\rho)Q_2'(\chi V)}}{2} \right\},$$

respectively. It is clear from the above calculations that the "purely" convective version of the systems (9)

$$\begin{pmatrix} \rho \\ U \\ V \end{pmatrix}_t + \begin{pmatrix} g(\rho)Q_1(\chi U) \\ -\gamma\rho \\ 0 \end{pmatrix}_x + \begin{pmatrix} g(\rho)Q_2(\chi U) \\ 0 \\ -\gamma\rho \end{pmatrix}_y = 0$$

is hyperbolic only if both

$$g'(\rho)Q_1(\chi U))^2 \geq 4\gamma g(\rho)Q_1'(\chi U) \quad \text{and} \quad g'(\rho)Q_2(\chi V))^2 \geq 4\gamma g(\rho)Q_2'(\chi V);$$

otherwise, it is elliptic. Unfortunately, the ellipticity condition is satisfied in generic cases, for example, when both g, Q_1 and Q_2 are linear functions (8), $U = V = 0$ and $\rho > 0$. Obviously, the complete chemotaxis system (9) contains stabilizing diffusion terms, but one has to be very careful since ellipticity of the convective part may still cause instabilities (especially if a fractional step approach is being implemented numerically).

It should also be emphasized that designing a positivity preserving second- and higher-order schemes is, in general, a nontrivial task and, to the best of our knowledge, the FVFD methods derived in [16] and presented below are among the first ones to achieve this goal for the chemotaxis system (7). In what follows we proceed with presenting these methods.

We consider the system (7) in a rectangular domain $\Omega \subset \mathbb{R}^2$, where we introduce a Cartesian mesh consisting of the cells $I_{j,k} := [x_{j-\frac{1}{2}}, x_{j+\frac{1}{2}}] \times [y_{k-\frac{1}{2}}, y_{k+\frac{1}{2}}]$, which, for the sake of simplicity, are assumed to be of the uniform size $|I_{j,k}| = \Delta x \Delta y$, that is, $x_{j+\frac{1}{2}} - x_{j-\frac{1}{2}} \equiv \Delta x$ for all j and $y_{k+\frac{1}{2}} - y_{k-\frac{1}{2}} \equiv \Delta y$ for all k. We assume that computed cell averages of ρ,

$$\overline{\rho}_{j,k}(t) := \frac{1}{|I_{j,k}|} \iint\limits_{I_{j,k}} \rho(x, y, t) \, \mathrm{d}x \, \mathrm{d}y,$$

and point values of the chemoattractant concentration, $c_{j,k}(t) \approx c(x_j, y_k, t)$, are available at a certain time level t (we will suppress the dependence of the indexed quantities on t in the subsequent text for the brevity of presentation unless it is required for clarity). These computed quantities are evolved in time according to a general semi-discrete hybrid FVFD scheme, which has the following form:

$$\begin{cases} \dfrac{\mathrm{d}\overline{\rho}_{j,k}}{\mathrm{d}t} = -\dfrac{\mathscr{F}_{j+\frac{1}{2},k} - \mathscr{F}_{j-\frac{1}{2},k}}{\Delta x} - \dfrac{\mathscr{G}_{j,k+\frac{1}{2}} - \mathscr{G}_{j,k-\frac{1}{2}}}{\Delta y}, \\[2mm] \tau \dfrac{\mathrm{d}c_{j,k}}{\mathrm{d}t} = \alpha \Delta_{j,k} c - \beta c_{j,k} + \gamma \rho_{j,k}. \end{cases} \tag{10}$$

Here, $\mathscr{F}_{j+\frac{1}{2},k}$ and $\mathscr{G}_{j,k+\frac{1}{2}}$ are the numerical fluxes in the x- and y-directions, respectively, $\Delta_{j,k}$ is a discrete Laplacian, and $\rho_{j,k}$ is an approximation of the point value of $\rho(x_j, y_k, t)$.

In the next sections, we will first provide the reader with a detailed description of both the second- and fourth-order versions of the FVFD scheme (10), including proofs of their positivity preserving property, and then present an AMM finite-volume method. These numerical schemes are obtained by different approximations of the numerical fluxes and discrete Laplacians in (10) and we will denote them by $\mathscr{F}^{\mathrm{II}}_{j+\frac{1}{2},k}$, $\mathscr{G}^{\mathrm{II}}_{j,k+\frac{1}{2}}$, $\Delta^{\mathrm{II}}_{j,k}$ and $\mathscr{F}^{\mathrm{IV}}_{j+\frac{1}{2},k}$, $\mathscr{G}^{\mathrm{IV}}_{j,k+\frac{1}{2}}$, $\Delta^{\mathrm{IV}}_{j,k}$ for the second- and fourth-order methods, respectively.

2.1 Second-Order Scheme

We approximate second-order numerical fluxes in (10),

$$\mathscr{F}^{\mathrm{II}}_{j+\frac{1}{2},k} = (gQ_1)^{\mathrm{II}}_{j+\frac{1}{2},k} - \mu(\rho_x)^{\mathrm{II}}_{j+\frac{1}{2},k} \quad \text{and} \quad \mathscr{G}^{\mathrm{II}}_{j,k+\frac{1}{2}} = (gQ_1)^{\mathrm{II}}_{j+\frac{1}{2},k} - \mu(\rho_y)^{\mathrm{II}}_{j,k+\frac{1}{2}},$$

(11)

using the central differences for the density derivatives:

$$(\rho_x)^{\mathrm{II}}_{j+\frac{1}{2},k} = \frac{\overline{\rho}_{j+1,k} - \overline{\rho}_{j,k}}{\Delta x}, \qquad (\rho_y)^{\mathrm{II}}_{j,k+\frac{1}{2}} = \frac{\overline{\rho}_{j,k+1} - \overline{\rho}_{j,k}}{\Delta y}, \qquad (12)$$

and an upwind differencing scheme for the chemotactic fluxes:

$$(gQ_1)^{\mathrm{II}}_{j+\frac{1}{2},k} = \begin{cases} g(\rho^{\mathrm{E}}_{j,k})\, Q_1\big(\chi U^{\mathrm{II}}_{j+\frac{1}{2},k}\big), & \text{if } Q_1\big(\chi U^{\mathrm{II}}_{j+\frac{1}{2},k}\big) > 0, \\[2mm] g(\rho^{\mathrm{W}}_{j+1,k})\, Q_1\big(\chi U^{\mathrm{II}}_{j+\frac{1}{2},k}\big), & \text{otherwise,} \end{cases}$$

(13)

$$(gQ_2)^{\mathrm{II}}_{j,k+\frac{1}{2}} = \begin{cases} g(\rho^{\mathrm{N}}_{j,k})\, Q_2\big(\chi V^{\mathrm{II}}_{j,k+\frac{1}{2}}\big), & \text{if } Q_2\big(\chi V^{\mathrm{II}}_{j,k+\frac{1}{2}}\big) > 0, \\[2mm] g(\rho^{\mathrm{S}}_{j,k+1})\, Q_2\big(\chi V^{\mathrm{II}}_{j,k+\frac{1}{2}}\big), & \text{otherwise.} \end{cases}$$

In (13), the point values of the velocities are obtained by the central differences:

$$U^{\mathrm{II}}_{j+\frac{1}{2},k} = \frac{c_{j+1,k} - c_{j,k}}{\Delta x}, \qquad V^{\mathrm{II}}_{j,k+\frac{1}{2}} = \frac{c_{j,k+1} - c_{j,k}}{\Delta y},$$

and the one-sided point values at the cell interfaces, $\rho^{\mathrm{E}}_{j,k}$, $\rho^{\mathrm{W}}_{j+1,k}$, $\rho^{\mathrm{N}}_{j,k}$, and $\rho^{\mathrm{S}}_{j,k+1}$, are calculated from a second-order piecewise linear reconstruction

$$\widetilde{\rho}(x,y) = \overline{\rho}_{j,k} + (\rho_x)_{j,k}(x - x_j) + (\rho_y)_{j,k}(y - y_k), \quad (x,y) \in I_{j,k}, \qquad (14)$$

as follows:

$$\rho^{\mathrm{E}}_{j,k} = \widetilde{\rho}(x_{j+\frac{1}{2}} - 0, y_k) = \overline{\rho}_{j,k} + \frac{\Delta x}{2}(\rho_x)^{\mathrm{II}}_{j,k},$$

$$\rho^{\mathrm{W}}_{j,k} = \widetilde{\rho}(x_{j-\frac{1}{2}} + 0, y_k) = \overline{\rho}_{j,k} - \frac{\Delta x}{2}(\rho_x)^{\mathrm{II}}_{j,k},$$

(15)

$$\rho^{\mathrm{N}}_{j,k} = \widetilde{\rho}(x_j, y_{k+\frac{1}{2}} - 0) = \overline{\rho}_{j,k} + \frac{\Delta y}{2}(\rho_y)^{\mathrm{II}}_{j,k},$$

$$\rho^{\mathrm{S}}_{j,k} = \widetilde{\rho}(x_j, y_{k-\frac{1}{2}} + 0) = \overline{\rho}_{j,k} - \frac{\Delta y}{2}(\rho_y)^{\mathrm{II}}_{j,k}.$$

It is important to guarantee nonnegativity of these reconstructed point values of ρ provided the computed cell averages $\overline{\rho}_{j,k}$ are nonnegative. One of the ways to achieve this goal is to use the following adaptive algorithm for computing the discrete slopes $(\rho_x)_{j,k}^{\mathrm{II}}$ and $(\rho_y)_{j,k}^{\mathrm{II}}$ in (15):

- Use central differences

$$(\rho_x)_{j,k}^{\mathrm{II}} = \frac{\overline{\rho}_{j+1,k} - \overline{\rho}_{j-1,k}}{2\Delta x}, \qquad (\rho_y)_{j,k}^{\mathrm{II}} = \frac{\overline{\rho}_{j,k+1} - \overline{\rho}_{j,k-1}}{2\Delta y}$$

to obtain the point values $\rho_{j,k}^{\mathrm{E,W,N,S}}$ in (15).
If either $\rho_{j,k}^{\mathrm{E}} < 0$ or $\rho_{j,k}^{\mathrm{W}} < 0$, then

Recompute these values by approximating the discrete derivative $(\rho_x)_{j,k}^{\mathrm{II}}$ in cell $I_{j,k}$ with the help of a positivity preserving nonlinear limiter. For instance, one can use the generalized minmod limiter [58, 60, 68, 86]:

$$(\rho_x)_{j,k}^{\mathrm{II}} = \mathrm{minmod}\left(2\frac{\overline{\rho}_{j+1,k} - \overline{\rho}_{j,k}}{\Delta x}, \frac{\overline{\rho}_{j+1,k} - \overline{\rho}_{j-1,k}}{2\Delta x}, 2\frac{\overline{\rho}_{j,k} - \overline{\rho}_{j-1,k}}{\Delta x}\right),$$

where

$$\mathrm{minmod}(z_1, z_2, \ldots) := \begin{cases} \min(z_1, z_2, \ldots), & \text{if } z_i > 0 \,\forall i, \\ \max(z_1, z_2, \ldots), & \text{if } z_i < 0 \,\forall i, \\ 0, & \text{otherwise.} \end{cases}$$

If either $\rho_{j,k}^{\mathrm{N}} < 0$ or $\rho_{j,k}^{\mathrm{S}} < 0$, then

Recompute these values by approximating the discrete derivatives $(\rho_y)_{j,k}^{\mathrm{II}}$ in cell $I_{j,k}$ with the help of a positivity preserving nonlinear limiter. Once again, one can use, for example, the generalized minmod limiter:

$$(\rho_y)_{j,k}^{\mathrm{II}} = \mathrm{minmod}\left(2\frac{\overline{\rho}_{j,k+1} - \overline{\rho}_{j,k}}{\Delta y}, \frac{\overline{\rho}_{j,k+1} - \overline{\rho}_{j,k-1}}{2\Delta y}, 2\frac{\overline{\rho}_{j,k} - \overline{\rho}_{j,k-1}}{\Delta y}\right).$$

This way, the positivity of reconstructed point values $\rho_{j,k}^{\mathrm{E,W,N,S}}$ will be guaranteed by the positivity preserving property of the chosen limiter. Besides the generalized minmod limiter, other positivity preserving limiters are available in the literature; see, e.g., [58, 60, 68, 86].

Finally, the Laplace operator in (10) is approximated using the standard five-point stencil:

$$\Delta_{j,k}^{\mathrm{II}} c = \frac{c_{j+1,k} - 2c_{j,k} + c_{j-1,k}}{(\Delta x)^2} + \frac{c_{j,k+1} - 2c_{j,k} + c_{j,k-1}}{(\Delta y)^2},$$

the point values $\rho_{j,k}$ are approximated by the corresponding cell averages, and then the resulting second-order semi-discrete hybrid FVFD scheme reads as

$$
\begin{cases}
\dfrac{d\overline{\rho}_{j,k}}{dt} = -\dfrac{\mathscr{F}^{\mathrm{II}}_{j+\frac{1}{2},k} - \mathscr{F}^{\mathrm{II}}_{j-\frac{1}{2},k}}{\Delta x} - \dfrac{\mathscr{G}^{\mathrm{II}}_{j,k+\frac{1}{2}} - \mathscr{G}^{\mathrm{II}}_{j,k-\frac{1}{2}}}{\Delta y}, \\[2mm]
\tau\dfrac{dc_{j,k}}{dt} = \alpha\Delta^{\mathrm{II}}_{j,k}c - \beta c_{j,k} + \gamma\overline{\rho}_{j,k}.
\end{cases}
\tag{16}
$$

The following theorems, proven in [16] for the PKS system (1), establish the positivity preserving property of the described numerical method in both the parabolic–parabolic ($\tau = 1$) and parabolic–elliptic ($\tau = 0$) cases. For the sake of completeness, we repeat the proof from [16] for a more general system (7).

Theorem 1 *Assume that the system of ODEs (16) with $\tau = 1$ is integrated using the forward Euler method:*

$$
\overline{\rho}_{j,k}(t + \Delta t) = \overline{\rho}_{j,k}(t) - \lambda\left(\mathscr{F}^{\mathrm{II}}_{j+\frac{1}{2},k}(t) - \mathscr{F}^{\mathrm{II}}_{j-\frac{1}{2},k}(t)\right)
$$
$$
- \sigma\left(\mathscr{G}^{\mathrm{II}}_{j,k+\frac{1}{2}}(t) - \mathscr{G}^{\mathrm{II}}_{j,k-\frac{1}{2}}(t)\right),
\tag{17}
$$

$$
c_{j,k}(t + \Delta t) = (1 - \beta\Delta t)c_{j,k}(t) + \alpha\Delta t\Delta^{\mathrm{II}}_{j,k}c_{j,k}(t) + \gamma\Delta t\overline{\rho}_{j,k}(t),
\tag{18}
$$

where $\lambda := \Delta t/\Delta x$ and $\sigma := \Delta t/\Delta y$. Then, the evolved cell densities $\overline{\rho}_{j,k}(t + \Delta t)$ and chemoattractant concentrations $c_{j,k}(t + \Delta t)$ will be nonnegative for all j, k provided $\overline{\rho}_{j,k}(t)$ and $c_{j,k}(t)$ are nonnegative for all j, k and the following CFL-like condition is satisfied:

$$
\Delta t \le \min\left\{\frac{\Delta x}{8a}, \frac{\Delta y}{8b}, \frac{\Delta x\Delta y}{4K\mu}, \frac{1}{\max\{K_1, \delta\}}\right\},
\tag{19}
$$

where

$$
a := \max_{j,k}|Q_1(\chi U^{\mathrm{II}}_{j+\frac{1}{2},k})| \cdot \max_{j,k}\left\{\frac{g(\rho^{\mathrm{E}}_{j,k})}{\rho^{\mathrm{E}}_{j,k}}, \frac{g(\rho^{\mathrm{W}}_{j,k})}{\rho^{\mathrm{W}}_{j,k}}\right\},
$$

$$
b := \max_{j,k}|Q_2(\chi V^{\mathrm{II}}_{j,k+\frac{1}{2}})| \cdot \max_{j,k}\left\{\frac{g(\rho^{\mathrm{N}}_{j,k})}{\rho^{\mathrm{N}}_{j,k}}, \frac{g(\rho^{\mathrm{S}}_{j,k})}{\rho^{\mathrm{S}}_{j,k}}\right\},
\tag{20}
$$

$$
K := \frac{\Delta x}{\Delta y} + \frac{\Delta y}{\Delta x}, \quad K_1 := \max_{j,k}\left(\beta + \frac{2K\alpha}{\Delta x\Delta y} - \gamma\overline{\rho}_{j,k}\right),
$$

and $\delta > 0$ is a small parameter.

Proof We follow the lines of the positivity proof in [16, 18] and begin with the cell density equation (17). Recall that the positivity preserving property of the interpolant (14) will guarantee that the reconstructed point values $\rho_{j,k}^E$, $\rho_{j,k}^W$, $\rho_{j,k}^N$, and $\rho_{j,k}^S$ will be nonnegative provided $\overline{\rho}_{j,k}(t) \geq 0$, $\forall j, k$. We then use (11)–(13) and the conservation property for the cell densities, $\overline{\rho}_{j,k} = \frac{1}{8}(\rho_{j,k}^E + \rho_{j,k}^W + \rho_{j,k}^S + \rho_{j,k}^N) + \frac{1}{2}\overline{\rho}_{j,k}$, to regroup the terms in (17) as follows:

$$
\begin{aligned}
\overline{\rho}_{j,k}(t+\Delta t) =\ & \left[\frac{1}{8}\rho_{j,k}^W - \frac{\lambda}{2}\Big(|Q_1(\chi U_{j-\frac{1}{2},k}^{II})| - Q_1(\chi U_{j-\frac{1}{2},k}^{II})\Big)g(\rho_{j,k}^W)\right] \\
& + \left[\frac{1}{8}\rho_{j,k}^E - \frac{\lambda}{2}\Big(|Q_1(\chi U_{j+\frac{1}{2},k}^{II})| + Q_1(\chi U_{j+\frac{1}{2},k}^{II})\Big)g(\rho_{j,k}^E)\right] \\
& + \frac{\lambda}{2}\left[|Q_1(\chi U_{j+\frac{1}{2},k}^{II})| - Q_1(\chi U_{j+\frac{1}{2},k}^{II})\right]g(\rho_{j+1,k}^W) \\
& + \frac{\lambda}{2}\left[|Q_1(\chi U_{j-\frac{1}{2},k}^{II})| + Q_1(\chi U_{j-\frac{1}{2},k}^{II})\right]g(\rho_{j-1,k}^E) \\
& + \left[\frac{1}{8}\rho_{j,k}^S - \frac{\sigma}{2}\Big(|Q_2(\chi V_{j,k-\frac{1}{2}}^{II})| - Q_2(\chi V_{j,k-\frac{1}{2}}^{II})\Big)g(\rho_{j,k}^S)\right] \\
& + \left[\frac{1}{8}\rho_{j,k}^N - \frac{\sigma}{2}\Big(|Q_2(\chi V_{j,k+\frac{1}{2}}^{II})| + Q_2(\chi V_{j,k+\frac{1}{2}}^{II})\Big)g(\rho_{j,k}^N)\right] \\
& + \frac{\sigma}{2}\left[|Q_2(\chi V_{j,k+\frac{1}{2}}^{II})| - Q_2(\chi V_{j,k+\frac{1}{2}}^{II})\right]g(\rho_{j,k+1}^S) \\
& + \frac{\sigma}{2}\left[|Q_2(\chi V_{j,k-\frac{1}{2}}^{II})| + Q_2(\chi V_{j,k-\frac{1}{2}}^{II})\right]g(\rho_{j,k-1}^N) + \left[\frac{1}{2} - \frac{2K\mu\Delta t}{\Delta x \Delta y}\right]\overline{\rho}_{j,k}(t) \\
& + \mu\Delta t\left[\frac{\overline{\rho}_{j+1,k}(t) + \overline{\rho}_{j-1,k}(t)}{(\Delta x)^2} + \frac{\overline{\rho}_{j,k+1}(t) + \overline{\rho}_{j,k-1}(t)}{(\Delta y)^2}\right].
\end{aligned}
$$

$$(21)$$

As one can see from (21), $\overline{\rho}_{j,k}(t + \Delta t)$ is a linear combination of the five cell averages $\overline{\rho}_{j,k}(t)$, $\overline{\rho}_{j\pm1,k}(t)$, $\overline{\rho}_{j,k\pm1}(t)$ and eight point values $g(\rho_{j,k}^W)$, $g(\rho_{j,k}^E)$, $g(\rho_{j+1,k}^W)$, $g(\rho_{j-1,k}^E)$, $g(\rho_{j,k}^S)$, $g(\rho_{j,k}^N)$, $g(\rho_{j,k+1}^S)$, $g(\rho_{j,k-1}^N)$, which are all nonnegative since $g > 0$ by assumption. The coefficients of this linear combination are also nonnegative under the CFL-like condition (19), which guarantees that $\overline{\rho}_{j,k}(t + \Delta t) \geq 0$ for all j, k.

Finally, the CFL-like condition (19) ensures that all of the terms on the right-hand side (RHS) of (18) are nonnegative and thus $c_{j,k}(t + \Delta t) \geq 0$ for all j, k, which completes the proof of the theorem. □

A similar theorem can be proven for the system of differential-algebraic equations (16) with $\tau = 0$.

Theorem 2 *Assume that the first equation of the system of (16) with $\tau = 0$ is integrated using the forward Euler method resulting in Eq. (17), while the system of linear algebraic equations for $c_{j,k}$ is solved exactly. Then, the evolved cell densities,*

$\bar{\rho}_{j,k}(t + \Delta t)$, and chemoattractant concentrations, $c_{j,k}(t + \Delta t)$, will be nonnegative for all j, k provided $\bar{\rho}_{j,k}(t)$ and $c_{j,k}(t)$ are nonnegative for all j, k and the following CFL-like condition is satisfied:

$$\Delta t \leq \min \left\{ \frac{\Delta x}{8a}, \frac{\Delta y}{8b}, \frac{\Delta x \Delta y}{4K\mu} \right\}, \tag{22}$$

where a, b, and K are given by (20).

Proof The proof of this theorem follows the lines of the proof of Theorem 1 and the positivity of ρ is enforced the same way as in the parabolic–parabolic case, but with a different CFL-like condition as stated in the theorem. The difference in the CFL conditions is due to the fact that for $\tau = 0$ the equation for the chemoattractant c reduces to a system of linear algebraic equations for $c_{j,k}$, which is to be solved by an accurate and efficient linear solver. It should be observed that the matrix of this linear system is diagonally dominant, which guarantees the positivity of c and no extra term is necessary in the CFL-like condition (22) compared to (19).   □

Remark 1 Theorems 1 and 2 are also valid if the forward Euler method is replaced by a higher-order strong-stability preserving (SSP) ODE solver, since a time step in such solver can be written as a convex combination of several forward Euler steps.

Remark 2 It is instructive to point out that the upper bounds on the time step in (19) and (22) are minima of either four or three terms: the first two terms are related to the chemotactic fluxes, while the third and fourth ones are due to the diffusion and source terms there. In the (near) blow-up regime, the system is convection-dominated, that is, the quantities a and b in (20) are large and thus the first two terms in (19) and (22) determine the size of time steps, in which case explicit methods are sufficiently efficient. However, when a and b are small, the third and fourth terms in (19) and (22) dominate and the efficiency of explicit methods is reduced. One of the ways to overcome this difficulty is to use implicit–explicit (IMEX) SSP methods (see, e.g., [4, 5, 47, 72]) as long as a and b remain relatively small. This does not affect the positivity preserving property of described method as was proven in [18].

2.1.1   Why Higher Resolution May Be Needed

In this section, we discuss the advantages and limitations of the second-order hybrid FVFD scheme described in Sect. 2.1 and also demonstrate why higher-order/higher-resolution methods may be required to accurately compute solutions of chemotaxis systems. To this end, we will consider several numerical examples.

Example 1 We start by considering the PKS system (1) with

$$\tau = 0, \quad \chi = 20, \quad \mu = \alpha = \beta = \gamma = 1$$

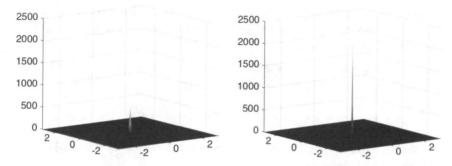

Fig. 1 Example 1: Cell density ρ computed using 101×101 (left) and 201×201 (right) meshes

and subject to the no-flux boundary conditions on $\Omega = [-3, 3] \times [-3, 3]$ and the following initial condition:

$$\rho(x, y, 0) = 100 \, e^{-100(x^2+y^2)}.$$

According to the analytical results (see, e.g., [12, 22, 36–39, 42, 44, 45, 49, 65, 75]), the initial mass is above the critical one and therefore the solution is expected to blow up in finite time. In order to illustrate the blow-up phenomenon, we compute the solution until the final time $t = 0.0038$ on a 101×101 uniform mesh. The obtained cell density is shown in Fig. 1 (left). As one can see,

$$\|\rho^{101}(\cdot, \cdot, 0.0038)\|_\infty \approx 559 \gg \|\rho(\cdot, \cdot, 0)\|_\infty = 100,$$

which suggests that the solution may have already blown up by the final computational time. In order to verify this, we refine the mesh to 201×201 and observe that the maximum of the computed cell density shown in Fig. 1 (right) is now

$$\|\rho^{201}(\cdot, \cdot, 0.0038)\|_\infty \approx 2248 \approx 4 \times \|\rho^{101}(\cdot, \cdot, 0.0038)\|_\infty.$$

This implies that the solution contains a δ-function, whose discrete maximum is supposed to be proportional to $1/(\Delta x \Delta y)$.

Comparing numerical solutions and its maximum norm on various meshes may allow one not only to confirm the solution blowup but also to numerically predict the blow-up time; see, e.g., [16]. The latter is an important piece of information, which may be required to be computed with a high precision and thus one may need to use a higher-order scheme and/or a certain adaptive strategy as an alternative to further mesh refinement study, which may become computationally unaffordable; see, e.g., [16, 20].

We would like to point out that even though using a higher-order or adaptive method in the initial-boundary value problem (IBVP) considered in Example 1 may enhance the achieved resolution, the second-order hybrid FVFD method seems to be sufficient to qualitatively understand the solution behavior (blowup). This may, however, not be the case in other situations like the one that will be considered in the second numerical example.

Example 2 We now consider the two-species chemotaxis system (4) with linear functions $g(\rho) = \rho$, $Q(\chi_1 \nabla c) = \chi_1 \nabla c$, and $Q(\chi_2 \nabla c) = \chi_2 \nabla c$, that is,

$$\tau = 0, \quad \chi_1 = 1 < \chi_2 = 20, \quad \mu_1 = \mu_2 = \alpha = \beta = \gamma_1 = \gamma_2 = 1,$$

and subject to the no-flux boundary conditions on $\Omega = [-3, 3] \times [-3, 3]$ and the following initial conditions:

$$\rho_1(x, y, 0) = \rho_2(x, y, 0) = 50 \, e^{-100(x^2 + y^2)}.$$

Although the scheme in Sect. 2.1 was described for the one-species system (2), it can be straightforwardly extended to the two-species system (4) since the equations for ρ_1 and ρ_2 are only coupled through the c-equation. We note that a detailed description of the second-order hybrid FVFD scheme for the two-species model can be found in [55].

We compute the solution of the studied IBVP on a 201×201 uniform mesh. The cell densities ρ_1 and ρ_2 computed at time $t = 0.0038$ are presented in Fig. 2 (left column). As in Example 1, one can observe that the second-order FVFD scheme captures the spiky structure of the solution. However, one can clearly see the difference between the maximum values of ρ_1 and ρ_2:

$$\|\rho_1^{201}(\cdot, \cdot, 0.0038)\|_\infty \approx 30.26 < \|\rho_1(\cdot, \cdot, 0)\|_\infty = 50,$$

$$\|\rho_2^{201}(\cdot, \cdot, 0.0038)\|_\infty \approx 646 \gg \|\rho_2(\cdot, \cdot, 0)\|_\infty = 50,$$

which suggests that ρ_2 had blown up, while ρ_1 remained bounded as its magnitude, in fact, had decreased. One may want to verify this conjecture by refining the mesh. The results computed on a 401×401 uniform mesh are shown in Fig. 2 (right column). The corresponding maximum values are now

$$\|\rho_1^{401}(\cdot, \cdot, 0.0038)\|_\infty \approx 34.05 < \|\rho_1(\cdot, \cdot, 0)\|_\infty = 50,$$

$$\|\rho_2^{401}(\cdot, \cdot, 0.0038)\|_\infty \approx 2365 \approx 4 \times \|\rho_2(\cdot, \cdot, 0.0038)\|_\infty.$$

These computations seem to confirm the above conjecture. However, the analytical results proven in [29, 31] state that only simultaneous blowup is possible, which means that ρ_1 should also have blown up by $t = 0.0038$ though at a much slower rate than ρ_2. This means that the above conjecture, which was made solely based on the above second-order numerical results, was wrong. Therefore, one would definitely need a higher-resolution method to obtain qualitatively more accurate results in this example.

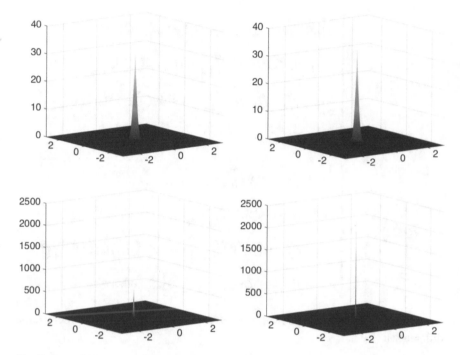

Fig. 2 Example 2: ρ_1 (top row) and ρ_2 (bottom row) computed using the second-order hybrid FVFD scheme on the 201×201 (left column) and 401×401 (right column) uniform meshes

2.2   Fourth-Order Scheme

In this section, we describe the fourth-order hybrid FVFD scheme, which is a slightly improved version of the scheme that was derived in [16]. As it was mentioned above, we denote the fourth-order fluxes by

$$\mathscr{F}^{\mathrm{IV}}_{j+\frac{1}{2},k} = (gQ_1)^{\mathrm{IV}}_{j+\frac{1}{2},k} - \mu(\rho_x)^{\mathrm{IV}}_{j+\frac{1}{2},k} \quad \text{and} \quad \mathscr{G}^{\mathrm{IV}}_{j,k+\frac{1}{2}} = (gQ_2)^{\mathrm{IV}}_{j,k+\frac{1}{2}} - \mu(\rho_y)^{\mathrm{IV}}_{j,k+\frac{1}{2}},$$
(23)

and approximate the density derivatives $(\rho_x)^{\mathrm{IV}}_{j+\frac{1}{2},k}$ and $(\rho_y)^{\mathrm{IV}}_{j,k+\frac{1}{2}}$ in (23) using the fourth-order central differences:

$$(\rho_x)^{\mathrm{IV}}_{j+\frac{1}{2},k} = \frac{\overline{\rho}_{j-1,k} - 15\overline{\rho}_{j,k} + 15\overline{\rho}_{j+1,k} - \overline{\rho}_{j+2,k}}{12\Delta x},$$

$$(\rho_y)^{\mathrm{IV}}_{j,k+\frac{1}{2}} = \frac{\overline{\rho}_{j,k-1} - 15\overline{\rho}_{j,k} + 15\overline{\rho}_{j,k+1} - \overline{\rho}_{j,k+2}}{12\Delta y}.$$

As in the second-order case, the chemotactic flux terms are computed in an upwind manner as follows:

$$
(gQ_1)^{\mathrm{IV}}_{j+\frac{1}{2},k} =
\begin{cases}
\dfrac{1}{6}\Big[g(\rho^{\mathrm{NE}}_{j,k})Q_1(\chi U^{\mathrm{IV}}_{j+\frac{1}{2},k+\frac{1}{2}}) + 4g(\rho^{\mathrm{E}}_{j,k})Q_1(\chi U^{\mathrm{IV}}_{j+\frac{1}{2},k}) \\
\qquad + g(\rho^{\mathrm{SE}}_{j,k})Q_1(\chi U^{\mathrm{IV}}_{j+\frac{1}{2},k-\frac{1}{2}})\Big], \quad \text{if } Q_1(\chi U^{\mathrm{IV}}_{j+\frac{1}{2},k}) > 0, \\[2mm]
\dfrac{1}{6}\Big[g(\rho^{\mathrm{NW}}_{j+1,k})Q_1(\chi U^{\mathrm{IV}}_{j+\frac{1}{2},k+\frac{1}{2}}) + 4g(\rho^{\mathrm{W}}_{j+1,k})Q_1(\chi U^{\mathrm{IV}}_{j+\frac{1}{2},k}) \\
\qquad + g(\rho^{\mathrm{SW}}_{j+1,k})Q_1(\chi U^{\mathrm{IV}}_{j+\frac{1}{2},k-\frac{1}{2}})\Big], \quad \text{otherwise,}
\end{cases}
$$

$$
(gQ_2)^{\mathrm{IV}}_{j,k+\frac{1}{2}} =
\begin{cases}
\dfrac{1}{6}\Big[g(\rho^{\mathrm{NW}}_{j,k})Q_2(\chi V^{\mathrm{IV}}_{j-\frac{1}{2},k+\frac{1}{2}}) + 4g(\rho^{\mathrm{N}}_{j,k})Q_2(\chi V^{\mathrm{IV}}_{j,k+\frac{1}{2}}) \\
\qquad + g(\rho^{\mathrm{NE}}_{j,k})Q_2(\chi V^{\mathrm{IV}}_{j+\frac{1}{2},k+\frac{1}{2}})\Big], \quad \text{if } Q_2(\chi V^{\mathrm{IV}}_{j,k+\frac{1}{2}}) > 0, \\[2mm]
\dfrac{1}{6}\Big[g(\rho^{\mathrm{SW}}_{j,k+1})Q_2(\chi V^{\mathrm{IV}}_{j-\frac{1}{2},k+\frac{1}{2}}) + 4g(\rho^{\mathrm{S}}_{j,k+1})Q_2(\chi V^{\mathrm{IV}}_{j,k+\frac{1}{2}}) \\
\qquad + g(\rho^{\mathrm{SE}}_{j,k+1})Q_2(\chi V^{\mathrm{IV}}_{j+\frac{1}{2},k+\frac{1}{2}})\Big], \quad \text{otherwise,}
\end{cases}
\tag{24}
$$

where the velocities at the cell interfaces are obtained using the fourth-order central differences:

$$
\begin{aligned}
U^{\mathrm{IV}}_{j+\frac{1}{2},k} &= \frac{c_{j-1,k} - 27c_{j,k} + 27c_{j+1,k} - c_{j+2,k}}{24\Delta x}, \\
V^{\mathrm{IV}}_{j,k+\frac{1}{2}} &= \frac{c_{j,k-1} - 27c_{j,k} + 27c_{j,k+1} - c_{j,k+2}}{24\Delta y},
\end{aligned}
\tag{25}
$$

and the velocities at the cell vertices are obtained using the fourth-order averaging of the cell interface velocities (25), which results in

$$
\begin{aligned}
U^{\mathrm{IV}}_{j+\frac{1}{2},k+\frac{1}{2}} &= \frac{-U^{\mathrm{IV}}_{j+\frac{1}{2},k-1} + 9U^{\mathrm{IV}}_{j+\frac{1}{2},k} + 9U^{\mathrm{IV}}_{j+\frac{1}{2},k+1} - U^{\mathrm{IV}}_{j+\frac{1}{2},k+2}}{16}, \\
V^{\mathrm{IV}}_{j+\frac{1}{2},k+\frac{1}{2}} &= \frac{-V^{\mathrm{IV}}_{j-1,k+\frac{1}{2}} + 9V^{\mathrm{IV}}_{j,k+\frac{1}{2}} + 9V^{\mathrm{IV}}_{j+1,k+\frac{1}{2}} - V^{\mathrm{IV}}_{j+2,k+\frac{1}{2}}}{16}.
\end{aligned}
\tag{26}
$$

Remark 3 We note that in [16], the velocities $U^{\mathrm{IV}}_{j+\frac{1}{2},k+\frac{1}{2}}$ and $V^{\mathrm{IV}}_{j+\frac{1}{2},k+\frac{1}{2}}$ were computed in a different manner. However, the formulae in (2.19) in [16] are not fully fourth-order and also computationally more expensive than (26).

In (24), the density point values along the cell interfaces, $\rho^{\mathrm{E}}_{j,k}$, $\rho^{\mathrm{W}}_{j,k}$, $\rho^{\mathrm{N}}_{j,k}$, $\rho^{\mathrm{S}}_{j,k}$, $\rho^{\mathrm{NE}}_{j,k}$, $\rho^{\mathrm{NW}}_{j,k}$, $\rho^{\mathrm{SE}}_{j,k}$, and $\rho^{\mathrm{SW}}_{j,k}$, are calculated using a conservative piecewise polynomial reconstruction

$$\mathscr{P}_{j,k}(x, y) = \overline{\rho}_{j,k} + (\rho_x)_{j,k}(x - x_j) + (\rho_y)_{j,k}(y - y_k)$$

$$\frac{1}{2}(\rho_{xx})_{j,k}(x - x_j)^2 + (\rho_{xy})_{j,k}(x - x_j)(y - y_k) + \frac{1}{2}(\rho_{yy})_{j,k}(y - y_k)^2$$

$$\frac{1}{6}(\rho_{xxx})_{j,k}(x - x_j)^3 + \frac{1}{2}(\rho_{xxy})_{j,k}(x - x_j)^2(y - y_k)$$

$$\frac{1}{2}(\rho_{xyy})_{j,k}(x - x_j)(y - y_k)^2 + \frac{1}{6}(\rho_{yyy})_{j,k}(y - y_k)^3$$

$$\frac{1}{24}(\rho_{xxxx})_{j,k}(x - x_j)^4 + \frac{1}{4}(\rho_{xxyy})_{j,k}(x - x_j)^2(y - y_k)^2$$

$$\frac{1}{24}(\rho_{yyyy})_{j,k}(y - y_k)^4, \qquad (x, y) \in I_{j,k},$$

which is almost fourth-order accurate provided its coefficients are obtained from the conservation requirements (see [54, Appendix B] for details):

$$\frac{1}{\Delta x \Delta y} \iint\limits_{I_{j+m,k+\ell}} \mathscr{P}_{j,k}(x, y) \, dx \, dy = \overline{\rho}_{j+m,k+\ell}, \quad \{m, \ell \in \mathbb{Z} : |m| + |\ell| \le 2\}.$$

The corresponding density point values are computed by (see [16] for detailed formulae and [54, Appendix B] for the algorithm of their efficient implementation)

$$\rho_{j,k}^{E} = \max\{\mathscr{P}_{j,k}(x_{j+\frac{1}{2}}, y_k), 0\}, \quad \rho_{j,k}^{W} = \max\{\mathscr{P}_{j,k}(x_{j-\frac{1}{2}}, y_k), 0\},$$

$$\rho_{j,k}^{N} = \max\{\mathscr{P}_{j,k}(x_j, y_{k+\frac{1}{2}}), 0\}, \quad \rho_{j,k}^{S} = \max\{\mathscr{P}_{j,k}(x_j, y_{k-\frac{1}{2}}), 0\},$$

$$\rho_{j,k}^{NE} = \max\{\mathscr{P}_{j,k}(x_{j+\frac{1}{2}}, y_{k+\frac{1}{2}}), 0\}, \quad \rho_{j,k}^{NW} = \max\{\mathscr{P}_{j,k}(x_{j-\frac{1}{2}}, y_{k+\frac{1}{2}}), 0\},$$

$$\rho_{j,k}^{SE} = \max\{\mathscr{P}_{j,k}(x_{j+\frac{1}{2}}, y_{k-\frac{1}{2}}), 0\}, \quad \rho_{j,k}^{SW} = \max\{\mathscr{P}_{j,k}(x_{j-\frac{1}{2}}, y_{k-\frac{1}{2}}), 0\},$$

whose nonnegativity is enforced in the most straightforward way.

Equipped with the above quantities, we obtain the following (almost) fourth-order semi-discrete hybrid FVFD scheme:

$$\begin{cases} \dfrac{d\overline{\rho}_{j,k}}{dt} = -\dfrac{\mathscr{F}_{j+\frac{1}{2},k}^{IV} - \mathscr{F}_{j-\frac{1}{2},k}^{IV}}{\Delta x} - \dfrac{\mathscr{G}_{j,k+\frac{1}{2}}^{IV} - \mathscr{G}_{j,k-\frac{1}{2}}^{IV}}{\Delta y}, \\[4mm] \tau \dfrac{dc_{j,k}}{dt} = \alpha \Delta_{j,k}^{IV} c - \beta c_{j,k} + \gamma \rho_{j,k}, \end{cases} \tag{27}$$

where a nine-point stencil is used to compute a fourth-order approximate Laplace operator,

$$\Delta^{IV}_{j,k} c = \frac{-c_{j-2,k} + 16c_{j-1,k} - 30c_{j,k} + 16c_{j+1,k} - c_{j+2,k}}{12(\Delta x)^2}$$
$$+ \frac{-c_{j,k-2} + 16c_{j,k-1} - 30c_{j,k} + 16c_{j,k+1} - c_{j,k+2}}{12(\Delta y)^2},$$

(28)

and $\rho_{j,k} = \max\{\mathscr{P}_{j,k}(x_j, y_k), 0\}$ are point values of ρ at the centers of cells $I_{j,k}$.

As in the case of the second-order method, the fourth-order FVFD scheme (27) is either a system of ODEs ($\tau = 1$) or differential-algebraic equations ($\tau = 0$) and should be integrated in time by a sufficiently accurate and stable ODE solver. We recall that the positivity preserving property of the second-order method was enforced by both the positivity of the reconstructed point values of the density and a proper choice of the ODE solver and its time step. Unfortunately, this is not the case here: even if one uses an SSP ODE solver, positivity of ρ and c cannot be guaranteed.

One of the ways to ensure that the computed solution will remain nonnegative at all times is to adapt the so-called draining time step strategy, which was originally proposed in [7] in the context of the Saint-Venant system of shallow water equations. This approach, which is described below for the forward Euler time discretization, is based on the idea of locally limiting the outgoing fluxes at cells where negative solution values appear.

We start by considering the parabolic–parabolic case ($\tau = 1$) and reformulate the fourth-order Laplacian in (28) in terms of diffusion fluxes as follows:

$$\alpha \Delta^{IV}_{j,k} c = -\frac{\mathscr{H}^{IV}_{j+\frac{1}{2},k} - \mathscr{H}^{IV}_{j-\frac{1}{2},k}}{\Delta x} - \frac{\mathscr{L}^{IV}_{j,k+\frac{1}{2}} - \mathscr{L}^{IV}_{j,k-\frac{1}{2}}}{\Delta y},$$

where

$$\mathscr{H}^{IV}_{j+\frac{1}{2},k} = \alpha \frac{-c_{j-1,k} + 15c_{j,k} - 15c_{j+1,k} + c_{j+2,k}}{12\Delta x},$$

$$\mathscr{L}^{IV}_{j,k+\frac{1}{2}} = \alpha \frac{-c_{j,k-1} + 15c_{j,k} - 15c_{j,k+1} + c_{j,k+2}}{12\Delta y}.$$

We then rewrite the fourth-order semi-discrete hybrid FVFD scheme (27) in the following flux form:

$$\begin{cases} \dfrac{d\bar{\rho}_{j,k}}{dt} = -\dfrac{\mathscr{F}^{IV}_{j+\frac{1}{2},k} - \mathscr{F}^{IV}_{j-\frac{1}{2},k}}{\Delta x} - \dfrac{\mathscr{G}^{IV}_{j,k+\frac{1}{2}} - \mathscr{G}^{IV}_{j,k-\frac{1}{2}}}{\Delta y}, \\[4mm] \dfrac{dc_{j,k}}{dt} = -\dfrac{\mathscr{H}^{IV}_{j+\frac{1}{2},k} - \mathscr{H}^{IV}_{j-\frac{1}{2},k}}{\Delta x} - \dfrac{\mathscr{L}^{IV}_{j,k+\frac{1}{2}} - \mathscr{L}^{IV}_{j,k-\frac{1}{2}}}{\Delta y} - \beta c_{j,k} + \gamma \rho_{j,k}, \end{cases}$$

(29)

and evolve the numerical solution in time using the forward Euler discretization of (29):

$$\bar{\rho}_{j,k}(t+\Delta t)=\bar{\rho}_{j,k}(t)-\lambda\Big(\widehat{\mathscr{F}}^{\,\mathrm{IV}}_{j+\frac{1}{2},k}(t)-\widehat{\mathscr{F}}^{\,\mathrm{IV}}_{j-\frac{1}{2},k}(t)\Big)-\sigma\Big(\widehat{\mathscr{G}}^{\,\mathrm{IV}}_{j,k+\frac{1}{2}}(t)-\widehat{\mathscr{G}}^{\,\mathrm{IV}}_{j,k-\frac{1}{2}}(t)\Big),$$
$$(30)$$

$$c_{j,k}(t+\Delta t)=(1-\beta\Delta t)c_{j,k}(t)+\gamma\Delta t\rho_{j,k}(t)$$
$$-\lambda\Big(\widehat{\mathscr{H}}^{\,\mathrm{IV}}_{j+\frac{1}{2},k}(t)-\widehat{\mathscr{H}}^{\,\mathrm{IV}}_{j-\frac{1}{2},k}(t)\Big)-\sigma\Big(\widehat{\mathscr{L}}^{\,\mathrm{IV}}_{j,k+\frac{1}{2}}(t)-\widehat{\mathscr{L}}^{\,\mathrm{IV}}_{j,k-\frac{1}{2}}(t)\Big),$$
$$(31)$$

where $\lambda:=\Delta t/\Delta x$, $\sigma:=\Delta t/\Delta y$, and $\widehat{\mathscr{F}}^{\,\mathrm{IV}}_{j+\frac{1}{2},k}$, $\widehat{\mathscr{G}}^{\,\mathrm{IV}}_{j,k+\frac{1}{2}}$, $\widehat{\mathscr{H}}^{\,\mathrm{IV}}_{j+\frac{1}{2},k}$, and $\widehat{\mathscr{L}}^{\,\mathrm{IV}}_{j,k+\frac{1}{2}}$ are modified numerical fluxes defined as follows:

$$\widehat{\mathscr{F}}^{\,\mathrm{IV}}_{j+\frac{1}{2},k}=\frac{\Delta t^{\rho}_{j+\frac{1}{2},k}}{\Delta t}\,\mathscr{F}^{\mathrm{IV}}_{j+\frac{1}{2},k},\qquad \widehat{\mathscr{G}}^{\,\mathrm{IV}}_{j,k+\frac{1}{2}}=\frac{\Delta t^{\rho}_{j,k+\frac{1}{2}}}{\Delta t}\,\mathscr{G}^{\mathrm{IV}}_{j,k+\frac{1}{2}}$$

$$\widehat{\mathscr{H}}^{\,\mathrm{IV}}_{j+\frac{1}{2},k}=\frac{\Delta t^{c}_{j+\frac{1}{2},k}}{\Delta t}\,\mathscr{H}^{\mathrm{IV}}_{j+\frac{1}{2},k},\qquad \widehat{\mathscr{L}}^{\,\mathrm{IV}}_{j,k+\frac{1}{2}}=\frac{\Delta t^{c}_{j,k+\frac{1}{2}}}{\Delta t}\,\mathscr{L}^{\mathrm{IV}}_{j,k+\frac{1}{2}}.$$
$$(32)$$

The time step Δt and the coefficients in front of the numerical fluxes in (32) are obtained according to the following algorithm:

- Compute Δt according to the CFL-like condition (19).
- Compute "draining" time steps:

$$\Delta t^{\rho}_{j,k}:=\frac{\Delta x\,\Delta y\,\bar{\rho}_{j,k}(t)}{f^{\rho}_{j,k}\Delta y+g^{\rho}_{j,k}\Delta x},$$

$$\Delta t^{c}_{j,k}:=\frac{\Delta x\,\Delta y[(1-\beta\Delta t)c_{j,k}(t)+\gamma\Delta t\rho_{j,k}(t)]}{f^{c}_{j,k}\Delta y+g^{c}_{j,k}\Delta x},$$
$$(33)$$

where

$$f^{\rho}_{j,k}:=\max(\mathscr{F}^{\mathrm{IV}}_{j+\frac{1}{2},k},0)+\max(-\mathscr{F}^{\mathrm{IV}}_{j-\frac{1}{2},k},0),$$

$$g^{\rho}_{j,k}:=\max(\mathscr{G}^{\mathrm{IV}}_{j,k+\frac{1}{2}},0)+\max(-\mathscr{G}^{\mathrm{IV}}_{j,k-\frac{1}{2}},0),$$

$$f^{c}_{j,k}:=\max(\mathscr{H}^{\mathrm{IV}}_{j+\frac{1}{2},k},0)+\max(-\mathscr{H}^{\mathrm{IV}}_{j-\frac{1}{2},k},0),$$
$$(34)$$

$$g^{c}_{j,k}:=\max(\mathscr{L}^{\mathrm{IV}}_{j,k+\frac{1}{2}},0)+\max(-\mathscr{L}^{\mathrm{IV}}_{j,k-\frac{1}{2}},0).$$

- Use the computed values of Δt and "draining" time steps to obtain

(continued)

$$\Delta t^\rho_{j+\frac{1}{2},k} := \min(\Delta t, \Delta t^\rho_{m,k}), \quad m = j + \frac{1}{2} - \frac{\operatorname{sgn}(\mathscr{F}^{\mathrm{IV}}_{j+\frac{1}{2},k})}{2},$$

$$\Delta t^\rho_{j,k+\frac{1}{2}} := \min(\Delta t, \Delta t^\rho_{j,\ell}), \quad \ell = k + \frac{1}{2} - \frac{\operatorname{sgn}(\mathscr{G}^{\mathrm{IV}}_{j,k+\frac{1}{2}})}{2},$$

$$\Delta t^c_{j+\frac{1}{2},k} := \min(\Delta t, \Delta t^c_{p,k}), \quad p = j + \frac{1}{2} - \frac{\operatorname{sgn}(\mathscr{H}^{\mathrm{IV}}_{j+\frac{1}{2},k})}{2}, \tag{35}$$

$$\Delta t^c_{j,k+\frac{1}{2}} := \min(\Delta t, \Delta t^c_{j,q}), \quad q = k + \frac{1}{2} - \frac{\operatorname{sgn}(\mathscr{L}^{\mathrm{IV}}_{j,k+\frac{1}{2}})}{2}.$$

In the parabolic–elliptic ($\tau = 0$) case, the second equation in (27) reduces to a system of linear (time-independent) algebraic equations for $c_{j,k}$, which, as in the second-order method, is to be solved by an accurate and efficient linear algebra solver. However, the matrix of this linear system is no longer diagonally dominant, and thus the positivity of c cannot be in general guaranteed. On the other hand, the following theorem from [16] establishes the nonnegativity of the computed ρ and c at all times in the parabolic–parabolic ($\tau = 1$) case and we repeat its proof here for the sake of completeness.

Theorem 3 *Assume that the system of ODEs (29) is integrated using the forward Euler method (30)–(35). Then, the evolved cell densities $\overline{\rho}_{j,k}(t + \Delta t)$ and chemoattractant concentrations $c_{j,k}(t + \Delta t)$ remain nonnegative for all j, k as long as $\overline{\rho}_{j,k}(t)$ and $c_{j,k}(t)$ are nonnegative for all j, k.*

Proof In order to prove the nonnegativity of ρ, one needs to consider different cases depending on the sign of the fluxes $\mathscr{F}^{\mathrm{IV}}_{j+\frac{1}{2},k}$ and $\mathscr{G}^{\mathrm{IV}}_{j,k+\frac{1}{2}}$ given by (23). We will only consider one of these cases, namely assuming that

$$\mathscr{F}^{\mathrm{IV}}_{j+\frac{1}{2},k} > 0, \quad \mathscr{F}^{\mathrm{IV}}_{j-\frac{1}{2},k} > 0 \quad \text{and} \quad \mathscr{G}^{\mathrm{IV}}_{j,k+\frac{1}{2}} < 0, \quad \mathscr{G}^{\mathrm{IV}}_{j,k-\frac{1}{2}} < 0, \tag{36}$$

in the cell $I_{j,k}$. All of the other cases can be analyzed in a similar way.

First, we use the definitions in (34) to obtain

$$f^\rho_{j,k} = \mathscr{F}^{\mathrm{IV}}_{j+\frac{1}{2},k}, \quad g^\rho_{j,k} = -\mathscr{G}^{\mathrm{IV}}_{j,k-\frac{1}{2}}, \tag{37}$$

and then substituting (37) into (33) results in

$$\Delta t^\rho_{j,k} = \frac{\Delta x \, \Delta y \, \overline{\rho}_{j,k}(t)}{\mathscr{F}^{\mathrm{IV}}_{j+\frac{1}{2},k} \Delta y - \mathscr{G}^{\mathrm{IV}}_{j,k-\frac{1}{2}} \Delta x} > 0. \tag{38}$$

It also follows from (36) and (35) that

$$\Delta t^{\rho}_{j+\frac{1}{2},k} = \min(\Delta t, \Delta t^{\rho}_{j,k}), \qquad \Delta t^{\rho}_{j-\frac{1}{2},k} = \min(\Delta t, \Delta t^{\rho}_{j-1,k}),$$

$$\Delta t^{\rho}_{j,k+\frac{1}{2}} = \min(\Delta t, \Delta t^{\rho}_{j,k+1}), \qquad \Delta t^{\rho}_{j,k-\frac{1}{2}} = \min(\Delta t, \Delta t^{\rho}_{j,k}).$$

We now rewrite Eq. (30) as

$$
\begin{aligned}
\overline{\rho}_{j,k}(t + \Delta t) = \overline{\rho}_{j,k}(t) &+ \frac{\Delta t^{\rho}_{j-\frac{1}{2},k}}{\Delta x} \mathscr{F}^{\mathrm{IV}}_{j-\frac{1}{2},k} - \frac{\Delta t^{\rho}_{j,k+\frac{1}{2}}}{\Delta y} \mathscr{G}^{\mathrm{IV}}_{j,k+\frac{1}{2}} \\
&+ \frac{\Delta t^{\rho}_{j,k-\frac{1}{2}}}{\Delta y} \mathscr{G}^{\mathrm{IV}}_{j,k-\frac{1}{2}} - \frac{\Delta t^{\rho}_{j+\frac{1}{2},k}}{\Delta x} \mathscr{F}^{\mathrm{IV}}_{j+\frac{1}{2},k},
\end{aligned}
\tag{39}
$$

and show that the RHS of (39) is nonnegative. To this end, we first note that (36) implies

$$\frac{\Delta t^{\rho}_{j-\frac{1}{2},k}}{\Delta x} \mathscr{F}^{\mathrm{IV}}_{j-\frac{1}{2},k} - \frac{\Delta t^{\rho}_{j,k+\frac{1}{2}}}{\Delta y} \mathscr{G}^{\mathrm{IV}}_{j,k+\frac{1}{2}} > 0. \tag{40}$$

We then note that $\Delta t^{\rho}_{j+\frac{1}{2},k} = \Delta t^{\rho}_{j,k-\frac{1}{2}} \le \Delta t^{\rho}_{j,k}$, and therefore using (38), (39), and (40), we conclude with

$$
\begin{aligned}
\overline{\rho}_{j,k}(t + \Delta t) &> \overline{\rho}_{j,k}(t) + \frac{\Delta t^{\rho}_{j,k-\frac{1}{2}}}{\Delta y} \mathscr{G}^{\mathrm{IV}}_{j,k-\frac{1}{2}} - \frac{\Delta t^{\rho}_{j+\frac{1}{2},k}}{\Delta x} \mathscr{F}^{\mathrm{IV}}_{j+\frac{1}{2},k} \\
&\ge \overline{\rho}_{j,k}(t) + \frac{\Delta x\, \overline{\rho}_{j,k}(t)}{\mathscr{F}^{\mathrm{IV}}_{j+\frac{1}{2},k} \Delta y - \mathscr{G}^{\mathrm{IV}}_{j,k-\frac{1}{2}} \Delta x} \mathscr{G}^{\mathrm{IV}}_{j,k-\frac{1}{2}} - \frac{\Delta y\, \overline{\rho}_{j,k}(t)}{\mathscr{F}^{\mathrm{IV}}_{j+\frac{1}{2},k} \Delta y - \mathscr{G}^{\mathrm{IV}}_{j,k-\frac{1}{2}} \Delta x} \mathscr{F}^{\mathrm{IV}}_{j+\frac{1}{2},k} = 0,
\end{aligned}
$$

which shows that $\overline{\rho}_{j,k}(t + \Delta t) \ge 0$ for all j, k, provided that $\overline{\rho}_{j,k}(\Delta t) \ge 0$ for all j, k.

The nonnegativity proof for the c component of the solution can be obtained similarly, and the proof of the theorem will be completed. □

Remark 4 Note that the positivity preserving property of the second-order scheme can be also enforced using the "draining" time step technique instead of the adaptive reconstruction approach implemented in Sect. 2.1.

Remark 5 As an alternative way of achieving the positivity preserving property of fourth- and higher-order methods, one can implement a maximum-principle-satisfying approach developed in [99] in the context of scalar conservation laws.

2.2.1 What Can Be Achieved with Higher-Resolution Methods

As it was mentioned in Sect. 2.1.1, capturing fast growing and/or singular solutions of chemotaxis systems is a challenging task. Example 2 and studies conducted in a series of works [14, 16, 20, 55] clearly illustrate that in some cases under-resolved numerical simulations may lead to very misleading results in terms of determining blow-up regimes of the computed solutions. In many of such cases, including the two-spices chemotaxis models like (4), where different blow-up time scales are exhibited by the two variables, a very fine mesh (often practically unaffordable) would typically be required to make a blow-up conjecture based on the numerical results even if the aforementioned fourth-order FVFD method is implemented as demonstrated in the following example; see also [16, 20, 55].

Example 3 We now consider the same IBVP for two-species chemotaxis system as in Example 2 and numerically solve it using the hybrid fourth-order FVFD scheme described in Sect. 2.2, which, similarly to its second-order counterpart, can be straightforwardly extended to the two-species system.

As in Example 2, we first compute the solution of the studied IBVP on a 201×201 uniform mesh. The cell densities ρ_1 and ρ_2 computed at time $t = 0.0038$ are presented in Fig. 3 (left column). Once again, we observe that the maximum values of the computed ρ_1 and ρ_2,

$$\|\rho_1^{201}(\cdot, \cdot, 0.0038)\|_\infty \approx 36.18 < \|\rho_1(\cdot, \cdot, 0)\|_\infty = 50,$$

$$\|\rho_2^{201}(\cdot, \cdot, 0.0038)\|_\infty \approx 2104 \gg \|\rho_2(\cdot, \cdot, 0)\|_\infty = 50,$$

clearly indicate that ρ_2 blows up. In fact, its maximum is substantially larger than the one obtained by the second-order scheme in Example 2. However, it is hard (or even impossible) to draw a blow-up conclusion about ρ_1, whose maximum is still below the initial one though it is larger than the maximum of ρ_1 computed using the 401×401 uniform meshes reported in Example 2. We then refine the mesh. The results computed on a 401×401 uniform mesh are shown in Fig. 3 (right column). The corresponding maximum values are now

$$\|\rho_1^{401}(\cdot, \cdot, 0.0038)\|_\infty \approx 42.25 < \|\rho_1(\cdot, \cdot, 0)\|_\infty = 50,$$

$$\|\rho_2^{401}(\cdot, \cdot, 0.0038)\|_\infty \approx 8284 \approx 4 \times \|\rho_2(\cdot, \cdot, 0.0038)\|_\infty.$$

These computations demonstrate that the required resolution has not been achieved yet. One therefore may want to further increase the order of the numerical method. This, however, may be quite cumbersome and also computationally expensive. In addition, implementing higher-order boundary conditions may also be challenging.

As an alternative approach of enhancing the resolution of spiky solutions in the concentration/blow-up regions, one may use an adaptive technique. For example, an adaptive moving mesh (AMM) method was proposed in [20] for the two-spices

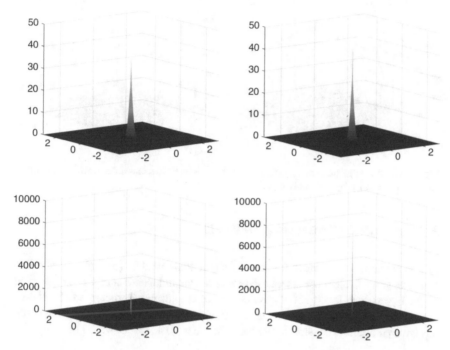

Fig. 3 Example 3: ρ_1 (top row) and ρ_2 (bottom row) computed using the fourth-order hybrid FVFD scheme on the 201×201 (left column) and 401×401 (right column) uniform meshes

parabolic–elliptic chemotaxis system (4) with $\tau = 0$ by combining the second-order positivity preserving finite-volume method with the AMM technique from [56]. A general algorithm of the AMM approach consists of the evolution, mesh adaptation, and projection steps. In particular, the solution is first evolved to the new time level on a given structured quadrilateral mesh. The mesh is then adapted to reflect the structure of the evolved solution and finally, the solution is projected onto the new mesh in a conservative manner. For more information about the AMM approach we refer the reader to [3, 46, 56, 87] and the references therein.

In order to illustrate the performance of the AMM method, we apply it to the aforementioned IBVP using 101×101 adaptively moving cells. The obtained results are reported in Fig. 4 and the maximum values of ρ_1 and ρ_2 are 62.38 and 622178. As one can see, the maximum value of ρ_2 is now much larger even than the one computed using the fourth-order FVFD scheme on the 401×401 uniform mesh. The maximum value of ρ_1 is now larger than the maximum if the initial datum, which suggests that ρ_1 may develop a sharp spiky structure even though one still cannot draw a definite conclusion whether the computed ρ_1 had blown up.

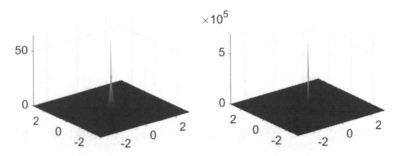

Fig. 4 Example 3: ρ_1 (left) and ρ_2 (right) computed using the second-order AMM finite-volume method 101×101 adaptively moving cells

3 Asymptotic Preserving Methods

In this section, we consider the kinetic chemotaxis models (5) and describe how consistent and stable methods whose properties are independent of ε can be derived.

As it was mentioned above, the choice of the turning kernel T in (5) plays a crucial role in chemotaxis modeling and we start by assuming that the turning kernel has an asymptotic expansion of the form (see, e.g., [14, 15, 35, 41, 71]):

$$T[c] = T_0[c] + \varepsilon T_1[c] + \mathcal{O}(\varepsilon^2). \tag{41}$$

Here, the leading term $T_0[c] = F(v) > 0$ is the bounded velocity distribution at the equilibrium, which satisfies the following assumptions:

$$\int_{\mathcal{V}} F(v)\, dv = 1 \quad \text{and} \quad F(v) = F(|v|). \tag{42}$$

The coefficient of the second term in (41), $T_1[c]$, describes the new favorable direction of the cells and following [14, 15] we assume that $T_1[c](v, v') = T_1[c](v)$ and consider the positive *taxis* towards the chemoattractant. We then substitute (6), (41), and (42) into the first equation of (5) and neglect $\mathcal{O}(\varepsilon^2)$ terms to obtain the following kinetic chemotaxis system:

$$\begin{cases} \varepsilon f_t + v \cdot \nabla_x f = 1 + \dfrac{1}{\varepsilon}[(F(v) + \varepsilon T_1)\rho - (1 + \mathscr{T}_1)f], \\[2mm] \tau c_t = \alpha \Delta c - \beta c + \gamma \rho, \end{cases} \tag{43}$$

where

$$\mathscr{T}_1(x, t) := \int_{\mathcal{V}} \varepsilon T_1[c](v)\, dv. \tag{44}$$

Specific models for turning kernels can be found in the literature. For instance, a group of the so-called local models, considered in [15, 71], suggests that $T_1[c]$ depends on point values of c and ∇c and thus (41) takes the form (up to high-order terms):

$$T[c](v) = F(v) + \varepsilon \max(v \cdot \nabla c, 0). \tag{45}$$

Other examples include nonlocal models, where the turning kernel is given by (see, e.g., [14, 15])

$$T[c](v) = \alpha_+ \psi(c(x, t), c(x + \varepsilon v, t)) + \alpha_- \psi(c(x, t), c(x - \varepsilon v, t)), \tag{46}$$

where ψ is a smooth, positive, nondecreasing function (in the second argument) defined on $\mathbb{R}^+ \times \mathbb{R}^+$ and such that $0 < \psi_{min} \leq \psi(c, \tilde{c}) \leq \alpha_1 \tilde{c} + \alpha_2$ with $\alpha_+, \alpha_-, \alpha_1$ and α_2 being some positive constants. This model implies that the cell is able to measure the chemoattractant concentration up to a distance εv_{max} away from its position, where v_{max} is the maximal speed in $\mathcal{V}$. A simplified version of (46) with $\alpha_+ = 1, \alpha_- = 0$, and $\psi(c, \tilde{c}) = \max(\tilde{c} - c, 0)$ was considered in [14] and reads as

$$T[c](v) = F(v) + \max(c(x + \varepsilon v, t) - c(x, t), 0). \tag{47}$$

In both models (45) and (47), the turning probability is higher for a change to a favorable direction and away from an unfavorable direction.

In what follows we describe an AP method for the system (43) following the idea of an odd–even formulation, which was first presented in [50] and further developed in [14, 19]. Such formulation allows one to efficiently implement the Strang splitting approach [83], by separating stiff and nonstiff parts of the system. In this setup, the nonstiff subsystem reduces to a system of linear transport equations, which can be solved by a second-order upwind method, and the stiff subsystem may either be solved exactly or by an implicit (uniformly stable in ε) method. The resulting numerical method becomes AP in the sense that it yields an accurate and uniformly stable in ε discretization, which stays consistent with the limiting system as $\varepsilon \to 0$.

3.1 Odd–Even Formulation

We restrict our attention to spherically symmetric sets $\mathcal{V} := \{v, |v| = v_0\}$ as a typical example and denote by $\mathcal{V}^+ := \{v = (u, v) \in \mathcal{V} \mid u > 0, v > 0\}$. From now on, we consider $v \in \mathcal{V}^+$ only and introduce new variables r_1, j_1, r_2, and j_2:

$$r_1(u, v) = R_1[f] := \frac{1}{2}[f(u, -v) + f(-u, v)],$$

$$r_2(u, v) = R_2[f] := \frac{1}{2}[f(u, v) + f(-u, -v)],$$

$$j_1(u, v) = J_1[f] := \frac{1}{2\varepsilon}[f(u, -v) - f(-u, v)],$$

$$j_2(u, v) = J_2[f] := \frac{1}{2\varepsilon}[f(u, v) - f(-u, -v)],$$

(48)

with a one-to-one correspondence between them and f:

$$f(u, v) = \begin{cases} r_2 + \varepsilon j_2, & u > 0, \ v > 0, \\ r_2 - \varepsilon j_2, & u < 0, \ v < 0, \\ r_1 + \varepsilon j_1, & u > 0, \ v < 0, \\ r_1 - \varepsilon j_1, & u < 0, \ v > 0. \end{cases}$$

It is instructive to point out that the macroscopic cell density ρ can be obtained from (6) and (48) in terms of the new variables r_1 and r_2:

$$\rho(\boldsymbol{x}, t) = 2 \int_{\mathcal{V}+} [r_1(\boldsymbol{x}, t, \boldsymbol{v}) + r_2(\boldsymbol{x}, t, \boldsymbol{v})] \, d\boldsymbol{v}.$$

(49)

Using for simplicity the notation $f(\pm u, \pm v)$ instead of $f(\boldsymbol{x}, t, \pm u, \pm v)$ and $T_1(\pm u, \pm v)$ instead of $T_1[c](\boldsymbol{x}, t, \pm u, \pm v)$ and taking into account that

$$\varepsilon f_t(u, v) + u f_x(u, v) + v f_y(u, v)$$

$$= \frac{\rho}{\varepsilon} [F(u, v) + \varepsilon T_1(u, v)] - \frac{1}{\varepsilon} (1 + \mathcal{T}_1) f(u, v),$$

$$\varepsilon f_t(-u, -v) - u f_x(-u, -v) - v f_y(-u, -v)$$

$$= \frac{\rho}{\varepsilon} [F(-u, -v) + \varepsilon T_1(-u, -v)] - \frac{1}{\varepsilon} (1 + \mathcal{T}_1) f(-u, -v),$$

$$\varepsilon f_t(u, -v) + u f_x(u, -v) - v f_y(u, -v)$$

$$= \frac{\rho}{\varepsilon} [F(u, -v) + \varepsilon T_1(u, -v)] - \frac{1}{\varepsilon} (1 + \mathcal{T}_1) f(u, -v),$$

$$\varepsilon f_t(-u, v) - u f_x(-u, v) + v f_y(-u, v)$$

$$= \frac{\rho}{\varepsilon} [F(-u, v) + \varepsilon T_1(-u, v)] - \frac{1}{\varepsilon} (1 + \mathcal{T}_1) f(-u, v),$$

we rewrite the first equation in (43) as a system of the following four coupled equations for r_1, j_1, r_2, and j_2:

$$(r_1)_t + u(j_1)_x - v(j_1)_x = \frac{\rho}{\varepsilon^2}(F(u,v) + \varepsilon R_1[T_1]) - \frac{1}{\varepsilon^2}(1 + \mathscr{T}_1)r_1,$$

$$(j_1)_t + \frac{1}{\varepsilon^2}u(r_1)_x - \frac{1}{\varepsilon^2}v(r_1)_y = \frac{\rho}{\varepsilon}J_1[T_1] - \frac{1}{\varepsilon^2}(1 + \mathscr{T}_1)j_1,$$

$$(r_2)_t + u(j_2)_x + v(j_2)_y = \frac{\rho}{\varepsilon^2}(F(u,v) + \varepsilon R_2[T_1]) - \frac{1}{\varepsilon^2}(1 + \mathscr{T}_1)r_2,$$

$$(j_2)_t + \frac{1}{\varepsilon^2}u(r_2)_x + \frac{1}{\varepsilon^2}v(r_2)_y = \frac{\rho}{\varepsilon}J_2[T_1] - \frac{1}{\varepsilon^2}(1 + \mathscr{T}_1)j_2.$$

(50)

Since the left-hand sides of the second and fourth equations in (50) include stiff terms with the $\frac{1}{\varepsilon^2}$ coefficients, we add and subtract $u(r_1)_x - v(r_1)_y$ and $u(r_2)_x + v(r_2)_y$ from the second and fourth equations, respectively, so that we finally replace the system (43) with the following system for r_1, j_1, r_2, j_2, and c:

$$\left\{ \begin{aligned} &(r_1)_t + u(j_1)_x - v(j_1)_x = \frac{\rho}{\varepsilon^2}(F(u,v) + \varepsilon R_1[T_1]) - \frac{1}{\varepsilon^2}(1 + \mathscr{T}_1)r_1, \\ &(j_1)_t + u(r_1)_x - v(r_1)_y = \frac{\rho}{\varepsilon}J_1[T_1] \\ &\qquad - \frac{1}{\varepsilon^2}\Big[(1 + \mathscr{T}_1)j_1 + (1 - \varepsilon^2)u\,(r_1)_x - (1 - \varepsilon^2)v\,(r_1)_y\Big], \\ &(r_2)_t + u(j_2)_x + v(j_2)_y = \frac{\rho}{\varepsilon^2}(F(u,v) + \varepsilon R_2[T_1]) - \frac{1}{\varepsilon^2}(1 + \mathscr{T}_1)r_2, \\ &(j_2)_t + u(r_2)_x + v(r_2)_y = \frac{\rho}{\varepsilon}J_2[T_1] \\ &\qquad - \frac{1}{\varepsilon^2}\Big[(1 + \mathscr{T}_1)j_2 + (1 - \varepsilon^2)u\,(r_2)_x + (1 - \varepsilon^2)v\,(r_2)_y\Big], \\ &\tau c_t = \alpha \Delta c - \beta c + \gamma \rho. \end{aligned} \right.$$

(51)

Notice that all of the stiff terms in the first four equations in (51) are moved to the RHS.

3.2 Strang Operator Splitting

The idea behind the operating splitting approach is to treat the stiff and nonstiff parts of the system (51) separately. To this end, we first introduce the vector $W := (r_1, j_1, r_2, j_2)^T$ and rewrite the system (51) in the following form:

$$\begin{cases} W_t + A_1 W_x + A_2 W_y = \mathscr{R}, \\ \tau c_t = \alpha \Delta c - \beta c + \gamma \rho, \end{cases}$$

where

$$A_1 = \begin{pmatrix} 0 & u & 0 & 0 \\ u & 0 & 0 & 0 \\ 0 & 0 & 0 & u \\ 0 & 0 & u & 0 \end{pmatrix}, \qquad A_2 = \begin{pmatrix} 0 & -v & 0 & 0 \\ -v & 0 & 0 & 0 \\ 0 & 0 & 0 & v \\ 0 & 0 & v & 0 \end{pmatrix},$$

and

$$\mathscr{R} = \begin{pmatrix} \frac{\rho}{\varepsilon^2}\left(F(u,v) + \varepsilon R_1[T_1]\right) - \frac{1}{\varepsilon^2}(1 + \mathscr{T}_1) r_1 \\ \frac{\rho}{\varepsilon} J_1[T_1] - \frac{1}{\varepsilon^2}\left[(1 + \mathscr{T}_1) j_1 + (1 - \varepsilon^2) u (r_1)_x - (1 - \varepsilon^2) v (r_1)_y\right] \\ \frac{\rho}{\varepsilon^2}\left(F(u,v) + \varepsilon R_2[T_1]\right) - \frac{1}{\varepsilon^2}(1 + \mathscr{T}_1) r_2 \\ \frac{\rho}{\varepsilon} J_2[T_1] - \frac{1}{\varepsilon^2}\left[(1 + \mathscr{T}_1) j_2 + (1 - \varepsilon^2) u (r_2)_x + (1 - \varepsilon^2) v (r_2)_y\right] \end{pmatrix}.$$

$$(52)$$

We then implement the splitting approach by considering the following two subsystems:

$$\begin{cases} W_t = \mathscr{R}, \\ \tau c_t = 0, \end{cases} \tag{53}$$

and

$$\begin{cases} W_t + A_1 W_x + A_2 W_y = 0, \\ \tau c_t = \alpha \Delta c - \beta c + \gamma \rho. \end{cases} \tag{54}$$

We note that in the subsystem (53), only the W variable is evolved in time, while c remains unchanged there. It is also instructive to point out that not only c, but also the macroscopic cell density ρ does not change in time in the subsystem (53). Indeed, it follows from (42), (49), (52), and (53) that

$$\rho_t = 2 \int_{\mathscr{V}+} [r_1 + r_2]_t \, dv = \frac{\rho}{\varepsilon^2}\left[\int_{\mathscr{V}+} 4F(v) \, dv \right.$$

$$+ \varepsilon \int_{\mathscr{V}+} \left\{ T_1(u, -v) + T_1(-u, v) + T_1(u, v) + T_1(-u, -v) \right\} dv \Bigg]$$

$$(55)$$

$$- \frac{2}{\varepsilon^2}(1 + \mathscr{T}_1) \int_{\mathscr{V}+} [r_1 + r_2] \, dv = 0.$$

Assuming that the solution at time t is available, we evolve it to the next time level using an operator splitting algorithm, [63, 83, 92], of either the first,

$$\begin{pmatrix} W(x, t + \Delta t, v) \\ c(x, t + \Delta t) \end{pmatrix} \approx \mathscr{L}_2(\Delta t)\mathscr{L}_1(\Delta t) \begin{pmatrix} W(x, t, v) \\ c(x, t) \end{pmatrix}, \tag{56}$$

or second,

$$\begin{pmatrix} W(x, t + \Delta t, v) \\ c(x, t + \Delta t) \end{pmatrix} \approx \mathscr{L}_1(\Delta t/2)\mathscr{L}_2(\Delta t)\mathscr{L}_1(\Delta t/2) \begin{pmatrix} W(x, t, v) \\ c(x, t) \end{pmatrix}, \tag{57}$$

order. Here, $\mathscr{L}_1$ and $\mathscr{L}_2$ stand for numerical solution operators for the stiff and nonstiff subsystems (53) and (54), respectively.

Remark 6 It should be observed that the order of the operators in (56) and (57) is interchangeable.

3.3 Time and Space Discretizations

We now proceed with the description of numerical methods for the subsystems (53) and (54). We consider a computational domain, $\Omega \times \mathscr{V}^+$, where the spatial domain Ω is assumed, as before, to be rectangular and partitioned into uniform Cartesian cells $I_{j,k} := [x_{j-\frac{1}{2}}, x_{j+\frac{1}{2}}] \times [y_{k-\frac{1}{2}}, y_{k+\frac{1}{2}}]$ of size $\Delta x \Delta y$ with the cell centers (x_j, y_k). We also introduce a uniform grid of size $\Delta \theta$ in the velocity domain $\mathscr{V}^+$:

$$v_i = (v_0 \cos \theta_i, v_0 \sin \theta_i), \quad \theta_i = (i - 1/2)\Delta\theta, \tag{58}$$

denoted by $F_i := F(v_i)$, $\rho_{j,k}(t) \approx \rho(x_k, y_k, t)$, $c_{j,k}(t) \approx c(x_j, y_k, t)$, and $W_{j,k,i}(t) \approx W(x_j, y_k, t, v_i)$, and assume that the solution $\rho_{j,k}^n = \rho_{j,k}(t^n)$, $c_{j,k}^n = c_{j,k}(t^n)$, and $W_{j,k,i}^n = W_{j,k}(t^n)$ is available at time level $t = t^n$. For simplicity, we will omit the dependence of all of the indexed quantities on time t in the rest of the text, unless it is required for clarity.

3.3.1 $\mathscr{L}_1$: Numerical Solution of the Stiff Subsystem (53)

We start by solving the equations for r_1 and r_2,

$$(r_1)_t = \frac{\rho}{\varepsilon^2}(F(u, v) + \varepsilon R_1[T_1]) - \frac{1}{\varepsilon^2}(1 + \mathscr{T}_1)r_1,$$

$$(r_2)_t = \frac{\rho}{\varepsilon^2}(F(u, v) + \varepsilon R_2[T_1]) - \frac{1}{\varepsilon^2}(1 + \mathscr{T}_1)r_2,$$

keeping in mind that both the chemoattractant concentration c and macroscopic density ρ do not change in time during this splitting step as shown in (55). The latter implies that $R_1[T_1]$ and $\mathscr{T}_1$ are also constants in time (as they depend on c and v only) and thus, the semi-discrete approximations for $(r_1)_{j,k,i}$ and $(r_2)_{j,k,i}$,

$$
\begin{aligned}
\frac{d}{dt}(r_1)_{j,k,i} + \frac{1}{\varepsilon^2}\left(1+(\mathscr{T}_1)_{j,k}\right)(r_1)_{j,k,i} &= \frac{\rho_{j,k}}{\varepsilon^2}(F_i + \varepsilon R_1[T_1]_{j,k,i}), \\
\frac{d}{dt}(r_2)_{j,k,i} + \frac{1}{\varepsilon^2}\left(1+(\mathscr{T}_1)_{j,k}\right)(r_2)_{j,k,i} &= \frac{\rho_{j,k}}{\varepsilon^2}(F_i + \varepsilon R_2[T_1]_{j,k,i}),
\end{aligned}
\tag{59}
$$

reduce to the system of linear ODEs, which can be solved exactly. The new point values of r_1 and r_2 are then used to solve the equations for j_1 and j_2:

$$
\begin{aligned}
(j_1)_t &= \frac{\rho}{\varepsilon}J_1[T_1] - \frac{1}{\varepsilon^2}\Big[(1+\mathscr{T}_1)\,j_1 + (1-\varepsilon^2)u\,(r_1)_x - (1-\varepsilon^2)v\,(r_1)_y\Big], \\
(j_2)_t &= \frac{\rho}{\varepsilon}J_2[T_1] - \frac{1}{\varepsilon^2}\Big[(1+\mathscr{T}_1)\,j_2 + (1-\varepsilon^2)u\,(r_2)_x + (1-\varepsilon^2)v\,(r_2)_y\Big].
\end{aligned}
$$

To this end, we use the central differences

$$
\begin{aligned}
((r_m)_x)_{j,k,i} &= \frac{(r_m)_{j+1,k,i} - (r_m)_{j-1,k,i}}{2\Delta x}, \\
((r_m)_y)_{j,k,i} &= \frac{(r_m)_{j,k+1,i} - (r_m)_{j,k-1,i}}{2\Delta y},
\end{aligned}
\qquad m=1,2,
\tag{60}
$$

and exactly solve the following linear ODEs obtained from a semi-discrete approximations for $(j_1)_{j,k,i}$ and $(j_2)_{j,k,i}$:

$$
\begin{aligned}
\frac{d}{dt}(j_1)_{j,k,i} &+ \frac{1}{\varepsilon^2}(1+(\mathscr{T}_1)_{j,k})(j_1)_{j,k,i} = \\
&-\frac{1}{\varepsilon^2}\Big[(1-\varepsilon^2)u_i((r_1)_x)_{j,k,i} - (1-\varepsilon^2)v_i((r_1)_y)_{j,k,i}\Big] + \frac{\rho_{j,k}}{\varepsilon}J_1[T_1]_{j,k,i}, \\
\frac{d}{dt}(j_2)_{j,k,i} &+ \frac{1}{\varepsilon^2}(1+(\mathscr{T}_1)_{j,k})(j_2)_{j,k,i} = \\
&-\frac{1}{\varepsilon^2}\Big[(1-\varepsilon^2)u_i((r_2)_x)_{j,k,i} + (1-\varepsilon^2)v_i((r_2)_y)_{j,k,i}\Big] + \frac{\rho_{j,k}}{\varepsilon}J_2[T_1]_{j,k,i}.
\end{aligned}
\tag{61}
$$

3.3.2 $\mathscr{L}_2$: Numerical Solution of the Nonstiff Subsystem (54)

We first solve the equation for W, from which we obtain the macroscopic density ρ and then use it to update the values of the chemoattractant concentration c.

The equation for W can be solved by the second-order upwind method (written here in the semi-discrete form):

$$\frac{d}{dt} W_{j,k,i} = - (A_1^+)_i (W_x^-)_{j,k,i} - (A_1^-)_i (W_x^+)_{j,k,i}$$
$$- (A_2^+)_i (W_y^-)_{j,k,i} - (A_2^-)_i (W_y^+)_{j,k,i}, \tag{62}$$

where

$$(A_1^+)_i = \frac{1}{2} \begin{pmatrix} u_i & u_i & 0 & 0 \\ u_i & u_i & 0 & 0 \\ 0 & 0 & u_i & u_i \\ 0 & 0 & u_i & u_i \end{pmatrix}, \quad (A_1^-)_i = \frac{1}{2} \begin{pmatrix} -u_i & u_i & 0 & 0 \\ u_i & -u_i & 0 & 0 \\ 0 & 0 & -u_i & u_i \\ 0 & 0 & u_i & -u_i \end{pmatrix},$$

$$(A_2^+)_i = \frac{1}{2} \begin{pmatrix} v_i & -v_i & 0 & 0 \\ -v_i & v_i & 0 & 0 \\ 0 & 0 & v_i & v_i \\ 0 & 0 & v_i & v_i \end{pmatrix}, \quad (A_2^-)_i = \frac{1}{2} \begin{pmatrix} -v_i & -v_i & 0 & 0 \\ -v_i & -v_i & 0 & 0 \\ 0 & 0 & -v_i & v_i \\ 0 & 0 & v_i & -v_i \end{pmatrix},$$

and

$$(W_x^+)_{j,k,i} = \frac{-W_{j+2,k,i} + 4W_{j+1,k,i} - 3W_{j,k,i}}{2\Delta x},$$

$$(W_x^-)_{j,k,i} = \frac{3W_{j,k,i} - 4W_{j-1,k,i} + W_{j-2,k,i}}{2\Delta x},$$

$$(W_y^+)_{j,k,i} = \frac{-W_{j,k+2,i} + 4W_{j,k+1,i} - 3W_{j,k,i}}{2\Delta y},$$

$$(W_y^-)_{j,k,i} = \frac{3W_{j,k,i} - 4W_{j,k-1,i} + W_{j,k-2,i}}{2\Delta y}$$

are the second-order forward and backward finite-difference approximations of the spatial derivatives. The system of time-dependent ODEs (62) should be numerically integrated in time using a stable and sufficiently accurate ODE solver. It is important to stress that according to the definitions in (48), both r_1 and r_2 (and hence ρ, see (49)) should be positive, which is not guaranteed unless the ODE solver is used with a very small (possibly impractical) time step Δt. Therefore, one may want to implement a "draining" time step technique described in Sect. 2.2; see [19] for details.

Once the new point values of $(r_1)_{j,k,i}$ and $(r_2)_{j,k,i}$ are obtained, they can be used to compute $\rho_{j,k}$ from (49), say, by the midpoint rule:

$$\rho_{j,k} = 2v_0 \sum_i \left[(r_1)_{j,k,i} + (r_2)_{j,k,i} \right] \Delta\theta, \quad \forall j, k. \tag{63}$$

Finally, the new point values of the chemoattractant concentration c can be computed by the spectral method using a fast Fourier transform (FFT); see [19] for details.

3.4 AP Property

As it was mentioned in the Introduction, the solutions of the studied kinetic chemotaxis model are expected to converge to the corresponding solutions of PKS system as $\varepsilon \to 0$. In this section, we repeat the arguments from [19] to show that the numerical method presented in Sects. 3.1–3.3 for (5) provides a consistent discretization of (1) in the limiting $\varepsilon \to 0$ case. This guarantees that the numerical method is AP as the uniform stability in ε is ensured by the fact that the stiff ODEs (59) and (61) are solved exactly.

For simplicity of presentation, we only consider the first-order splitting (56) and either local or nonlocal turning kernels described in (45) and (47), respectively. We denote by

$$\begin{pmatrix} W^* \\ c^* \end{pmatrix} := \mathcal{L}_1(\Delta t) \begin{pmatrix} W^n \\ c^n \end{pmatrix},$$

so that the first-order splitting (56) yields

$$\begin{pmatrix} W^{n+1} \\ c^{n+1} \end{pmatrix} := \mathcal{L}_2(\Delta t) \begin{pmatrix} W^* \\ c^* \end{pmatrix} = \mathcal{L}_2(\Delta t)\mathcal{L}_1(\Delta t) \begin{pmatrix} W^n \\ c^n \end{pmatrix}.$$

We recall that ρ and c do not change in time during the first splitting step and thus $\rho_{j,k}^* = \rho_{j,k}^n$ and $c_{j,k}^* = c_{j,k}^n$. Also, formulae (45), (47), and (48) imply that $(\mathscr{T}_1)_{j,k} = \mathcal{O}(\varepsilon)$, $\varepsilon J_1[T_1]_{j,k,i} = \mathcal{O}(1)$, and $\varepsilon J_2[T_1]_{j,k,i} = \mathcal{O}(1)$. Using these facts, we obtain that as $\varepsilon \to 0$ the leading terms in (59) and (61) are equal to (see also [14, 19]):

$$(r_1)_{j,k,i}^* = \rho_{j,k}^n F_i, \qquad (r_2)_{j,k,i}^* = \rho_{j,k}^n F_i \tag{64}$$

and

$$(j_1)_{j,k,i}^* = \frac{\rho_{j,k}^n(\varepsilon J_1[T_1]_{j,k,i}^n) - u_i((r_1)_x)_{j,k,i}^* + v_i((r_1)_y)_{j,k,i}^*}{1 + (\mathscr{T}_1)_{j,k}^n}$$

$$= \frac{\rho_{j,k}^n}{2}\left[u_i(c_x)_{j,k}^n - v_i(c_y)_{j,k}^n\right] - u_i F_i(\rho_x)_{j,k}^n + v_i F_i(\rho_y)_{j,k}^n, \tag{65}$$

$$(j_2)_{j,k,i}^* = \frac{\rho_{j,k}^n(\varepsilon J_2[T_1]_{j,k,i}^n) - u_i((r_2)_x)_{j,k,i}^* - v_i((r_2)_y)_{j,k,i}^*}{1 + (\mathscr{T}_1)_{j,k}^n}$$

$$= \frac{\rho_{j,k}^n}{2}\left[u_i(c_x)_{j,k} + v_i(c_y)_{j,k}\right] - u_i F_i(\rho_x)_{j,k}^n - v_i F_i(\rho_y)_{j,k}^n,$$

respectively, where $(\rho_x)_{j,k}^n$ and $(\rho_y)_{j,k}^n$ are obtained from (60) and (63) and equal to

$$(\rho_x)_{j,k}^n = \frac{\rho_{j+1,k}^n - \rho_{j-1,k}^n}{2\Delta x}, \qquad (\rho_y)_{j,k}^n = \frac{\rho_{j,k+1}^n - \rho_{j,k-1}^n}{2\Delta y}.$$

Next, we consider the first and third equations in the semi-discrete upwind scheme (62), which after the forward Euler time discretization read as

$$\frac{(r_1)_{j,k,i}^{n+1} - (r_1)_{j,k,i}^*}{\Delta t} = -\frac{1}{2}\Big[u_i((r_1)_x^-)_{j,k,i}^* + u_i((j_1)_x^-)_{j,k,i}^* - u_i((r_1)_x^+)_{j,k,i}^*$$

$$+ u_i((j_1)_x^+)_{j,k,i}^* + v_i((r_1)_y^-)_{j,k,i}^* - v_i((j_1)_y^-)_{j,k,i}^*$$

$$- v_i((r_1)_y^+)_{j,k,i}^* - v_i((j_1)_y^+)_{j,k,i}^* \Big],$$

$$\frac{(r_2)_{j,k,i}^{n+1} - (r_2)_{j,k,i}^*}{\Delta t} = -\frac{1}{2}\Big[u_i((r_2)_x^-)_{j,k,i}^* + u_i((j_2)_x^-)_{j,k,i}^*) - u_i((r_2)_x^+)_{j,k,i}^*$$

$$+ u_i((j_2)_x^+)_{j,k,i}^* + v_i((r_2)_y^-)_{j,k,i}^* + v_i((j_2)_y^-)_{j,k,i}^*$$

$$- v_i((r_2)_y^+)_{j,k,i}^* + v_i((j_2)_y^+)_{j,k,i}^* \Big].$$

$$\text{(66)}$$

Substituting (64) and (65) into (66), adding the above two equations, multiplying by $2v_0\Delta\theta$, summing up over all i, and using (63) yield (see [19] for details)

$$\rho_{j,k}^{n+1} = \rho_{j,k}^n v_0 \sum_i 4F_i\Delta\theta$$

$$- v_0\Delta t\Big[(\Delta x)^3 (\rho_{xxxx})_{j,k}^n \sum_i u_i F_i\Delta\theta + (\Delta y)^3 (\rho_{yyyy})_{j,k}^n \sum_i v_i F_i\Delta\theta$$

$$+ 2((\rho c_x)_x)_{j,k}^n \sum_i u_i^2 \Delta\theta + 2((\rho c_y)_y)_{j,k}^n \sum_i v_i^2 \Delta\theta$$

$$- 4(\rho_{xx})_{j,k}^n \sum_i u_i^2 F_i\Delta\theta - 4(\rho_{yy})_{j,k}^n \sum_i v_i^2 F_i\Delta\theta \Big].$$

$$\text{(67)}$$

We finally use (42), (58), and the approximation property of the midpoint rule to establish the following estimates and identities:

$$v_0 \sum_i 4F_i\Delta\theta = 1 + \mathcal{O}((\Delta\theta)^2),$$

$$v_0 \sum_i u_i F_i\Delta\theta \le \frac{v_0^2}{4} + \mathcal{O}((\Delta\theta)^2), \qquad v_0 \sum_i v_i F_i\Delta\theta \le \frac{v_0^2}{4} + \mathcal{O}((\Delta\theta)^2),$$

$$2v_0 \sum_i u_i^2 \Delta\theta = 2v_0 \sum_i v_i^2 \Delta\theta = v_0 \sum_i (u_i^2 + v_i^2)\Delta\theta = \sum_i v_0^4 \Delta\theta \approx \chi,$$

$$4v_0 \sum_i u_i^2 F_i \Delta\theta = 4v_0 \sum_i v_i^2 F_i \Delta\theta = 2v_0 \sum_i (u_i^2 + v_i^2) F_i \Delta\theta = 2\sum_i v_0^4 F_i \Delta\theta \approx \mu,$$

which can be used to show that (67) provides a consistent approximation of the first equation in (1).

Acknowledgements A large portion of the material covered in this review is based on the work of the authors with Yekaterina Epshteyn, Hengrui Hu, Mária Lukáčová-Medvidŏvá, Mario Ricchiuto, Şeyma Nur Özcan, and Tong Wu, whose valuable contribution we would like to acknowledge here. The work of A. Chertock was supported in part by NSF grants DMS-1521051 and DMS-1818684. The work of A. Kurganov was supported in part by NSFC grant 11771201 and NSF grants DMS-1521009 and DMS-1818666.

References

1. Adler, A.: Chemotaxis in bacteria. Ann. Rev. Biochem. **44**, 341–356 (1975)
2. Alt, W.: Biased random walk models for chemotaxis and related diffusion approximations. J. Math. Biol. **9**(2), 147–177 (1980)
3. Arpaia, L., Ricchiuto, M.: r-adaptation for shallow water flows: conservation, well balancedness, efficiency. Comput. & Fluids **160**, 175–203 (2018)
4. Ascher, U.M., Ruuth, S.J., Spiteri, R.J.: Implicit-explicit Runge-Kutta methods for time-dependent partial differential equations. Appl. Numer. Math. **25**(2-3), 151–167 (1997). Special issue on time integration (Amsterdam, 1996)
5. Ascher, U.M., Ruuth, S.J., Wetton, B.T.R.: Implicit-explicit methods for time-dependent partial differential equations. SIAM J. Numer. Anal. **32**(3), 797–823 (1995)
6. Bialké, J., Löwen, H., Speck, T.: Microscopic theory for the phase separation of self-propelled repulsive disks. EPL (Europhysics Letters) **103**(3), 30,008 (2013)
7. Bollermann, A., Noelle, S., Lukáčová-Medvid'ová, M.: Finite volume evolution Galerkin methods for the shallow water equations with dry beds. Commun. Comput. Phys. **10**(2), 371–404 (2011)
8. Bonner, J.T.: The cellular slime molds, 2nd edn. Princeton University Press, Princeton, New Jersey (1967)
9. Bournaveas, N., Calvez, V.: Critical mass phenomenon for a chemotaxis kinetic model with spherically symmetric initial data **26**(5), 1871–1895 (2009)
10. Budrene, E.O., Berg, H.C.: Complex patterns formed by motile cells of Escherichia coli. Nature **349**, 630–633 (1991)
11. Budrene, E.O., Berg, H.C.: Dynamics of formation of symmetrical patterns by chemotactic bacteria. Nature **376**, 49–53 (1995)
12. Calvez, V., Carrillo, J.A.: Volume effects in the Keller-Segel model: energy estimates preventing blow-up. J. Math. Pures Appl. (9) **86**(2), 155–175 (2006)
13. Calvez, V., Perthame, B., Sharifi Tabar, M.: Modified Keller-Segel system and critical mass for the log interaction kernel. In: Stochastic analysis and partial differential equations, *Contemp. Math.*, vol. 429, pp. 45–62. Amer. Math. Soc., Providence, RI (2007)
14. Carrillo, J.A., Yan, B.: An asymptotic preserving scheme for the diffusive limit of kinetic systems for chemotaxis. Multiscale Model. Simul. **11**(1), 336–361 (2013)

15. Chalub, F.A.C.C., Markowich, P.A., Perthame, B., Schmeiser, C.: Kinetic models for chemotaxis and their drift-diffusion limits. Monatsh. Math. **142**(1-2), 123–141 (2004)
16. Chertock, A., Epshteyn, Y., Hu, H., Kurganov, A.: High-order positivity-preserving hybrid finite-volume-finite-difference methods for chemotaxis systems. Adv. Comput. Math. **44**(1), 327–350 (2018)
17. Chertock, A., Fellner, K., Kurganov, A., Lorz, A., Markowich, P.A.: Sinking, merging and stationary plumes in a coupled chemotaxis-fluid model: a high-resolution numerical approach. J. Fluid Mech. **694**, 155–190 (2012)
18. Chertock, A., Kurganov, A.: A positivity preserving central-upwind scheme for chemotaxis and haptotaxis models. Numer. Math. **111**, 169–205 (2008)
19. Chertock, A., Kurganov, A., Lukáčová-Medvid'ová, M., Özcan, c.N.: An asymptotic preserving scheme for kinetic chemotaxis models in two space dimensions. Kinet. Relat. Models **12**, 195–216 (2019)
20. Chertock, A., Kurganov, A., Ricchiuto, M., Wu, T.: Adaptive moving mesh upwind scheme for the two-species chemotaxis model. Comput. Math. Appl. **77**, 3172–3185 (2019)
21. Chertock, A., Kurganov, A., Wang, X., Wu, Y.: On a chemotaxis model with saturated chemotactic flux. Kinet. Relat. Models **5**(1), 51–95 (2012)
22. Childress, S., Percus, J.K.: Nonlinear aspects of chemotaxis. Math. Biosc. **56**, 217–237 (1981)
23. Cohen, M.H., Robertson, A.: Wave propagation in the early stages of aggregation of cellular slime molds. J. Theor. Biol. **31**, 101–118 (1971)
24. Conca, C., Espejo, E., Vilches, K.: Remarks on the blowup and global existence for a two species chemotactic Keller-Segel system in $\mathbb{R}^2$. European J. Appl. Math. **22**(6), 553–580 (2011)
25. Eisenbach, M., Lengeler, J.W., Varon, M., Gutnick, D., Meili, R., Firtel, R.A., Segall, J.E., Omann, G.M., Tamada, A., Murakami, F.: Chemotaxis. Imperial College Press (2004)
26. Epshteyn, Y.: Upwind-difference potentials method for Patlak-Keller-Segel chemotaxis model. J. Sci. Comput. **53**(3), 689–713 (2012)
27. Epshteyn, Y., Izmirlioglu, A.: Fully discrete analysis of a discontinuous finite element method for the Keller-Segel chemotaxis model. J. Sci. Comput. **40**(1-3), 211–256 (2009)
28. Epshteyn, Y., Kurganov, A.: New interior penalty discontinuous Galerkin methods for the Keller-Segel chemotaxis model. SIAM J. Numer. Anal. **47**, 386–408 (2008)
29. Espejo, E.E., Stevens, A., Suzuki, T.: Simultaneous blowup and mass separation during collapse in an interacting system of chemotactic species. Differential Integral Equations **25**(3-4), 251–288 (2012)
30. Espejo, E.E., Stevens, A., Velázquez, J.J.L.: A note on non-simultaneous blow-up for a drift-diffusion model. Differential Integral Equations **23**(5-6), 451–462 (2010)
31. Espejo, E.E., Vilches, K., Conca, C.: Sharp condition for blow-up and global existence in a two species chemotactic Keller-Segel system in $\mathbb{R}^2$. European J. Appl. Math. **24**, 297–313 (2013)
32. Espejo Arenas, E.E., Stevens, A., Velázquez, J.J.L.: Simultaneous finite time blow-up in a two-species model for chemotaxis. Analysis (Munich) **29**(3), 317–338 (2009)
33. Fasano, A., Mancini, A., Primicerio, M.: Equilibrium of two populations subject to chemotaxis. Math. Models Methods Appl. Sci. **14**, 503–533 (2004)
34. Filbet, F.: A finite volume scheme for the Patlak-Keller-Segel chemotaxis model. Numer. Math. **104**, 457–488 (2006)
35. Filbet, F., Yang, C.: Numerical simulations of kinetic models for chemotaxis. SIAM J. Sci. Comput. **36**(3), B348–B366 (2014)
36. Gajewski, H., Zacharias, K., Gröger, K.: Global behaviour of a reaction-diffusion system modelling chemotaxis. Mathematische Nachrichten **195**(1), 77–114 (1998)
37. Herrero, M., Velázquez, J.: A blow-up mechanism for a chemotaxis model. Ann. Scuola Normale Superiore **24**, 633–683 (1997)
38. Herrero, M.A., Medina, E., Velázquez, J.: Finite-time aggregation into a single point in a reaction-diffusion system. Nonlinearity **10**(6), 1739 (1997)
39. Herrero, M.A., Velázquez, J.J.: Chemotactic collapse for the iKeller-Segel model. J. Math. Biol. **35**(2), 177–194 (1996)

40. Herrero, M.A., Velázquez, J.J.L.: A blow-up mechanism for a chemotaxis model. Ann. Scuola Normale Superiore **24**, 633–683 (1997)
41. Hillen, T., Othmer, H.G.: The diffusion limit of transport equations derived from velocity-jump processes. SIAM J. Appl. Math. **61**(3), 751–775 (electronic) (2000)
42. Hillen, T., Painter, K.: Global existence for a parabolic chemotaxis model with prevention of overcrowding. Adv. in Appl. Math. **26**(4), 280–301 (2001)
43. Hillen, T., Painter, K.J.: A user's guide to PDE models for chemotaxis. J. Math. Biol. **58**(1-2), 183–217 (2009)
44. Horstmann, D.: From 1970 until now: The Keller-Segel model in chemotaxis and its consequences I. Jahresber. DMV **105**, 103–165 (2003)
45. Horstmann, D.: From 1970 until now: The Keller-Segel model in chemotaxis and its consequences II. Jahresber. DMV **106**, 51–69 (2004)
46. Huang, W., Russell, R.D.: Adaptive moving mesh methods, *Applied Mathematical Sciences*, vol. 174. Springer, New York (2011)
47. Hundsdorfer, W., Ruuth, S.J.: IMEX extensions of linear multistep methods with general monotonicity and boundedness properties. J. Comput. Phys. **225**(2), 2016–2042 (2007)
48. Hwang, H.J., Kang, K., Stevens, A.: Drift-diffusion limits of kinetic models for chemotaxis: a generalization. Discrete Contin. Dyn. Syst. Ser. B **5**(2), 319–334 (2005)
49. Jäger, W., Luckhaus, S.: On explosions of solutions to a system of partial differential equations modelling chemotaxis. Trans. Amer. Math. Soc. **329**(2), 819–824 (1992)
50. Jin, S., Pareschi, L., Toscani, G.: Diffusive relaxation schemes for multiscale discrete-velocity kinetic equations. SIAM J. Numer. Anal. **35**(6), 2405–2439 (electronic) (1998)
51. Keller, E.F., Segel, L.A.: Initiation of slime mold aggregation viewed as an instability. J. Theor. Biol. **26**, 399–415 (1970)
52. Keller, E.F., Segel, L.A.: Model for chemotaxis. J. Theor. Biol. **30**, 225–234 (1971)
53. Keller, E.F., Segel, L.A.: Traveling bands of chemotactic bacteria: A theoretical analysis. J. Theor. Biol. **30**, 235–248 (1971)
54. Kurganov, A., Liu, Y.: New adaptive artificial viscosity method for hyperbolic systems of conservation laws. J. Comput. Phys. **231**, 8114–8132 (2012)
55. Kurganov, A., Lukáčová-Medviďová, M.: Numerical study of two-species chemotaxis models. Discrete Contin. Dyn. Syst. Ser. B **19**, 131–152 (2014)
56. Kurganov, A., Qu, Z., Rozanova, O., Wu, T.: Adaptive moving mesh central-upwind schemes for hyperbolic system of PDEs. Applications to compressible Euler equations and granular hydrodynamics Submitted
57. Kurokiba, M., Ogawa, T.: Finite time blow-up of the solution for a nonlinear parabolic equation of drift-diffusion type. Diff. Integral Eqns **4**, 427–452 (2003)
58. van Leer, B.: Towards the ultimate conservative difference scheme. V. A second-order sequel to Godunov's method. J. Comput. Phys. **32**(1), 101–136 (1979)
59. Levy, D., Requeijo, T.: Modeling group dynamics of phototaxis: from particle systems to PDEs. Discrete Contin. Dyn. Syst. Ser. B **9**(1), 103–128 (electronic) (2008)
60. Lie, K.A., Noelle, S.: On the artificial compression method for second-order nonoscillatory central difference schemes for systems of conservation laws. SIAM J. Sci. Comput. **24**(4), 1157–1174 (2003)
61. Liebchen, B., Löwen, H.: Modelling chemotaxis of microswimmers: from individual to collective behavior. arXiv preprint arXiv:1802.07933 (2018)
62. Lin, C.S., Ni, W.M., Takagi, I.: Large amplitude stationary solutions to a chemotaxis system. J. Differential Equations **72**(1), 1–27 (1988)
63. Marchuk, G.I.: Splitting and alternating direction methods. In: Handbook of numerical analysis, Vol. I, Handb. Numer. Anal., I, pp. 197–462. North-Holland, Amsterdam (1990)
64. Marrocco, A.: 2d simulation of chemotaxis bacteria aggregation. M2AN Math. Model. Numer. Anal. **37**, 617–630 (2003)
65. Nagai, T.: Blowup of nonradial solutions to parabolic-elliptic systems modeling chemotaxis in two-dimensional domains. J. Inequal. Appl. pp. 37–55

66. Nagai, T., Senba, T., Yoshida, K.: Application of the Trudinger-Moser inequality to a parabolic system of chemotaxis. Funkcial. Ekvac. **40**(3), 411–433 (1997)
67. Nanjundiah, V.: Chemotaxis, signal relaying and aggregation morphology. J. Theor. Biol. **42**, 63–105 (1973)
68. Nessyahu, H., Tadmor, E.: Nonoscillatory central differencing for hyperbolic conservation laws. J. Comput. Phys. **87**(2), 408–463 (1990)
69. Ni, W.M.: Diffusion, cross-diffusion, and their spike-layer steady states. Notices Amer. Math. Soc. **45**(1), 9–18 (1998)
70. Othmer, H.G., Dunbar, S.R., Alt, W.: Models of dispersal in biological systems. J. Math. Biol. **26**(3), 263–298 (1988)
71. Othmer, H.G., Hillen, T.: The diffusion limit of transport equations. II. Chemotaxis equations. SIAM J. Appl. Math. **62**(4), 1222–1250 (electronic) (2002)
72. Pareschi, L., Russo, G.: Implicit-Explicit Runge-Kutta schemes and applications to hyperbolic systems with relaxation. J. Sci. Comput. **25**(1-2), 129–155 (2005)
73. Patlak, C.S.: Random walk with persistence and external bias. Bull. Math: Biophys. **15**, 311–338 (1953)
74. Pedley, T.J., Kessler, J.O.: Hydrodynamic phenomena in suspensions of swimming microorganisms. Annu. Rev. Fluid Mech. **24**(1), 313–358 (1992)
75. Perthame, B.: PDE models for chemotactic movements: parabolic, hyperbolic and kinetic. Appl. Math. **49**, 539–564 (2004)
76. Perthame, B.: Transport equations in biology. Frontiers in Mathematics. Birkhäuser Verlag, Basel (2007)
77. Pohl, O., Stark, H.: Dynamic clustering and chemotactic collapse of self-phoretic active particles. Phys. Rev. Lett. **112**(23), 238,303 (2014)
78. Prescott, L.M., Harley, J.P., Klein, D.A.: Microbiology, 3rd edn. Wm. C. Brown Publishers, Chicago, London (1996)
79. Saito, N.: Conservative upwind finite-element method for a simplified Keller-Segel system modelling chemotaxis. IMA J. Numer. Anal. **27**(2), 332–365 (2007)
80. Sleeman, B.D., Ward, M.J., Wei, J.C.: The existence and stability of spike patterns in a chemotaxis model. SIAM J. Appl. Math. **65**(3), 790–817 (electronic) (2005)
81. Stevens, A.: The derivation of chemotaxis equations as limit dynamics of moderately interacting stochastic many-particle systems. SIAM J. Appl. Math. **61**(1), 183–212 (electronic) (2000)
82. Stevens, A., Othmer, H.G.: Aggregation, blowup, and collapse: the ABC's of taxis in reinforced random walks. SIAM J. Appl. Math. **57**(4), 1044–1081 (1997)
83. Strang, G.: On the construction and comparison of difference schemes. SIAM J. Numer. Anal. **5**, 506–517 (1968)
84. Strehl, R., Sokolov, A., Kuzmin, D., Turek, S.: A flux-corrected finite element method for chemotaxis problems. Computational Methods in Applied Mathematics **10**(2), 219–232 (2010)
85. Stroock, D.W.: Some stochastic processes which arise from a model of the motion of a bacterium. Probab. Theory Relat. Fields **28**(4), 305–315 (1974)
86. Sweby, P.K.: High resolution schemes using flux limiters for hyperbolic conservation laws. SIAM J. Numer. Anal. **21**(5), 995–1011 (1984)
87. Tang, H., Tang, T.: Adaptive mesh methods for one- and two-dimensional hyperbolic conservation laws. SIAM J. Numer. Anal. **41**(2), 487–515 (electronic) (2003)
88. Tuval, I., Cisneros, L., Dombrowski, C., Wolgemuth, C.W., Kessler, J.O., Goldstein, R.E.: Bacterial swimming and oxygen transport near contact lines. PNAS **102**, 2277–2282 (2005)
89. Tyson, R., Lubkin, S.R., Murray, J.D.: A minimal mechanism for bacterial pattern formation. Proc. Roy. Soc. Lond. B **266**, 299–304 (1999)
90. Tyson, R., Lubkin, S.R., Murray, J.D.: Model and analysis of chemotactic bacterial patterns in a liquid medium. J. Math. Biol. **38**(4), 359–375 (1999)
91. Tyson, R., Stern, L.G., LeVeque, R.J.: Fractional step methods applied to a chemotaxis model. J. Math. Biol. **41**, 455–475 (2000)

92. Vabishchevich, P.N.: Additive operator-difference schemes. De Gruyter, Berlin (2014). Splitting schemes
93. Velázquez, J.J.L.: Point dynamics in a singular limit of the Keller-Segel model. I. Motion of the concentration regions. SIAM J. Appl. Math. **64**(4), 1198–1223 (electronic) (2004)
94. Velázquez, J.J.L.: Point dynamics in a singular limit of the Keller-Segel model. II. Formation of the concentration regions. SIAM J. Appl. Math. **64**(4), 1224–1248 (electronic) (2004)
95. Wang, X.: Qualitative behavior of solutions of chemotactic diffusion systems: effects of motility and chemotaxis and dynamics. SIAM J. Math. Anal. **31**(3), 535–560 (electronic) (2000)
96. Wolansky, G.: Multi-components chemotactic system in the absence of conflicts. European J. Appl. Math. **13**, 641–661 (2002)
97. Woodward, D., Tyson, R., Myerscough, M., Murray, J., Budrene, E., Berg, H.: Spatio-temporal patterns generated by S. typhimurium. Biophys. J. **68**, 2181–2189 (1995)
98. Yeomans, J.: The hydrodynamics of active systems. In: C.N. Likas, F. Sciortino, E. Zaccarelli, P. Ziherl (eds.) Proceedings of the International School of Physics "Enrico Fermi", pp. 383–415. IOS, Amsterdam, SIF, Bologna (2016)
99. Zhang, X., Shu, C.W.: On maximum-principle-satisfying high order schemes for scalar conservation laws. J. Comput. Phys. **229**(9), 3091–3120 (2010)

Control Strategies for the Dynamics of Large Particle Systems

Michael Herty, Lorenzo Pareschi, and Sonja Steffensen

Abstract We survey some recent approaches to control problems for large particle systems. Particle systems are transversal to many applications, ranging from classical physics to social sciences. The temporal evolution of the particles is determined by deterministic or stochastic dynamics and they are additionally able to optimize their trajectory over a large time. In particular, we investigate the limit of infinitely many particles leading to control of kinetic partial differential equations. To this goal a different notion of differentiability of the meanfield equation is introduced. Different mathematical methods based on meanfield games, model predictive control, and optimal control techniques will be discussed.

1 Introduction

Many particle dynamical systems have been recently investigated in different context compared to classical particle physics, including life sciences, social sciences, and economy, see, for example, the recent reviews [9–11, 18, 31–33]. In some of those systems control actions may be applied in order to drive the system towards a desired state using either open, closed loop control or a game theoretic setting. For dynamical system allowing for suitable growth conditions in measure space and symmetry the meanfield limit for infinitely many particles can be analyzed (see [24], for example). The interplay of this limit and the control actions has been subject of recent investigations leading to results on meanfield games and control [13, 16, 23, 29], feedback controls [4, 19], and results for control under further constraints, as, e.g., sparsity [14, 17, 22] or Stackelberg type controls [1, 12, 15].

M. Herty (✉) · S. Steffensen
RWTH Aachen University, IGPM, Aachen, Germany
e-mail: herty@igpm.rwth-aachen.de; steffensen@igpm.rwth-aachen.de

L. Pareschi
Department of Mathematics and Computer Science, University of Ferrara, Ferrara, Italy
e-mail: lorenzo.pareschi@unife.it

© Springer Nature Switzerland AG 2019
N. Bellomo et al. (eds.), *Active Particles, Volume 2*, Modeling
and Simulation in Science, Engineering and Technology,
https://doi.org/10.1007/978-3-030-20297-2_5

The dynamical particle systems consist of possible nonlinear coupled systems of ordinary differential equations. Hence, classical tools from control theory for ordinary differential equations as, for example, dynamic programing and Pontryagin's maximum principle can be applied. However, a major obstacle has been the fact the size of the system of associated Hamilton–Jacobi equations is proportional to the number of particles. Except for particular situations, like the case of linear dynamics and quadratic cost functionals [27], the computation of solutions to the Hamilton–Jacobi equations or Pontryagin's maximum principle is therefore prohibitive for increasingly number of particles. A concept to circumvent this approach has been discussed, e.g., in [3–5, 7, 21]. It is based on model predictive control and aims to derive a (suboptimal) closed loop control. The approach is based on a notion of differentiability of the underlying dynamics. For model predictive control with short time horizon this notion provides an efficiently computable and suboptimal control. However, for large control horizons the used notion of differentiability may lead to inconsistencies with Pontryagin's maximum principle. Therefore, we propose in a separate Appendix a different notion of differentiability of the meanfield equation.

In the rest of the manuscript we shortly survey some of these control strategies. In order to focus on the major ideas we will illustrate the results on a simple model used, for example, for opinion formation and wealth distributions [26, 32]. The control problem for the prototype model and its meanfield limit are presented in Sect. 2 together with some related problems. Section 3 is devoted to the case of linear dynamics and quadratic cost functionals. The case of general nonlinear systems and the introduction of the model predictive control are discussed in more detail in Sect. 4 where we also apply the notion of differentiability presented in the Appendix.

2 Controlled Particle Dynamics and Meanfield Control Problem

In order to illustrate the different strategies we will discuss the derivation and control concepts on a prototype particle model (see [33] for applications to market dynamics and opinion formation). The particle model will be introduced in Sect. 2.1 with the problem be given by Eq. (3). Its corresponding meanfield model is introduced in Sect. 2.2 and the equations are given by Eq. (5b). Other results from the literature on related problems are reviewed in Sect. 2.3.

2.1 Exemplary Particle Problem

Consider N particles with state space $y_i \in I \subset \mathbb{R}$. The dynamics of each particle y_i is driven by an alignment process due to pairwise interaction and weighted by a function $P(\cdot, \cdot)$. For given initial conditions $y_{i,0}$ the dynamics reads

$$y_i' = \frac{1}{N} \sum_{j \neq i}^{N} P(y_i, y_j)(y_j - y_i) + u_i, \qquad y_i(0) = y_{i,0}. \tag{1}$$

Depending on the choice of the interval I different applications are possible: For example, in the case of wealth models we have $I = \mathbb{R}_+$ denoting the available money for each agent. In the case of opinion formation we have $I = [-1, 1]$ denoting extreme opinions $\{-1, 1\}$. For traffic flow applications the interval I is $I = [0, v_{\max}]$, where $v_{\max}$ is the maximal speed allowed on a road. Other examples are $I = [0, 2\pi]$ in the case of one-dimensional Kuramoto-type models or $I = \mathbb{R}$ in the case of the one-dimensional Cucker–Smale model.

In the previous model each particle is subject to a control $u_i = u_i(t)$ modeling either an individual strategy or if $u_i = u(t)$ a global external influence. We assume that the control u_i is obtained as solution to optimal control problems, $i = 1, \ldots, N$,

$$u_i = \operatorname{argmin}_v \frac{1}{2} \int_0^T \left(\frac{1}{N} \sum_{j=1}^{N} g(y_i) \right)^2 + \frac{\beta}{2} v^2 dt, \tag{2}$$

where $\beta > 0$ is a weight to balance the cost of control and the desired state $\frac{1}{N} \sum_{j=1}^{N} g(y_i)$ for some differentiable function $g : \mathbb{R} \to \mathbb{R}$. For simplicity we chose a desired state as moment of the particle distribution. Clearly, the desired state may be dependent on further parameters and costs at terminal time $t = T$ may be added to problem (2). The terminal time $T > 0$ is fixed but can be replaced by $T = +\infty$ provided a suitable weight $\exp(-t)$ is added towards the cost. At this point it is important to note that there is a major difference in discussing the case of particles having each an individual control u_i or particles having a single control u. The problem (1)–(2) yields a system of N coupled optimization problems. One possibility to define a notion of optimality is to consider dynamic Nash equilibria. This approach has been investigated, e.g., in [19] and will be discussed also briefly in Sect. 2.3. The case of $u_i = u$ leads to a high dimensional optimal control problem given by (3)

$$u = \operatorname{argmin}_v \frac{1}{2} \int_0^T \left(\frac{1}{N} \sum_{j=1}^{N} g(y_i) \right)^2 + \frac{\beta}{2} v^2 dt \tag{3a}$$

$$\text{subject to } y_i' = \frac{1}{N} \sum_{j \neq i}^{N} P(y_i, y_j)(y_j - y_i) + u, \qquad y_i(0) = y_{i,0}. \tag{3b}$$

2.2 Meanfield Control Problem

In order to state the meanfield problem associated with Eq. (3) we introduce the following notation. We denote by $\mu_Y(\cdot) \in \mathcal{P}(\mathbb{R})$ the discrete measure on $\mathbb{R}$ located at points $Y \in \mathbb{R}^N$, i.e., $\mu_Y(y) = \frac{1}{N} \sum_{i=1}^{N} \delta(y - y_i)$, where $Y = (y_i)_{i=1}^{N}$. By $f = f(t, \cdot) \in \mathcal{P}(\mathbb{R})$ we denote the probability measure at time $t > 0$. We denote by $U = U(t)$ the control on the meanfield level. Then, we obtain as limit of the particle dynamics (3b) the following strong form of the meanfield equation:

$$\partial_t f(t, y) + \partial_y \left(\int P(y, z)(z - y) f(t, z) dz f(t, y) + U(t) f(t, y) \right) = 0, \qquad (4)$$

$$f(0, y) = f_0(y).$$

In the case $f(0, y) = \mu_Y(y)$ we recover the dynamics (3b) provided $u \equiv U$ and the initial data $f_0(y) = \mu_{Y_0}(y)$. The cost functional on meanfield level and the optimization problem are obtained as

$$U = \mathrm{argmin}_v \frac{1}{2} \int_0^T \left(\int g(y) f(t, y) dy \right)^2 + \frac{\beta}{2} v^2 ds, \qquad (5a)$$

$$\partial_t f(t, y) + \partial_y \left(\int P(y, z)(z - y) f(t, z) dz f(t, y) + U(t) f(t, y) \right) = 0, \qquad (5b)$$

$$f(0, y) = f_0(y).$$

The relation between U and u will be investigated in more detail in Sect. 3. This requires an analysis of the adjoint equation and a suitable notion of differentiability. Prior to this approach we discuss a possibility to derive a closed loop formula for u and U, respectively.

2.3 Related Problems from Literature

Problems of the structure (3) have been widely studied in the literature and we review some extension below without discussing models having high dimensional phase space, e.g., Cucker–Smale type models.

- In [2] white noise dW is added towards the dynamics leading to the problem

$$u = \mathrm{argmin}_{v \in \mathbb{R}} \frac{1}{2} \int_0^T \mathbb{E} \left(\left(\frac{1}{N} \sum_{j=1}^{N} g(y_i) \right)^2 + \frac{\beta}{2} v^2 \right) dt$$

$$\text{subject to } y_i = \frac{1}{N} \sum_{j \neq i}^{N} P(y_i, y_j)(y_j - y_i) + u\,dt + dW, \qquad y_i(0) = y_{i,0}.$$

The problem is studied on the meanfield level but not on the particle based level.

- Another possibility to include stochasticity in the dynamics has been discussed in [7]. Therein, a stochastic parameter W is added in the collision operator P to account for modeling errors. Then, the resulting problem reads

$$u = \text{argmin}_{v \in \mathbb{R}} \frac{1}{2} \int_0^T \mathbb{E} \left(\left(\frac{1}{N} \sum_{j=1}^{N} g(y_i) \right)^2 + \frac{\beta}{2} v^2 \right) dt$$

$$\text{subject to } y_i' = \frac{1}{N} \sum_{j \neq i}^{N} P(y_i, y_j, W)(y_j - y_i) + u, \qquad y_i(0) = y_{i,0}.$$

The solution $y_i = y_i(t, W)$ is then formally expanded in a series using polynomial chaos expansion in the scalar unknown W, i.e., $y_i(t, W) = \sum_k y_{i,k}(t)\psi_k(W)$ for suitable basis functions ψ_k. This results in an extended optimal control problem for the coefficients $y_{i,k}$. In order to obtain a computable control again model predictive control concepts are applied for the resulting problem.

- In specialized setting an explicit comparison of the optimal strategies is possible. For the problem (3) this involves taking $P(y_j, y_i) = P$ constant and modifying the cost functional to

$$u = \text{argmin}_{v \in \mathbb{R}} \frac{1}{2} \int_0^T \frac{1}{N} \sum_{j=1}^{N} y_i^2 + \frac{\beta}{2} v^2 dt.$$

For an open loop control problem of linear dynamical systems and quadratic cost the solution can be expressed by a matrix Riccati equation. It can be shown that the solution to the Riccati equation inherits the symmetry properties required to obtain the meanfield limit in (1). The associated Riccati equation therefore allows also for a meanfield limit and we refer to Sect. 3 and [27] for more details.

- A possible extension to include requirements of sparsity in the control approximation has been introduced in [22]. Adapted to the setup of problem (3) this amounts to modify the cost functional as

$$u = \text{argmin}_{v \in \mathbb{R}} \frac{1}{2} \int_0^T \frac{1}{N} \sum_{j=1}^{N} y_i^2 dt + \beta \|v\|_L,$$

where $L = L_p(0, T)$ for $0 < p \leq 1$. This term promotes sparsity in time of the control. Another type of sparsity is introduced in [1] where sparsity is introduced in the number of controllable particles. In the notation of problem (3) this leads to a modified dynamic as

$$y_i' = \frac{1}{N} \sum_{j \neq i}^{N} P(y_i, y_j, W)(y_j - y_i) + b_i u, \qquad y_i(0) = y_{i,0},$$

where $b_i \in \{0, 1\}$ fixed and with $\|b\|_{l^1}$ sufficiently small.

- The model predictive control concept is applied in settings where multiple different species of particles are considered, e.g., [6]. In this case not all particles are subject to control, but there is a possibly small class of particles that are controlled, the others remain independent. Typically, this leads to adding further ordinary differential equations to the dynamics (1).
- The game theoretic setting (1)–(2) using model predictive control with single time horizon has been discussed in [19] using a meanfield game setting as in [29]. Due to the restriction to pursue model predictive control on a single time horizon the game theoretic aspect does not appear in the derived closed loop control formula. For the studied case of distributed cost the feedback is proportional to the negative gradient of the cost functional. Therefore, model predictive control using the given approximation is equivalent to the control concept "best-reply-strategy" proposed for meanfield dynamics, e.g., in [20, 21].

3   Riccati Control Approach for Particle and Meanfield Model

The problem (3) in the case of linear dynamics and quadratic cost functionals allows for a closed solution based on the Riccati equation. The solution is optimal and it can be computed provided that the matrix Riccati equation could be solved—even in the meanfield limit, see [27]. We exemplify the results for problem (3) where we assume that

$$P(y_i, y_j) = P \text{ and } u = \underset{v}{\operatorname{argmin}} \int_0^T \frac{1}{2N} \sum_{i=1}^{N} y_j^2 + \frac{\beta}{2} v^2 ds, \qquad (6)$$

for some fixed $\beta > 0$. As in [27] this allows to rewrite the problem as

$$\min_u \int_0^T \frac{1}{2} \mathbf{y}^T M \mathbf{y} + \frac{\beta}{2} u^2 ds \quad \text{subject to} \quad \frac{d}{dt} \mathbf{y} = A\mathbf{y} + Bu, \quad \mathbf{y}(0) = \mathbf{y}_0, \qquad (7)$$

where $\mathbf{y} = (y_1, \ldots, y_N)$, $B \in \mathbb{R}^{N \times 1}$ with $B_{i,1} = 1$ and

$$A_{ij} = \begin{pmatrix} a_o & i \neq j \\ a_d & i = j \end{pmatrix}, a_o = \frac{P}{N}, a_d = P\frac{1-N}{N}, \text{ and } M_{ij} = \frac{1}{N}\delta_{ij}.$$

The minimization problem (7) is a convex quadratic problem since M is positive definite. Therefore, Pontryagin's maximum principle is necessary and sufficient for optimality. Under suitable regularity assumptions on the solution it is known that the optimal control is given by

$$u(t) = -\frac{1}{\beta}B^T K(t)\mathbf{y}(t), \tag{8}$$

where $K(t) \in \mathbb{R}^{N \times N}$ fulfills the Riccati equation

$$-\frac{d}{dt}K = KA + A^T K - \frac{1}{\beta}KBB^T K + M, \quad K(T) = 0 \in \mathbb{R}^{N \times N}. \tag{9}$$

It has been proven that the particular structure of A and M translates to the structure of $K = K(t)$, i.e., if $K(t) \in \mathbb{R}^{N \times N}$ is a solution to equation (9), then we have for (i, j) with $i, j = 1, \ldots, N$

$$(BB^T K(t))_{i,j} = \mathscr{K}(t),$$

where $\mathscr{K}(t) \in \mathbb{R}$ is the solution to equation

$$-\frac{d}{dt}\mathscr{K}(t) = \frac{1}{N} - \frac{N}{\nu}\mathscr{K}^2(t), \quad \mathscr{K}(T) = 0. \tag{10}$$

This equation is the key factor to derive a meanfield limit for infinitely many particles in the Riccati equation. More precisely, consider the following new variable:

$$k(t) := N\mathscr{K}(t) \tag{11}$$

and the resulting controlled dynamics where we use (11) and (8) to obtain

$$\frac{dy_i(t)}{dt} = \frac{1}{N}\sum_{j=1}^{N}P(y_j(t) - y_i(t)) - \frac{1}{\beta N}k(t)\sum_{j=1}^{N}y_j(t),$$

$$-\frac{dk(t)}{dt} = 1 - \frac{1}{\beta}k^2(t).$$

The previous system allows for a meanfield limit representing the optimal solution for the problem (3) under assumption (6) for infinitely many particles. The meanfield equation for the probability density $f(t, x)$ is given by the coupled system

$$\partial_t f(x, t) + \partial_x \left(\int_I P(y - x) f(x, t) f(y, t) dy \right)$$

$$- \partial_x \left(k(t) \int_I y f(y, t) f(x, t) dy \right) = 0,$$

$$- \frac{dk(t)}{dt} = 1 - \frac{1}{\beta} k^2(t).$$

The previous closed system allows to analyze moments of the probability density. For example, the first moment is given by $m(t) = \int y f(y, t) dy$ and we obtain

$$\frac{d}{dt} m(t) = -\frac{k(t)}{\nu} m(t) = -\frac{1}{\sqrt{\beta}} \tanh \left(\frac{T - t}{\sqrt{\beta}} \right) m(t). \tag{12}$$

Here, we used the explicitly analytical solution for $k(\cdot)$. The ordinary differential equation (12) has the solution

$$m(t) = \frac{m(0)}{\cosh \left(\frac{T}{\sqrt{\beta}} \right)} \cosh \left(\frac{T - t}{\sqrt{\beta}} \right), \tag{13}$$

for some initial data $m(0) = \int y f_0(y) dy$. For $T \gg 1$ sufficiently large we observe that $m(t)$ decays exponentially fast to $m(0)/\cosh(T/\sqrt{\beta})$ and tends to zero for $T \to \infty$. The convergence rate is $1/\sqrt{\beta}$ where small values of β correspond to a strong influence of the cost functional. The second-order moment $E(t) = \int \frac{1}{2} y^2 f(y, t) dy$ fulfills the equation

$$\frac{d}{dt} E(t) = P \left(m^2(t) - 2E(t) \right) - \frac{k(t)}{\sqrt{\beta}} m^2(t).$$

Hence, $E(t)$ tends to zero as t tends to infinity at rate $2P$ provided that $P > 0$ and T is sufficiently large. Combining the results on the moments we obtain that f in fact converges to $f_\infty = \delta(y)$ for $t \to \infty$ exponentially fast.

4   Model Predictive Control Approach for Particle and Meanfield Model

We now consider again the problem (3) in the general nonlinear case. Necessary optimality conditions for the problem are given by the application of Pontryagin's maximum principle. This yields a system of $2N + 1$ coupled differential-algebraic equations. For an increasing number of particles solving this system is challenging numerically. Therefore, we aim for a suboptimal control using model predictive control also called receding horizon control [30]. In the case of a short receding horizon it allows also to derive closed loop control. Estimates on optimality are available for the model predictive control [25].

The closed loop control can be derived as following. Consider $\Delta t > 0$ and assume the control $u(t) = \sum_{k=1}^{K} u_k \chi_{[t_k,t_{k+1}]}(t)$ is piecewise constant on $I_k :=$ $[t_k, t_{k+1}]$ for $t_{k+1} = t_k + \Delta t$. The time horizon for optimization is also split into optimal control problems with horizon Δt. Hence, on the kth time interval we solve the problem for a control $v \in \mathbb{R}$

$$u_k = \mathrm{argmin}_{v \in \mathbb{R}} \frac{1}{2} \int_{I_k} \left(\frac{1}{N} \sum_{j=1}^{N} g(y_i) \right)^2 + \frac{\beta \Delta t}{2} v^2 dt \tag{14a}$$

$$\text{subject to } y_i' = \frac{1}{N} \sum_{\substack{j=1 \\ j \neq i}}^{N} P(y_i, y_j)(y_j - y_i) + v, \qquad y_i(t_k) = y_i^k, \tag{14b}$$

where $y_i^0 = y_{i,0}$ and where we scale the weight of the control. Having solved the problem (14) for u_k new initial conditions are computed by solving

$$y_i' = \frac{1}{N} \sum_{\substack{j=1 \\ j \neq i}}^{N} P(y_i, y_j)(y_j - y_i) + u_k, \qquad y_i(t_k) = y_i^k.$$

We define $y_i^{k+1} := y_i(t_{k+1})$. Then, we repeat the optimization step (14) on the time interval I_{k+1}. In [4] it has been shown that using an explicit Euler discretization of (14b) the previous steps can be solved explicitly: Denote by $y_i^k = y_i(t_{k+1})$ and consider the following discretization of (14):

$$u_k = \mathrm{argmin}_{v \in \mathbb{R}} \frac{\Delta t}{2} \left(\left(\frac{1}{N} \sum_{j=1}^{N} g(y_i^{k+1}) \right)^2 + \frac{\beta \Delta t}{2} v^2 \right) \tag{15a}$$

$$\text{subject to } y_i^{k+1} - y_i^k = \frac{\Delta t}{N} \sum_{\substack{j=1 \\ j \neq i}}^{N} P(y_i^k, y_j^k)(y_j^k - y_i^k) + \Delta t v. \tag{15b}$$

Using $\frac{\partial y_i^{k+1}}{\partial v} = \Delta t$ we have

$$u_k = -\frac{1}{\beta} \left(\frac{1}{N} \sum_{j=1}^{N} g(y_i^{k+1}) \right) \frac{1}{N} \sum_{j=1}^{N} g'(y_j^{k+1})$$

$$= \frac{1}{\beta} \left(\frac{1}{N} \sum_{j=1}^{N} g(y_i^k) \right) \frac{1}{N} \sum_{j=1}^{N} g'(y_j^k) + O(\Delta t).$$

The last term does only dependent on the known state of the system at time t_k. Up to $O(\Delta t)$ we obtain a closed loop representation of the control. Since the discretized dynamics are only accurate up to order $O(\Delta t^2)$ we substitute u_k and obtain discretized controlled dynamics and in the limit $\Delta t \to 0$ for all t :

$$
y_i' = \frac{1}{N} \sum_{j\neq i}^{N} P(y_i, y_j)(y_j - y_i) - \frac{1}{\beta} \left(\frac{1}{N} \sum_{j=1}^{N} g(y_i) \right) \frac{1}{N} \sum_{j=1}^{N} g'(y_j), \quad y_i(0) = y_{i,0}.
$$
(16)

The corresponding meanfield limit for the controlled equation (16) is given by

$$
\partial_t f(t, y) + \partial_y \left(\int P(y, z)(z - y) f(t, z) f(t, y) dz \right.
$$
(17)

$$
\left. - \frac{1}{\beta} \int g(z) f(t, z) g'(w) f(t, w) f(t, y) dz dw \right) = 0.
$$
(18)

The relation between (17) and the optimal control problem (17) will be investigated in detail in Sect. 3. It is important to note that *only* the cost function enters in the previously derived formulas for the closed loop control. In fact, the obtained control is the gradient of the distributed cost weighted by the inverse of the regularization parameter. This is due to, first, the structure of the problem having a quadratic penalization term and, second, the control enters linearly in the dynamics. Hence, the closed loop control is *independent* of the dynamics. Using a different motivation this formula also been proposed in [20, 21] as "best-reply-strategy." Further, in the case of a game theoretic setting (1)–(2) a similar computation has been carried out in [19].

The control obtained previously is suboptimal since only a short horizon $\Delta t > 0$ is considered for optimization. A possible measure for optimality is the value function. For problem (3) and initial conditions $X = (x_i)_{i=1}^{N}$ at time $t = \tau$ it is defined by

$$
V(\tau, X) = \min_v \int_\tau^T \frac{1}{2} \left(\frac{1}{N} \sum_{j=1}^{N} g(y_i) \right)^2 + \frac{\beta}{2} v^2 dt,
$$

where y_i solves (3b) with initial data $y_i(\tau) = x_i$. This value can be compared with the value $V^{MPC}(\tau, X)$ defined also as the future costs, but where y_i solves (16) for initial data $y_i(\tau) = x_i$. An estimate for the value function V/V^{MPC} and the corresponding value function for the meanfield limit has been studied in [28].

The meanfield equation (17) may also be obtained as limit of a binary particle game. This approach has been discussed, e.g., in [4, 6] and it is particularly interesting for numerical computations. We derive a particle game that after suitable scaling results in the meanfield equation (17). Consider $i = 1, \ldots, N$ particles where two particles i and j interact according to the binary interaction:

$$y_i^* = y_i + \frac{\tau}{2} P(y_i, y_j)(y_j - y_i) - \frac{\tau}{4\beta} \left(g(y_j) + g(y_i) \right) \left(g'(y_j) + g'(y_i) \right),$$

$$y_j^* = y_j + \frac{\tau}{2} P(y_j, y_i)(y_i - y_j) - \frac{\tau}{4\beta} \left(g(y_j) + g(y_i) \right) \left(g'(y_j) + g'(y_i) \right).$$

The states y_i and y_j are pre-interaction, y_i^*, y_j^* are post-interaction states, and $\tau > 0$ is the interaction rate.

4.1 Application of the Notion of Differentiability to the Example

Details on the used notion of differentiability and an example in simplified setting are provided in the Appendix. The particle optimization problem (3) is given by

$$u = \mathrm{argmin}_v \frac{1}{2} \int_0^T \left(\frac{1}{N} \sum_{j=1}^{N} g(y_i) \right)^2 + \frac{\beta}{2} v^2 dt \qquad (19a)$$

$$\text{subject to } y_i' = \frac{1}{N} \sum_{j \neq i}^{N} P(y_i, y_j)(y_j - y_i) + u, \qquad y_i(0) = y_{i,0}. \qquad (19b)$$

Under suitable assumptions Pontryagin's maximum principle gives necessary conditions for optimality [34]. In order to simplify the result we assume that

$$\textbf{Assumption. } P(y, z) = P(z, y) \qquad (20)$$

and denote by $P_1(y, z) = \partial_y P(y, z)$. Denote by $\eta_i = \eta_i(t)$ the adjoint variable for state y_i. Then, the necessary conditions for optimality are (19b) and (21).

$$-\eta_i' = \frac{1}{N} \sum_{j \neq i} P_1(y_i, y_j)(y_j - y_i)(\eta_i - \eta_j) + \frac{1}{N} \sum_{j \neq i} P(y_i, y_j)(\eta_j - \eta_i) \qquad (21a)$$

$$+ \frac{1}{N} \sum_{j=1}^{N} g(y_j) g'(y_i), \qquad \eta_i(T) = 0$$

$$0 = \beta u + \frac{1}{N} \sum_{i=1}^{N} \eta_i. \qquad (21b)$$

Remark 1 Equations (21) allow to recover the model predictive control strategy of the previous section. A backward Euler discretization yields for time step $\Delta t > 0$, zero terminal data of Eq. (21a), and constant u on $(t - \Delta t, t)$:

$$\eta_i(t - \Delta t) = \frac{\Delta t}{N} \sum_j g(y_j(t)) g'(y_i(t)), \qquad u = -\frac{1}{\beta} \frac{\Delta t}{N^2} \sum_{i,j} g(y_j(t)) g'(y_i(t)).$$

Only the gradient of the cost functional enters and after scaling of β we obtain (16).

4.2 Formal Karush–Kuhn–Tucker Conditions for the Kinetic Equation

We recall the kinetic equation of problem (5) as

$$\partial_t f(t, y) + \partial_y \left(\int P(y, z)(z - y) f(t, z) dz f(t, y) + U(t) f(t, y) \right) = 0. \qquad (22)$$

Equation (21) should be obtained from the necessary optimality conditions of problem (5) when considering

$$f(t, x) = \frac{1}{N} \sum_{i=1}^{N} \delta(x - y_i(t)) =: \mu_{Y(t)}(x)$$

for some distinct trajectories $Y(t) = (y_i(t))_{i=1}^{N}$. The correct space for f is $\mathscr{P}(\mathbb{R})$ but this is not a vector space and we define a geometric derivative using the following motivation based on the empirical measure. Consider variations of the cost and the constraint in $\mathscr{P}(\mathbb{R})$ for a fixed, smooth vector function $c = c(t, y)$. Further, we introduce a function $\psi(s, t, y)$ such that $\psi(0, t, y) = f(t, y)$, where f fulfills (22). In case $\mu_{Y(t)}(x)$ we move all particles with speed c to obtain a family of trajectories $\mathscr{Y}(s, t) \in \mathbb{R}^N$ as

$$\mathscr{Y}_i(s, t) = c(t, \mathscr{Y}_i(s, t)), \qquad \mathscr{Y}_i(0, t) = y_i(t).$$

The kinetic description of this equation with the corresponding probability density ψ evolves according to

$$\partial_s \psi(s, t, y) + \partial_y (c(t, y) \psi(s, t, y)) = 0, \qquad \psi(0, t, y) = f(t, y).$$

Then, for a functional $J : \mathscr{P}(\mathbb{R}) \to \mathbb{R}$ depending on $f(t, y)$ we consider the differential

$$\frac{d}{df} J(f(t, y))c := \lim_{s \to 0} \frac{J(\psi(s, t, y)) - J(\psi(0, t, y))}{s}.$$

In the case that J is Frechet-differentiable with $J'(\cdot)$ we obtain formally

$$\frac{d}{df} J(f(t, y))c = J'(f(t, y))\partial_s \psi(0, t, y) = -J'(f(t, y))\partial_y(c(t, y) f(t, y)).$$

Hence, the derivative might be computed using the usual differential applied to $-\partial_y(c(t, y) f(t, y))$. Denote by

$$\mathbf{P}(f)(t, y) := \int P(y, z)(z - y) f(t, z)dz f(t, y), \quad \mathbf{R}(f) := \frac{1}{2}\left(\int g(y) f dy\right)^2,$$

and then the Lagrangian $\mathscr{L}$ by

$$\mathscr{L}(f, \lambda, U) := \int \int \lambda_t f + \lambda_y \left(\mathbf{P}(f) + U(t)f\right) dt dy + \mathbf{R}(f) + \frac{\beta}{2} U^2(t) dt.$$

Applying the previous calculus to obtain the stationary points of $\mathscr{L}$ for some given c leads to

$$0 = \iint \lambda_t (c(t, y) f)_y + \lambda_y \mathbf{P}'(f)(c(t, y) f)_y$$

$$+ U(t)\lambda_y(c(t, y) f)_y dy dt - \int \mathbf{R}'(f) dt, \tag{23a}$$

$$0 = \int \lambda_y(t, y) f(t, y) dy + \beta U(t). \tag{23b}$$

The terminal conditions for λ are equal to zero and omitted. The derivatives of $\mathbf{R}$ are given by

$$\mathbf{R}(f)' = \left(\int g(y) f dy\right)\left(\int g(y)(-c(t, y) f)_y dy\right)$$

$$= \left(\int g(y) f dy\right)\left(\int g'(y) f c(t, y) dy\right),$$

$$-\iint U(t)\lambda_y(c(t, y) f)_y dy dt = \iint \lambda_{yy} f U c(t, y) dy dt,$$

and for **P** we obtain

$$\mathbf{P}(f)'(-c(t, y)f)_y = \left(\int P(y, z)(z - y)f \, dz\right)(-c(t, y)f)_y +$$

$$\int P(y, z)(z - y)(-c(t, z)f)_z dz f(t, y).$$

In (23a) we integrate by parts with respect to y to simplify. Further, we denote by $f^* = f(t, z)$ and similarly for λ_y and c.

$$\int \mathbf{P}(f)'(-c(t, y)f)_y \lambda_y dy = -\int \int P(y, z)(z - y)\lambda_y f^* f_y c \, dy dz$$

$$-\int \int P(z, y)(y - z)\lambda_y^* f^* c f_y \, dy dz$$

$$= \int \int (P(y, z)(z - y))_y \, \lambda_y f^* f$$

$$+ (P(z, y)(y - z))_y \, \lambda_y^* f^* f \, dy dz$$

$$+ \int \int P(y, z)(z - y)\lambda_{yy} f^* f \, dy dz.$$

Due to the assumption on the symmetry of P we have

$$(P(y, z)(z - y))_y = -(P(z, y)(y - z))_y = P_1(y, z)(z - y) - P(y, z),$$

and therefore

$$\int \mathbf{P}(f)'(-c(t, y)f)_y \lambda_y dy = \int \int P(y, z)(z - y)\lambda_{yy} f^* f \, dy dz +$$

$$\int \int P_1(y, z)(z - y)(\lambda_y - \lambda_y^*)f^* f \, dy dz + \int \int P(y, z)(\lambda_y - \lambda_y^*)f^* f \, dy dz.$$

Summarizing, we obtain from (23) for all c

$$0 = \int \int \int \lambda_{ty} f c + g(z)f^* g'(y)f c \, dy dz dt + \int \int \lambda_{yy} U(t)f c \, dy dt \tag{24}$$

$$+ \int \int \mathbf{P}(f)'(-cf)_y \lambda_y dy dt y,$$

$$U(t) = -\frac{1}{\beta} \int \lambda_y f \, dy. \tag{25}$$

4.3 Relation of the Kinetic Equation and Particle System

The derived equation includes mixed derivatives λ_{ty} as well as second-order derivatives λ_{yy}. We observe that λ_y is always multiplied by f and λ_y^* by f^*, respectively. In order to investigate the relation of (24) and (21a)–(21b) we therefore have only to consider products of $\lambda_y f$. Further, we have from standard kinetic theory

$$f(t, y) = \frac{1}{N} \sum_{j=1}^{N} \delta(y - y_j(t)).$$

For any given $Y(t) \in \mathbb{R}^N$ we recover then the particle dynamics (19b) provided f solves (22). For the adjoint we expect to have the relation $\lambda_y(t, y_j) = \lambda_y(t, y_j(t)) = \eta_j(t)$, as seen, for example, by Eq. (25). This relation can also be formulated as $\lambda_y(t, y) f(t, y) = \frac{1}{N} \sum_j \eta_j(t) \delta(y - y_j(t))$ or for all test functions ψ

$$\int \lambda_y(t, y) f(t, y) \psi(y) dy = \frac{1}{N} \sum_{j=1}^{N} \eta_j \psi(y_j). \tag{26}$$

Using this relation we recover from (25) the form

$$U(t) = -\frac{1}{\beta N} \sum_{j=1}^{N} \eta_j(t)$$

that coincides with Eq. (21b). Finally, we show the relation on the adjoint equation (24). For notational simplicity we discuss the terms separately. On the kinetic level the equations have to hold for all functions $c(t, y)$, in particular for c being constant we have

$$\int \int \lambda_{ty} c f \, dy dt = \int \int (\lambda_y f)_t c \, dy dt + \int \int \lambda_y c \, (\mathbf{P}(f) + Uf)_y \, dy dt$$

$$= \frac{1}{N} \sum_j \eta_j'(t) c - \int \int \lambda_{yy} c \left(\int P(y, z)(z - y) f^* f + Uf \right) dy dt,$$

where we used that f fulfills the kinetic equation. Further, we obtain

$$\int \int g(z) f^* g'(y) f c \, dy dz = \frac{1}{N^2} \sum_{i,j} g(y_j) g'(y_j) c,$$

$$\int \int P(y, z)(\lambda_y - \lambda_y^*) f^* f c \, dy dz = \frac{1}{N^2} \sum_{i,j} P(y_i, y_j)(\eta_i - \eta_j) c,$$

$$\int \int P_1(y,z)(z-y)(\lambda_y - \lambda_y^*)f^* f c \, dy \, dz = \frac{1}{N^2} \sum_{i,j} P_1(y_i, y_j)(y_j - y_i)(\eta_i - \eta_j)c.$$

The full form of the adjoint is now obtained by summation. Comparing with the adjoint equation (24) we observe that all terms with second derivative in λ all add up to zero. Hence, we obtain

$$0 = \frac{1}{N} \sum_{j} \eta_j'(t) + \frac{1}{N^2} \sum_{i,j} g(y_j) g'(y_j) + \frac{1}{N^2} \sum_{i,j} P(y_i, y_j)(\eta_i - \eta_j)$$

$$+ \frac{1}{N^2} \sum_{i,j} P_1(y_i, y_j)(y_j - y_i)(\eta_i - \eta_j).$$

This is precisely the sum over i of the discrete adjoints (21a). The terminal conditions also coincide, since $\lambda_y(T, y) = 0$. This shows that with the introduced notion of differentiability we may derive a consistent kinetic approximation of the particle dynamics.

4.4 A Numerical Comparison

In order to evaluate the quality of the model predictive control we compare the optimal control with the model predictive control. In the setting of Sect. 3 the optimal control is known analytically due to the explicit solution of the Riccati equation. We consider precisely this setting with the following parameters: We take a time horizon of $T = 2$, $N = 100$ agents, and a penalization of the control with $\beta \in \{\frac{1}{10}, 1\}$. Due to assumption (6) we have constant interaction kernel P taken to be equal to $P = \frac{1}{2}$. Since the control objective is to drive $m(t) := \frac{1}{N} \sum_{j=1}^{N} y_j(t)$ to zero we plot m over time in logarithmic scale. We run two examples. In the first we apply as control the optimal control u^* obtained in Sect. 3. In the second test we apply the model predictive control u^{MPC} obtained in Sect. 4. Clearly, u^{MPC} is suboptimal and we observe the difference in the order of magnitude for both approaches. Depending on the weight on the control the decay during the given time horizon leads to values of $m(T)$ between $10^{-1.2}$ and 10^{-10} (Fig. 1).

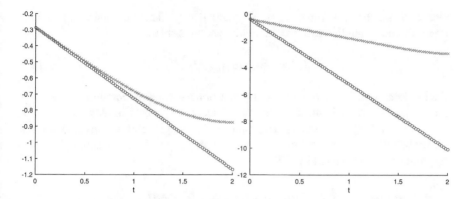

Fig. 1 Numerical simulation of the controlled particle system with optimal control u^* in blue of Sect. 3 and model predictive control in red of Sect. 4. The time horizon is fixed to $T = 2$ and we consider linear dynamics with quadratic objective. Shown is the value of the mean of the particles that is supposed to tend to zero under the control action. Left: Time evolution of the mean $m(t)$ for a penalization of the control of $\beta = 1$. Right: Time evolution of the mean $m(t)$ for a penalization of the control of $\beta = \frac{1}{10}$. All plots are in log scale

Acknowledgements This work has been supported by DFG HE5386/13,14,15-1, BMBF 05M18PAA, and the DAAD-MIUR project.

Appendix: Notion of Differentiability and Calculus for Consistent Derivation

This section is devoted to more details on the notion of differentiability. In order to keep the presentation simple on a technical level we consider the following particle system. Consider N particles i with state space $x_i \in \mathbb{R}$, a sufficiently smooth function $\phi, \psi : \mathbb{R} \to \mathbb{R}$ and assume each particle fulfills

$$x_i'(t) = \frac{1}{N} \sum_{j=1}^{N} \phi(x_j(t)), \, x_i(0) = u_i. \tag{27}$$

Here, u_i is the unknown control applied as initial datum and we assume that the previous dynamics is solved on a time interval $t \in (0, T)$. The control $U = (u_i)_{i=1}^{N} \in \mathbb{R}^N$ is chosen to minimize the cost $J_N : \mathbb{R}^N \to \mathbb{R}$

$$J_N(U) = \int_0^T \frac{1}{N} \sum_{i=1}^{N} \psi(x_i(t)) dt, \tag{28}$$

where x_i are the solution to (27). Standard theory like Pontryagin's maximum principle can be applied to solve the minimization problem

$$U_N^* = \text{argmin}_{U \in \mathbb{R}^N} J_N(U). \tag{29}$$

We are interested in the relation of the first-order optimality conditions to (29) in the case $N \to \infty$. We introduce for each fixed t the probability density $\mu(t, \cdot) \in \mathscr{P}(\mathbb{R})$, i.e., $\mu(t, \cdot) \geq 0$ and $\int_{\mathbb{R}} \mu(t, x)dx = 1$. The probability density describes the probability to have particles at time t with property x. In the meanfield limit the dynamics of μ is obtained by (27) as

$$\partial_t \mu(t, x) + \partial_x \int \phi(y)\mu(t, y)\mu(t, x)dy = 0, \quad \mu(0, x) = v(x), \tag{30}$$

where $\int v(x)dx = 1$ is the corresponding control. We introduce the empirical measure on $\mathbb{R}$ as $v_X(x) = \frac{1}{N} \sum_i \delta(x - x_i)$ with $X = (x_i)_{i=1}^N$. Then, for $\mu(t, x) = v_{X(t)} = \frac{1}{N} \sum_i \delta(x - x_i(t))$ and $v = v_U$ we recover (27) from the weak form of Eq. (30) and vice versa. The meanfield limit of the sequence $(J_N)_N$ of cost functional for $N \to \infty$ is obtained as $\mathscr{J} : \mathscr{P}(\mathbb{R}) \to \mathbb{R}$, where

$$\mathscr{J}(v) = \int_0^T \int \psi(x)\mu(t, x)dxdt, \tag{31}$$

and where μ solves Eq. (30). Hence, on the level of the meanfield limit we may formulate the problem

$$v^* = \text{argmin}_{v \in \mathscr{P}(\mathbb{R})} \mathscr{J}(v). \tag{32}$$

The latter is an optimal control problem with differential equations as constraints and may be solved using techniques from infinite-dimensional optimal control theory. The main result of this section is to establish the relation between the meanfield limit in the Pontryagin's maximum principle and the formal adjoint calculus applied to the optimal control problem (32).

Remark 2 A simple application of formal Lagrange calculus to problem (32) yields as the adjoint equation

$$-\partial_t \lambda - \partial_x \lambda \left(\int \phi \mu dy \right) - \int \mu \partial_x \lambda dy \phi = \psi(x), \quad \lambda(T, 0) = 0. \tag{33}$$

From the previous equation we observe that $\lambda = v_L$ with $L = (\ell_i)_{i=1}^N$ is not a solution since $\int \lambda dx$ is not conserved. Further, from Pontryagin's maximum principle we expect the N adjoints $\ell_i(t)$ to fulfill

$$-\ell_i'(t) = \psi'(x_i(t)) + \frac{1}{N}\sum_{j=1}^{N}\ell_j(t)\phi'(x_i(t)), \qquad \ell_i(T) = 0, \tag{34}$$

involving the derivatives of ψ and ϕ that are not present in (33).

The problem with the formal Lagrange calculus is the fact that Eq. (30) is posed on the space $\mathscr{P}(\mathbb{R})$ that is *not* a vector space. Therefore, we consider geometric derivatives also used, e.g., in [8]. As preliminary discussion note that if

$$\frac{d}{ds}y_i(s) = c, \qquad y_i(0) = x_{i,0}$$

holds for $i = 1 : N$, then the empirical measure fulfills in weak form

$$\partial_s \nu_{Y(s)}(z) + \partial_z c, \qquad \nu_{Y(s)}(z) = 0$$

for initial condition

$$\nu_{Y(0)}(z) = \nu_{X_0}.$$

Clearly, ν is a probability measure. It also serves as first-order variation of the points $x_{i,0}, i = 1 : N$, in the direction of the given velocity field c in the following sense: We define the derivative of $\mathscr{J}$ at measure $\mu = \nu_{X_0}$ in direction c by considering the measure associated with moving all points X_0 with velocity c. This measure is $\eta(t, \cdot) := \nu_{Y(t)} \in \mathscr{P}(\mathbb{R})$. The derivative of $\mathscr{J}$ is then infinitesimal difference at $t = 0$ between the limit

$$\frac{d\mathscr{J}}{d\mu}(\mu)c := \lim_{t\to 0}\frac{1}{t}\left(\mathscr{J}(\eta(t,\cdot)) - \mathscr{J}(\eta(0,\cdot))\right),$$

where $\partial_t \eta + \partial_x(c\eta) = 0, \eta(0) = \mu$. For the example $\mathscr{J}(\mu) = \int \psi(x)\mu(x)dx$ we obtain

$$\frac{dJ}{d\mu}(\mu)c = \lim_{t\to 0}\frac{J(\eta(t,\cdot)) - J(\mu)}{t} =$$

$$\lim_{t\to 0}\frac{1}{t}\int \psi(\eta(t,x) - \eta(0,x))dx = \int \psi\eta_t(t,x)dx =$$

$$-\int \psi(\eta(t,x)c)_x dx = \int \psi'(x)c(x)\mu(x)dx.$$

A similar formalism is applied to the weak form of Eq. (30). We write for a function $\rho \in C^2(\mathbb{R})$ and probability measures $\mu(t,x)$ and $\mu(t,y)$ the weak form as operator A as

$$< A\mu, \rho >:= \int_0^T \int \rho_t + \rho_x \int \phi d\mu(t, y)d\mu(t, x)dt.$$

Applying a similar calculus we obtain

$$\frac{d}{d\mu} < A\mu, \rho > c = \int_0^T \int \rho_{tx}c(x)d\mu(t, x)dt$$

$$+ \int_0^T \int \int \rho_x \phi'(y)c(y)d\mu(t, y)d\mu(t, x)dt$$

$$+ \int_0^T \int \int \rho_{xx}\phi(y)c(x)d\mu(t, y)d\mu(t, x)dt.$$

In order to derive first-order optimality conditions for the problem (32) we consider the Lagrangian $\mathscr{L} : \mathscr{P}(\mathbb{R}) \times \mathscr{P}(\mathbb{R}) \times (\mathscr{P}(\mathbb{R}))' \to \mathbb{R}$ given by

$$\mathscr{L}(v, \mu, \rho) = \int_0^T \int \psi(x)d\mu(t, x)dt+ < A\mu, \rho > + \int \rho(0, x)dv(x).$$

Then, the formal optimality conditions are obtained as saddle point of the Lagrangian $\mathscr{L}$ for everey field c in weak form and are given by

$$\int_0^T \int \rho_t + \rho_x \int \phi d\mu(t, y)d\mu(t, x)dt + \int \rho dv(x) = 0, \tag{35a}$$

$$\int_0^T \int \rho_{tx}c(x)d\mu(t, x)dt + \int_0^T \int \int \rho_x \phi'(y)c(y)d\mu(t, y)d\mu(t, x)dt \tag{35b}$$

$$+ \int_0^T \int \int \rho_{xx}\phi(y)c(x)d\mu(t, y)d\mu(t, x)dt + \int_0^T \int \psi'(x)c(x)d\mu(t, x)dt = 0,$$

$$\int \rho_x(0, x)c(x)dv(x) = 0. \tag{35c}$$

Finally, we show that Eqs. (34) and (27) are recovered from (35) using the empirical measure. Fix N and $U = (u_i)_{i=1}^N$. Let $v(x) := \frac{1}{N} \sum_{i=1}^N \delta(x - u_i)$. Then, a weak solution to (30) is given by $\mu(t, x) := \frac{1}{N} \sum_i \delta(x - x_i(t))$ provided that x_i fulfills (27). Further, we denote by $c_i(t) = \int c(x)\delta(x - x_i(t))dx$ for any c to get

$$\frac{1}{N} \int_0^T \sum_i c_i(t) \left(\psi'(x_i) + \rho_{tx}(t, x_i) + \rho_{xx}(t, x_i) \left(\frac{1}{N} \sum_j \phi(x_j) \right) \right.$$

$$\left. + \phi'(x_i) \left(\frac{1}{N} \sum_j \rho_x(t, x_j) \right) \right) dt = 0.$$

Since the previous equation has to hold true for any function c this yields

$$\psi'(x_i) + \rho_{tx}(t, x_i) + \rho_{xx}(t, x_i) \left(\frac{1}{N} \sum \phi(x_j) \right) + \phi'(x_i) \left(\frac{1}{N} \sum \rho_x(t, x_j) \right) = 0.$$

Define now

$$\lambda_i(t) := \rho_x(t, x_i(t)), i = 1, \ldots, N.$$

Then,

$$\lambda_i'(t) = \rho_{tx}(t, x_i) + \rho_{xx}(t, x_i) x_i'(t)$$

and therefore

$$\psi'(x_i) + \lambda_i'(t) + \rho_{xx}(t, x_i) \left(-x_i' + \frac{1}{N} \sum_j \phi(x_j) \right) + \phi'(x_i) \frac{1}{N} \sum \lambda_j = 0.$$

Since x_i' fulfills (27) we obtain that $\lambda_i \equiv \ell_i$ for all i and fulfills (34). Summarizing, the first-order optimality conditions for the meanfield problem (32) coincide with the first-order optimality conditions for (29) provided the previous notion of differentiability in $\mathscr{P}(\mathbb{R})$ is used. The following relation holds true for the corresponding Lagrange multipliers:

$$\ell_i(t) := \partial_x \rho(t, x_i(t)).$$

References

1. G. ALBI, M. BONGINI, E. CRISTIANI, AND D. KALISE, *Invisible control of self-organizing agents leaving unknown environments*, SIAM J. Appl. Math., 76 (2016), pp. 1683–1710.
2. G. ALBI, Y.-P. CHOI, M. FORNASIER, AND D. KALISE, *Mean field control hierarchy*, Appl. Math. Optim., 76 (2017), pp. 93–135.
3. G. ALBI, M. FORNASIER, AND D. KALISE, *A Boltzmann approach to mean-field sparse feedback control*, IFAC-PapersOnLine, 50 (2017), pp. 2898–2903. 20th IFAC World Congress.
4. G. ALBI, M. HERTY, AND L. PARESCHI, *Kinetic description of optimal control problems and applications to opinion consensus*, Comm. Math. Sci., 13 (2015), pp. 1407–1429.
5. G. ALBI AND L. PARESCHI, *Modeling of self-organized systems interacting with a few individuals: from microscopic to macroscopic dynamics*, Appl. Math. Lett., 26 (2013), pp. 397–401.
6. G. ALBI, L. PARESCHI, AND M. ZANELLA, *Boltzmann type control of opinion consensus through leaders*, Philos. Trans. R. Soc. Lond. Ser. A, 372 (2014).
7. G. ALBI, L. PARESCHI, AND M. ZANELLA, *Uncertainty quantification in control problems for flocking models*, Math. Probl. Eng., (2015), pp. Art. ID 850124, 14.
8. L. AMBROSIO, N. GIGLI, AND G. SAVARE, *Gradient Flows in Metric Spaces of Probability Measures*, Lectures in Mathematics, Birkhäuser Verlag, Basel, Boston, Berlin, 2008.

9. N. BELLOMO, G. AJMONE MARSAN, AND A. TOSIN, *Complex Systems and Society. Modeling and Simulation*, SpringerBriefs in Mathematics, Springer, 2013.

10. N. BELLOMO, P. DEGOND, AND E. TADMOR, *Active Particles, Volume 1 : Advances in Theory, Models, and Applications*, Modeling and Simulation in Science, Engineering and Technology, Birkhäuser Verlag, 2017.

11. N. BELLOMO AND J. SOLER, *On the mathematical theory of the dynamics of swarms viewed as complex systems*, Math. Models Methods Appl. Sci., 22 (2012), p. 1140006.

12. A. BENSOUSSAN, M. H. M. CHAU, Y. LAI, AND S. C. P. YAM, *Linear-quadratic mean field Stackelberg games with state and control delays*, SIAM J. Control Optim., 55 (2017), pp. 2748–2781.

13. A. BENSOUSSAN, J. FREHSE, AND P. YAM, *Mean Field Games and Mean Field Type Control Theory*, Series: SpringerBriefs in Mathematics, New York, 2013.

14. M. BONGINI, M. FORNASIER, O. JUNGE, AND B. SCHARF, *Sparse control of alignment models in high dimension*, Netw. Heterog. Media, 10 (2015), pp. 647–697.

15. A. BORZÌ AND S. WONGKAEW, *Modeling and control through leadership of a refined flocking system*, Math. Models Methods Appl. Sci., 25 (2015), pp. 255–282.

16. M. BURGER, M. DI FRANCESCO, P. A. MARKOWICH, AND M.-T. WOLFRAM, *Mean field games with nonlinear mobilities in pedestrian dynamics*, Discrete Contin. Dyn. Syst. Ser. B, 19 (2014), pp. 1311–1333.

17. M. CAPONIGRO, M. FORNASIER, B. PICCOLI, AND E. TRÉLAT, *Sparse stabilization and optimal control of the Cucker-Smale model*, Math. Control Relat. Fields, 3 (2013), pp. 447–466.

18. E. CRISTIANI, B. PICCOLI, AND A. TOSIN, *Multiscale modeling of pedestrian dynamics*, vol. 12 of MS&A. Modeling, Simulation and Applications, Springer, Cham, 2014.

19. P. DEGOND, M. HERTY, AND J.-G. LIU, *Meanfield games and model predictive control*, Comm. Math. Sci., 5 (2017), pp. 1403–1422.

20. P. DEGOND, J.-G. LIU, AND C. RINGHOFER, *Evolution of the distribution of wealth in an economic environment driven by local Nash equilibria*, Journal of Statistical Physics, 154 (2014), pp. 751–780.

21. P. DEGOND, J.-G. LIU, AND C. RINGHOFER, *Evolution of wealth in a nonconservative economy driven by local Nash equilibria*, Phl. Trans. Roy. Soc. A, 372 (2014).

22. M. FORNASIER, B. PICCOLI, AND F. ROSSI, *Mean-field sparse optimal control*, Philos. Trans. R. Soc. Lond. Ser. A Math. Phys. Eng. Sci., 372 (2014), pp. 20130400, 21.

23. M. FORNASIER AND F. SOLOMBRINO, *Mean-field optimal control*, ESAIM Control Optim. Calc. Var., 20 (2014), pp. 1123–1152.

24. F. GOLSE, *On the dynamics of large particle systems in the mean field limit*, in Macroscopic and large scale phenomena: coarse graining, mean field limits and ergodicity, vol. 3 of Lect. Notes Appl. Math. Mech., Springer, [Cham], 2016, pp. 1–144.

25. L. GRÜNE, *Analysis and design of unconstrained nonlinear MPC schemes for finite and infinite dimensional systems*, SIAM J. Control Optim., 48 (2009), pp. 1206–1228.

26. R. HEGSELMANN AND U. KRAUSE, *Opinion dynamics and bounded confidence, models, analysis and simulation*, Journal of Artificial Societies and Social Simulation, 5 (2002), p. 2.

27. M. HERTY, S. STEFFENSEN, AND L. PARESCHI, *Mean-field control and Riccati equations*, Netw. Heterog. Media, 10 (2015).

28. M. HERTY AND M. ZANELLA, *Performance bounds for the mean-field limit of constrained dynamics*, Discrete Contin. Dyn. Syst., 37 (2017), pp. 2023–2043.

29. J.-M. LASRY AND P.-L. LIONS, *Mean field games*, Japanese Journal of Mathematics, 2 (2007), pp. 229–260.

30. D. Q. MAYNE AND H. MICHALSKA, *Receding horizon control of nonlinear systems*, IEEE Trans. Automat. Control, 35 (1990), pp. 814–824.

31. S. MOTSCH AND E. TADMOR, *Heterophilious dynamics enhances consensus*, SIAM Rev., 4 (2014), pp. 577–621.

32. G. NALDI, L. PARESCHI, AND G. TOSCANI, *Mathematical Modeling of Collective Behavior in Socio-Economic and Life Sciences*, Series: Modeling and Simulation in Science, Engineering and Technology, Birkhauser, Boston, 2010.

33. L. PARESCHI AND G. TOSCANI, *Interacting Multiagent Systems: Kinetic Equations and Monte Carlo Methods*, Oxford University Press, 2013.

34. E. D. SONTAG, *Mathematical control theory*, vol. 6 of Texts in Applied Mathematics, Springer-Verlag, New York, second ed., 1998. Deterministic finite-dimensional systems.

Kinetic Equations and Self-organized Band Formations

Quentin Griette and Sebastien Motsch

Abstract Self-organization is a ubiquitous phenomenon in nature which can be observed in a variety of different contexts and scales, with examples ranging from schools of fish, swarms of birds or locusts to flocks of bacteria. The observation of such global patterns can often be reproduced in models based on simple interactions between neighboring particles. In this paper we focus on two particular interaction dynamics closely related to the one described in the seminal paper of Vicsek and collaborators. After reviewing the current state of the art in the subject, we study a numerical scheme for the kinetic equation associated with the Vicsek models which has the specificity of reproducing many physical properties of the continuous models, like the preservation of energy and positivity and the diminution of an entropy functional. We describe a stable pattern of bands emerging in the dynamics proposed by Degond–Frouvelle–Liu dynamics and give some insights about their formation.

1 Introduction

Swarming dynamics have attracted a lot of attention in recent years raising the question how simple interaction rules could lead to complex pattern formation [23]. One of the main difficulties is to link the individual behaviors of agents and the pattern formations observed at a larger scale. Fortunately the framework of kinetic equations allows such transition between microscopic and macroscopic dynamics. Among the many existing swarming models [3], the Vicsek model [22] is one of the most popular since it describes a rather simple dynamics (there is only alignment)

Q. Griette
Université de Bordeaux - Institut de Mathématiques, Talence, France
e-mail: quentin.griette@math.u-bordeaux.fr

S. Motsch (✉)
Arizona State University - School of Mathematics and Statistical Sciences, Tempe, AZ, USA
e-mail: smotsch@asu.edu

© Springer Nature Switzerland AG 2019
N. Bellomo et al. (eds.), *Active Particles, Volume 2*, Modeling
and Simulation in Science, Engineering and Technology,
https://doi.org/10.1007/978-3-030-20297-2_6

with few parameters but it is however able to generate complex pattern which are challenging to predict analytically. The Vicsek model has been well studied both numerically [6, 18, 19] and analytically [2, 12, 16] and the derivation of its kinetic and macroscopic equation is well-understood [9, 10]. However, as noted first by Chaté and Grégoire[18], there exists a certain regime where the Vicsek model leads to the formation of traveling bands. Many numerical studies have been conducted to better analyze the formation of these bands at the particle level but none have been proposed so far to study the bands using a kinetic or macroscopic framework. This manuscript aims at proposing a first study on such band formation from the angle of kinetic equations.

After the discovery of band formation in the Vicsek dynamics by Chaté and co-worker [18], there has been a debate [1, 5] about the order of the phase transition in the Vicsek model (continuous or discontinuous). As there was no analytic framework available, the conjecture could be only based on (particle) numerical simulations. However, the derivation of kinetic and macroscopic equation for the Vicsek model [9, 10] indicated that in a dense regime of particles (the so-called moderately interacting particle [20]), the Vicsek model has a continuous transition from order to disorder. In this regime of high density, no phase transition or band formation could be observed. A major discovery was then provided by Degond, Frouvelle, and Liu [7, 8, 14] where a modification of the (continuous) Vicsek was considered: alignment is proportional to the local density rather than to the mean direction. In their dynamics, a phase transition occurs: at low density, the velocity distribution becomes uniform, whereas at large density, the dynamics converge to a so-called von Mises distribution. This analytic result was only proven in a homogeneous setting (no spatial variable). Thus, it is still unknown what effect a transport term would have on the dynamics. This is however a very challenging question as the transport term breaks the entropy dissipation. In this manuscript, we propose to investigate numerically the Degond–Frouvelle–Liu (DFL) dynamics in a non-homogeneous setting.

Starting from the kinetic equation associated with the Vicsek model, we first review some properties of the collisional operator (entropy dissipation) that will be central for the building of our numerical scheme. Most of these estimates are built on the Fokker–Planck structure of the operator. We do take advantage of this formulation in the design of our numerical scheme. The key properties of the collision operator (positivity preserving, entropy dissipation) are also satisfied by the discrete operator. Since we aim at analyzing the long-time behavior of the solution, it is essential to preserve these properties. For instance, several papers have already proposed to solve numerically the kinetic equation associated with the Vicsek model using other methods (spectral method [15], particle method [11], discontinuous Galerkin [13]). But we would rather have lower accuracy and a preserving numerical scheme to study the long-time behavior of the solution (even though our scheme is still second order accurate in the velocity-variable). We then explore the dynamics of the kinetic equation in various regimes. In the original Vicsek model, no band formations are observed, the spatial density becomes homogeneous while the velocity distribution becomes distributed according to a (global) von Mises distribution. In the Degond–Frouvelle–Liu (DFL) dynamics, however, when the density is above a threshold, band formation occurs starting from random initial configuration. As far as the authors know, this is the first time such band formations are observed at the kinetic level. Band were also observed in [13] but there were only "transient," the density profile would become flat after a long time. Here, the density profile is not flattening out, but instead is becoming more and more concentrated. Numerically, we have to introduce an adaptive time step to deal with a demanding CFL condition because of this very phenomenon.

Although our numerical investigation suggests that band formation emerges from the DFL dynamics, it would be crucial to also develop an analytic framework to further understand this phenomenon. Our results indicate that the transport operator could further amplify the concentration originating from the alignment operator. From these observations, it seems unlikely that there exists an analytic profile for these bands. But the question remains open. Similarly, we could perform simulation in dimension 3, but the discretization of the unit sphere $\mathbb{S}^2$ is more delicate than $\mathbb{S}^1$ (there is no "uniform grid" on $\mathbb{S}^2$) and thus having discrete entropy dissipation or symmetry preserving would be more challenging. Finally, higher order accuracy in time should also be investigated using, for instance, [4, 17].

The structure of the paper is as follows. In Sect. 2 we discuss the microscopic Vicsek model and the DFL dynamics with an eye toward band formation. In Sect. 3 we precise the associated kinetic models and discuss related analytical properties. Section 4 is devoted to a precise description of the numerical scheme used to approximate the kinetic models. Finally in Sect. 5 we present our numerical results and in particular our observations concerning self-organized band formation.

2 Microscopic Description

2.1 Vicsek Model

The Vicsek model [10, 22] at the particle level describes the motion of N particles with position $\mathbf{x}_i \in \mathbb{R}^d$ (with $d = 2, 3$) and a direction $\omega_i \in \mathbb{S}^{d-1}$ (i.e., $|\omega_i| = 1$). The evolution of the particles is given by the following system:

$$\begin{aligned} \mathbf{x}'_i &= c\,\omega_i \\ \mathrm{d}\omega_i &= P_{\omega_i^\perp}(\mu\,\Omega_i \mathrm{d}t + \sqrt{2\sigma} \circ \mathrm{d}B_i^t), \end{aligned} \tag{1}$$

where $c > 0$ is the speed of the particle, μ is the strength of the alignment interaction, σ is the intensity of the noise (note that μ and σ are both scalar), $\mathrm{d}B_i^t$ are independent white noises (which must be understood in the Stratonovich sense as indicated by the symbol $\circ$), $P_{\omega_i^\perp}$ is the orthogonal projection on the orthogonal of ω_i defined as:

$$P_{\omega_i^\perp} = \mathrm{Id} - \omega_i \otimes \omega_i, \qquad (2)$$

which ensures that $|\omega_i(t)| = 1$ over time, and Ω_i is the average direction of the particle i:

$$\Omega_i = \frac{\mathbf{j}_i}{|\mathbf{j}_i|}, \quad \mathbf{j}_i = \sum_{j, |\mathbf{x}_j - \mathbf{x}_i| \leq R} \omega_j,$$

where R is the radius of interaction.

2.2 Degond–Frouvelle–Liu (DFL) Dynamics

Degond, Frouvelle, and Liu [8, 14] proposed a modification of the dynamics where the alignment interaction μ is proportional to the norm of the flux $\mathbf{j}_i$:

$$\begin{aligned}
\mathbf{x}_i' &= c\omega_i \\
\mathrm{d}\omega_i &= P_{\omega_i^\perp}(\mu\,\mathbf{j}_i\mathrm{d}t + \sqrt{2\sigma} \circ \mathrm{d}B_i^t).
\end{aligned} \qquad (3)$$

This modification has several consequences: (1) unlike the Vicsek dynamics (1) which is ill-defined when the flux $\mathbf{j}_i$ equal zero (Ω_i not defined), the DFL dynamics does not have any singularity, (2) there is a phase transition in the dynamics (3) as the number of particles increases (or similarly as μ increases). The kinetic description of this dynamics will allow to better explain this phase transition (see Sect. 3.2).

2.3 Band Formation

Band formations have been first analyzed by Grégoire and Chaté [18] in the case of the original *discrete* Vicsek model and several numerical studies have been conducted since [1, 6, 19]. To motivate our study, we present numerically an example of such band formation in the context of the *continuous* dynamics (1).

The numerical simulation presented in this subsection is performed with $N = 30{,}000$ particles on a square domain with length $L = 4$ and periodic boundary condition. Initially, particles are distributed at random in space and velocity. Table 1 gives the list of values for the parameters. We observe in Fig. 1 the formation of a traveling wave moving in the x-direction. To further quantify this formation, we estimate the average density ρ and velocity u in the x-direction:

$$\overline{\rho}(x, t)\Delta x = \sum_{i=1}^{N} \mathbb{1}_{[-\frac{\Delta x}{2}, \frac{\Delta x}{2})}(x_i(t) - x)$$

Table 1 Parameters used in the simulations for Figs. 1 and 2

Description	Notation	Value
Number of particles	N	30,000
Strength alignment	μ	100
Noise intensity	σ	20
Radius interaction	R	0.02
Length domain	L	4
Time step	Δt	10^{-2}

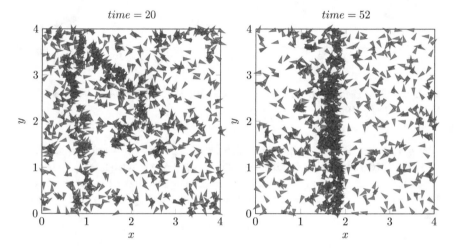

Fig. 1 Illustration of the simulation of the Vicsek model (1) at two different time. We observe the formation of a vertical *band*. See Table 1 for the parameters used

$$\overline{\rho}(x,t)\overline{u}(x,t) = \sum_{i=1}^{N} \mathbb{1}_{[-\frac{\Delta x}{2}, \frac{\Delta x}{2})}(x_i(t) - x)\cos\theta_i(t).$$

where x_i and $\cos\theta_i$ are, respectively, the x-component of the position vector $\mathbf{x}_i$ and velocity ω_i. We give an example of such ρ and u in Fig. 2.

We notice that the regime in which the band formation occurs is far from being *dense*. In other words, each particle has few neighbors (in average) in this regime: assuming a uniform distribution of particles on all the domain, the average number of neighbors is given by

$$\text{Average number of neighbors} \approx \frac{|B(0, R)|}{L^2} \times N = 2.36,$$

therefore we are far from being in kinetic regime (let alone macroscopic region). Thus, the validity of the kinetic equation associated with the dynamics (described in the next section) is questionable in this regime. Particles are not necessarily "moderately interacting" [20]. Numerically experiments suggest that band formation only

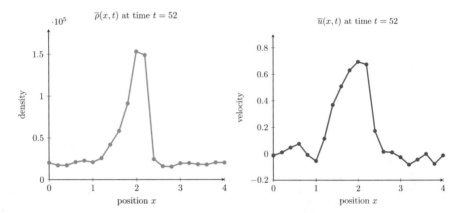

Fig. 2 Density ρ and average velocity u in the x-direction at $t = 52$. Where the density ρ is larger, the speed u increases

occurs in this low density regime which could explain why band formation had been first observed 10 years after the Vicsek model was introduced. Preliminary numerical investigations of the DFL dynamics suggest that band formations can be observed for a larger spectrum of density within this dynamics compare to the Vicsek model.

3   Kinetic Description

3.1   Introduction

The kinetic equation associated with the Vicsek dynamics (1) is described through the density distribution $f(\mathbf{x}, \omega, t)$. As the number of particles N tends to infinity, the particle dynamics converge to the solution of the deterministic equation:

$$\partial_t f + c\,\omega \cdot \nabla_{\mathbf{x}} f = -\mu \nabla_\omega \cdot (P_{\omega^\perp}(\Omega_f)\,f) + \sigma \Delta_\omega f, \tag{4}$$

where $c > 0$ is the speed of the particles, $\mu > 0$ is the intensity of the relaxation toward the mean velocity, $\sigma > 0$ is the diffusion coefficient, $P_{\omega^\perp}$ the projection operator (2), Ω is the mean velocity at the point $\mathbf{x}$:

$$\Omega_f(\mathbf{x}, t) = \frac{\mathbf{j}_f(\mathbf{x}, t)}{|\mathbf{j}_f(\mathbf{x}, t)|} \quad \text{with} \quad \mathbf{j}_f(\mathbf{x}, t) = \int_{y \in B(\mathbf{x}, R), \omega \in \mathbb{S}^{d-1}} \omega f(\mathbf{y}, \omega, t)\, \mathrm{d}y\mathrm{d}\omega,$$

$R > 0$ being the radius of interaction.

The DFL dynamics lead to a similar kinetic equation except for the transport term in ω:

$$\partial_t f + c\,\omega \cdot \nabla_{\mathbf{x}} f = -\mu \nabla_\omega \cdot (P_{\omega^\perp}(\mathbf{j}_f)\,f) + \sigma \Delta_\omega f. \tag{5}$$

Notice that (5) can be formally obtained from (4) by allowing the strength of interaction μ to depend linearly on the local flux $|\mathbf{j}_f|$.

A rigorous derivation of Eq. (5) from the microscopic model can be found in [2]. There is an additional difficulty to derive rigorously the Vicsek kinetic equation (4) due to a singularity in the dynamics: the mean velocity Ω_f is undefined when the local flux $\mathbf{j}_f$ is zero.

3.2   Homogeneous Case

To investigate kinetic equations, we study the homogeneous case, assuming that f is independent of $\mathbf{x}$. Thus, the kinetic equations (4) and (5) become

$$\partial_t f = Q(f), \tag{6}$$

with:

$$Q(f) = -\mu_f \nabla_\omega \cdot (P_{\omega^\perp}(\Omega_f) f) + \sigma \Delta_\omega f, \tag{7}$$

and $\Omega_f = \frac{\mathbf{j}_f}{|\mathbf{j}_f|}$ with $\mathbf{j}_f = \int_{\omega \in \mathbb{S}^{d-1}} \omega f(\omega) \, d\omega$ and

$$\mu_f = \begin{cases} \mu & \text{Vicsek dynamics} \\ \mu|\mathbf{j}| & \text{DFL dynamics.} \end{cases}$$

The operator Q defined by (7) can be written as a Fokker–Planck type operator. Introducing:

$$\phi(\omega) = \begin{cases} \langle \Omega_f, \omega \rangle & \text{Vicsek dynamics} \\ \langle \mathbf{j}_f, \omega \rangle & \text{DFL dynamics,} \end{cases} \tag{8}$$

with $\langle , \rangle$ the usual scalar product in $\mathbb{R}^n$, we find

$$Q(f) = \sigma \nabla_\omega \cdot \left(M_f \nabla_\omega \left(\frac{f}{M_f} \right) \right), \quad \text{with} \quad M_f(\omega) = e^{\frac{\mu}{\sigma} \phi(\omega)}, \tag{9}$$

using the identity $\nabla_\omega \langle \mathbf{u}, \omega \rangle = P_{\omega^\perp}(\mathbf{u})$, where M_f is the *von Mises distribution* associated with Q. We deduce a first identity:

$$\int_\omega \partial_t f \frac{f}{M_f} \, d\omega = -\sigma \int_\omega M_f \left| \nabla_\omega \left(\frac{f}{M_f} \right) \right|^2 \, d\omega \leq 0. \tag{10}$$

Unfortunately, the left-hand side of (10) cannot be written as a total time derivative and thus we cannot deduce any *entropy decay*. The *trick* is to notice the following:

$$Q(f) = \sigma \nabla_\omega \cdot \left(f \frac{M_f}{f} \nabla_\omega \left(\frac{f}{M_f} \right) \right)$$

$$= \sigma \nabla_\omega \cdot \left(f \nabla_\omega \ln \left(\frac{f}{M_f} \right) \right).$$

Therefore,

$$\int_\omega \partial_t f \ln \left(\frac{f}{M_f} \right) d\omega = -\sigma \int_\omega f \left| \nabla_\omega \ln \left(\frac{f}{M_f} \right) \right|^2 d\omega \leq 0. \tag{11}$$

Thanks to the property of the logarithm, the left-hand side can now be written as a total time derivative and we deduce the following proposition.

Proposition 1 *Suppose f is a solution to the homogeneous kinetic equation* (6) *and consider the free energy:*

$$\mathcal{F}[f] = \int_\omega f \ln f \, d\omega - \frac{\mu}{\sigma} \Phi_f, \tag{12}$$

with:

$$\Phi_f = \begin{cases} |\mathbf{j}_f| & \textit{Vicsek dynamics} \\[2mm] \frac{1}{2} |\mathbf{j}_f|^2 & \textit{DFL dynamics.} \end{cases} \tag{13}$$

It satisfies:

$$\frac{d}{dt} \mathcal{F} = -\sigma \int_\omega f \left| \nabla_\omega \ln \left(\frac{f}{M_f} \right) \right|^2 d\omega \leq 0.$$

Proof We remark that left-hand side of (11) is a total derivative:

$$\int_\omega \partial_t f \ln \left(\frac{f}{M_f} \right) d\omega = \int_\omega \partial_t f \left(\ln f - \frac{\mu}{\sigma} \phi \right) d\omega$$

$$= \int_\omega \partial_t (f \ln f) - \frac{\mu}{\sigma} (\partial_t f) \phi \, d\omega,$$

using the conservation of mass $\int_\omega \partial_t f \, d\omega = 0$. Notice moreover that ϕ (8) can be expressed as gradient (making the dynamics (7) a gradient flow as noted in [12]). Indeed, taking $f + h$ a small perturbation of f, we have

$$|\mathbf{j}_{f+h}|^2 = \left| \int_\omega (f+h)\omega \, \mathrm{d}\omega \right|^2 = |\mathbf{j}_f|^2 + 2\langle \int_\omega f\omega \, \mathrm{d}\omega, \int_\omega h\omega \, \mathrm{d}\omega \rangle + \mathcal{O}(h^2)$$

$$= |\mathbf{j}_f|^2 + 2\int_\omega \langle \mathbf{j}_f, \omega \rangle h \, \mathrm{d}\omega + \mathcal{O}(h^2),$$

thus $\frac{\delta |\mathbf{j}_f|^2}{\delta f}(\omega) = 2\langle \mathbf{j}_f, \omega \rangle$. In particular $\phi = \frac{\delta \Phi}{\delta f}$ with Φ given by (13). We deduce

$$\int_\omega (\partial_t f)\phi(\omega) \, \mathrm{d}\omega = \int_\omega (\partial_t f)\frac{\delta \Phi}{\delta f}(\omega) \, \mathrm{d}\omega = \frac{\mathrm{d}}{\mathrm{d}t}\Phi(f(t)).$$

Therefore, we obtain

$$\int_\omega \partial_t f \ln\left(\frac{f}{M_f}\right) \mathrm{d}\omega = \frac{\mathrm{d}}{\mathrm{d}t}\mathcal{F},$$

with $\mathcal{F}$ given by (12).

Finally, for the sake of completeness, we prove that the stationary states of the homogeneous equation (6) are the minimizers to the free energy defined by (12), in the case of the Vicsek dynamics.

Proposition 2 (Minimizers for the Vicsek Dynamics) *Let $f \in H^1(\mathbb{S}^1)$ be such that $\int_{\mathbb{S}^1} f(\omega)\mathrm{d}\omega = 1$. Then f is a stationary solution to Eq. (6) for the Vicsek dynamics (i.e., $\phi = \Omega_f$) if, and only if, f is a minimizer to $\mathcal{F}$ in the sense that*

$$\mathcal{F}(f) = \min_{g \in H^1, \int g = 1} \mathcal{F}(g).$$

Proof We remark that $\mathcal{F}$ can be written as a generalized entropy:

$$\mathcal{F}(f) = \int_{\mathbb{S}^1} f \ln f \mathrm{d}\omega - \frac{\mu}{\sigma}\langle \Omega_f, \mathbf{j}_f \rangle = \int_{\mathbb{S}^1} f \ln f \mathrm{d}\omega - \frac{\mu}{\sigma}\langle \Omega_f, \int_{\mathbb{S}^1} \omega f(\omega)\mathrm{d}\omega \rangle$$

$$= \int_{\mathbb{S}^1} f\left(\ln f - \frac{\mu}{\sigma}\langle \Omega_f, \omega \rangle\right) \mathrm{d}\omega = \int_{\mathbb{S}^1} f \ln\left(\frac{f}{M_f}\right) \mathrm{d}\omega.$$

Jensen's inequality then yields

$$\left(\int M_f\right)\int \frac{f}{M_f} \ln\left(\frac{f}{M_f}\right)\frac{M_f}{\int M_f} \mathrm{d}\omega \geq \left(\int M_f\right)\int \frac{f}{\int M_f}\mathrm{d}\omega \ln\left(\int \frac{f}{\int M_f}\mathrm{d}\omega\right)$$

$$= -\ln\left(\int_{\mathbb{S}^1} M_f(\omega)\mathrm{d}\omega\right),$$

with equality if, and only if, $f = \frac{M_f}{\int M_f}$ (remark that $\ln\left(\int_{\mathbb{S}^1} M_f(\omega)d\omega\right)$ is independent from f). In view of (10), this is equivalent to $Q(f) = 0$.

The case of the DFL dynamics is more complex and has been investigated in [14]. It has been shown that the stationary solutions are indeed the minimizers of $\mathcal{F}$.

3.3 Phase Transition

Since the dynamics (7) have entropy, one can study the long-time behavior and deduce the convergence toward an equilibrium given as the minimizer of the *free energy* $\mathcal{F}$ (12). But first we need to identify the minimizers of $\mathcal{F}$. To do so, we notice that once we fix the flux $\mathbf{j}_f$, the minimizer would be given by von Mises.

Lemma 1 *Fix* $\mathbf{j}$ *with* $0 < |\mathbf{j}| < 1$ *and consider the affine space:*

$$\mathcal{A} = \left\{ f \in L^2(\mathbb{S}^{n-1}) \mid \int_{\mathbb{S}^{n-1}} \omega f(\omega)\,d\omega = \mathbf{j} \quad and \quad \int_{\mathbb{S}^{n-1}} f(\omega)\,d\omega = 1 \right\}.$$

Then:

$$\inf_{\mathcal{A}} \left\{ \int_\omega f \ln f \, d\omega \right\} = \int_\omega M_* \ln M_* \, d\omega,$$

where M_* *is the von Mises* (9) *satisfying* $\int_\omega \omega M_*(\omega)\,d\omega = \mathbf{j}$.

Proof Assume there exists a minimizer f_* and rewrite the constraint as

$$H[f] = \int f \ln f \quad , \quad \alpha[f] = \left| \int_{\mathbb{S}^{n-1}} \omega f - \mathbf{j} \right|^2 - 1 \quad , \quad \beta[f] = \left(\int_{\mathbb{S}^{n-1}} f - 1 \right)^2.$$

Denote λ_1 and λ_2 the Lagrange multiplier associated with f_*:

$$\left. \frac{\delta H}{\delta f} \right|_{f_*} = \lambda_1 \left. \frac{\delta \alpha}{\delta f} \right|_{f_*} + \lambda_2 \left. \frac{\delta \beta}{\delta f} \right|_{f_*}$$

$$\Rightarrow \ln f_* + 1 = \lambda_1 \, 2\omega \cdot (-\mathbf{j}) + \lambda_2 \cdot 0.$$

since $\int f_* = 1$. Thus, taking the exponential leads to:

$$f_*(\omega) = C e^{-2\lambda_1 \omega \cdot \mathbf{j}},$$

and therefore f_* is a von Mises distribution.

As a consequence of Lemma 1, we can restrict the search of minimizers of the free energy $\mathcal{F}$ on von Mises distributions. In Fig. 3, we estimate numerically the

Fig. 3 Entropy $\int M \log M$
for M von Mises distribution
as a function of the length
$|\int \omega M|$. The curve increases
quadratically near $|\mathbf{j}| = 0$

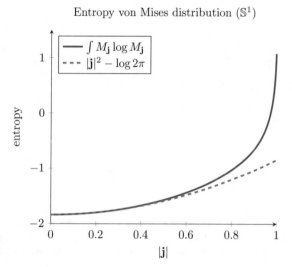

Entropy von Mises distribution ($\mathbb{S}^1$)

entropy $\int_\omega M \ln M$ of von Mises distributions depending on their average velocity
$|\mathbf{j}| = |\int_\omega \omega M|$ along with its approximation near $\mathbf{j} \approx 0$:

$$\int_\omega M \log M = -\log 2\pi + |\mathbf{j}|^2 + \mathcal{O}(|\mathbf{j}|^3).$$

We deduce that the free energy $\mathcal{F}$ for the Vicsek model will never have a minimum
at $\mathbf{j} = 0$ meaning that the uniform distribution is never stable. However, for the
DFL dynamics, when the diffusion σ is large, the free entropy $\mathcal{F}$ will be minimum
at $\mathbf{j} = 0$ and therefore the uniform distribution will become the stable equilibrium.
These two situations are depicted in Fig. 4.

The dynamics of the homogeneous DFL kinetic model has been investigated
in much detail [14]. In particular, the well-posedness and convergence to the
equilibrium are already known; estimates of the rate of convergence have been
established. To the extent of our knowledge, however, these results have no
equivalent in the spatial model.

4 Numerical Scheme

Several schemes have already been proposed to study the kinetic equation (4)
using spectral method [15], discontinuous Galerkin [13], or semi-Lagrangian [11].
However, we are now interested in the long-time behavior of the solution, thus we
would like to design a numerical scheme with several properties:

– conservative, preserve positivity (under some CFL condition)
– satisfy a discrete version of inequalities (10) and (11).

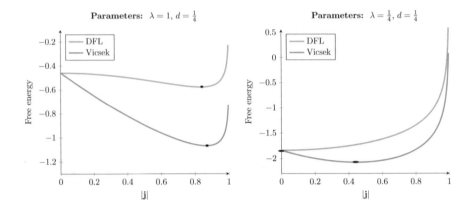

Fig. 4 **Left**: for low value of the diffusion coefficient σ, the minimizers for both free energies are von Mises distribution. **Right**: when the diffusion coefficient σ exceeds a certain threshold, the uniform distribution, i.e., $\mathbf{j} = 0$, becomes the minimizer for the DFL dynamics

In the following, we study the 2D scenario, taking advantage of the fact that the velocity space $\omega \in \mathbb{S}^1$ can be parametrized using polar coordinates by $\theta \in \mathbb{R}/2\pi\mathbb{Z}$ with $\omega = \begin{pmatrix} \cos\theta \\ \sin\theta \end{pmatrix}$. Thus, the kinetic equation (4) becomes

$$\partial_t f + c\,\omega \cdot \nabla_{\mathbf{x}} f = -\mu_f\,\partial_\theta(\sin(\bar\theta - \theta)f) + \sigma\partial_\theta^2 f,$$

where $\bar\theta$ is such that:

$$\Omega = \begin{pmatrix} \cos\bar\theta \\ \sin\bar\theta \end{pmatrix},$$

and μ_f is either a constant (Vicsek model) or proportional to $|\mathbf{j}|$ (DFL dynamics). Our numerical scheme is then based on a splitting method solving separately:

– the *transport part*

$$\partial_t f + c\,\omega \cdot \nabla_{\mathbf{x}} f = 0, \tag{14}$$

– the *collision part* :

$$\partial_t f = Q(f), \tag{15}$$

where $Q(f) = -\mu_f\,\partial_\theta(\sin(\bar\theta - \theta)f) + \sigma\partial_\theta^2 f$.

4.1 Collision Operator

In this section we focus on numerically solving Eq. (15). We write

$$Q(f) = -\mu_f \partial_\theta (\sin(\bar{\theta} - \theta) f) + \sigma \partial_\theta^2 f = \sigma \partial_\theta \left(M_f \partial_\theta \left(\frac{f}{M_f} \right) \right),$$

where M_f is the *von Mises distribution* :

$$M_f(\theta) = C_0 \exp^{\frac{\mu_f}{\sigma} \cos(\theta - \bar{\theta})},$$

and C_0 is a normalization constant. Here, we simply take $C_0 = 1$.

4.1.1 Discretization in θ

Fix $N > 0$ and consider a uniform discretization of the interval $[0, 2\pi)$ with $\theta_k = i \Delta\theta$ and $\Delta\theta = \frac{2\pi}{N}$. Denote: $f_k = f(\theta_k)$ and $f_{k+\frac{1}{2}} = f(\theta_k + \Delta\theta/2)$ and similarly $M_k = M_f(\theta_k)$. To approximate Q (which is a differential operator), we use the second order approximation:

$$\partial_\theta \left(\frac{f}{M_f} \right) \Bigg|_{\theta_k} = \frac{1}{\Delta\theta} \left(\frac{f_{k+\frac{1}{2}}}{M_{k+\frac{1}{2}}} - \frac{f_{k-\frac{1}{2}}}{M_{k-\frac{1}{2}}} \right) + \mathcal{O}(\Delta\theta^2),$$

which gives

$$Q(f)(\theta_k) = Q_N(f)(\theta_k) + \mathcal{O}(\Delta\theta^2),$$

with

$$Q_N(f)(\theta_k) = \frac{\sigma}{\Delta\theta^2} \left[M_{k+\frac{1}{2}} \left(\frac{f_{k+1}}{M_{k+1}} - \frac{f_k}{M_k} \right) - M_{k-\frac{1}{2}} \left(\frac{f_k}{M_k} - \frac{f_{k-1}}{M_{k-1}} \right) \right] \quad (16)$$

$$= \frac{\sigma}{\Delta\theta^2} \left[\frac{M_{k+\frac{1}{2}}}{M_{k+1}} f_{k+1} - \left(\frac{M_{k+\frac{1}{2}} + M_{k-\frac{1}{2}}}{M_k} \right) f_k + \frac{M_{k-\frac{1}{2}}}{M_{k-1}} f_{k-1} \right].$$

The discrete operator Q_N can be identified with a square $N \times N$ matrix:

$$Q_N := \frac{\sigma}{\Delta\theta^2} \begin{pmatrix} b_1 & c_1 & 0 & \cdots & 0 & a_1 \\ a_2 & b_2 & c_2 & \cdots & & 0 \\ 0 & a_3 & b_3 & & & \vdots \\ \vdots & & & \ddots & & \\ 0 & & & & b_{n-1} & c_{n-1} \\ c_n & 0 & \cdots & & a_n & b_n \end{pmatrix},$$

and

$$a_k = \frac{M_{k-\frac{1}{2}}}{M_{k-1}}, \quad b_k = -\frac{M_{k+\frac{1}{2}} + M_{k-\frac{1}{2}}}{M_k}, \quad c_k = \frac{M_{k+\frac{1}{2}}}{M_{k+1}},$$

with a slight abuse of notation such as $M_{-\frac{1}{2}} = M_{N-\frac{1}{2}}$ (by the periodicity of M_f).

The discrete operator Q_N reproduces many features of the differential operator Q. Define the scalar product:

$$\langle u, v \rangle_{M_f^{-1}} = \sum_{i=1}^{N} \frac{u_k v_k}{M_k},$$

the operator Q_N then satisfies the discrete equivalent to (10) and (11).

Proposition 3 *The operator Q_N (16) is symmetric with respect to this scalar product:*

$$\langle Q_N(u), v \rangle_{M^{-1}} = \langle u, Q_N(v) \rangle_{M^{-1}},$$

and satisfies

$$\langle Q_N(u), u \rangle_{M^{-1}} = -\frac{\sigma}{\Delta\theta^2} \sum_k M_{k+\frac{1}{2}} \left(\frac{u_{k+1}}{M_{k+1}} - \frac{u_k}{M_k} \right)^2 \leq 0, \qquad (17)$$

$$\langle Q_N(u), \ln \frac{u}{M} \rangle \leq 0. \qquad (18)$$

Proof Take any vectors u and v, then the Abel formula (discrete integration by parts) gives

$$\langle Q_N(u), v \rangle_{M^{-1}} = \frac{\sigma}{\Delta\theta^2} \sum_k M_{k+\frac{1}{2}} \left(\frac{u_{k+1}}{M_{k+1}} - \frac{u_k}{M_k} \right) \left(\frac{v_k}{M_k} - \frac{v_{k+1}}{M_{k+1}} \right)$$

$$= \langle u, Q_N(v) \rangle_{M^{-1}}.$$

From (16), we also deduce (17).

Moreover, using once again the Abel formula:

$$\langle Q_N(u), \ln \frac{u}{M} \rangle = \frac{\sigma}{\Delta\theta^2} \sum_k M_{k+\frac{1}{2}} \left(\frac{u_{k+1}}{M_{k+1}} - \frac{u_k}{M_k} \right) \left(\ln \frac{u_k}{M_k} - \ln \frac{u_{k+1}}{M_{k+1}} \right).$$

Thus, denoting $x = \frac{u_{k+1}}{M_{k+1}}$ and $y = \frac{u_k}{M_k}$, we have an expression of the form:

$$(x - y)(\ln y - \ln x) = (x - y) \ln \frac{y}{x} \leq 0$$

for any $x, y > 0$. We deduce (18).

Remark 1 In particular, the computations from Proposition 1 can be reproduced for the discrete free energy:

$$\mathcal{F}_N(f) := \Delta\theta \sum_k f_k \ln f_k - \frac{\mu}{\sigma} \Phi_f$$

where

$$\Phi_f := \begin{cases} |\mathbf{j}_f| & \text{Vicsek dynamics} \\ \frac{1}{2}|\mathbf{j}_f|^2 & \text{DFL dynamics} \end{cases}$$

and $\mathbf{j}_f := \sum_k \begin{pmatrix} \cos\theta_k \\ \sin\theta_k \end{pmatrix} f(\theta_k)\Delta\theta$. Any solution to the ODE $\frac{df}{dt} = Q_N(f)$ will then satisfy

$$\frac{d}{dt}\mathcal{F}_N(f) \leq 0.$$

4.1.2 Explicit Euler

The Euler method can be used to discretize in time the collisional part of the kinetic equation (15):

$$f^{n+1} = f^n + \Delta t Q_N(f^n) = (\mathrm{Id} + \Delta t Q_N)f^n.$$

This method readily preserves the mass. A sufficient condition for the L^∞ stability of the scheme is to have the matrix $\mathrm{Id} + \Delta t\, Q_N$ positive (i.e., all coefficients positive). This sufficient condition leads to the following CFL condition:

$$\max_k\{|b_k|\} \frac{\sigma \Delta t}{\Delta\theta^2} < 1, \tag{19}$$

which is usual for diffusion type operator. Moreover, if the CFL condition is met, then positivity is preserved.

Remark 2 We can find an explicit sufficient condition to guarantee the CFL condition (19). Indeed, writing:

$$\frac{M_{k+\frac{1}{2}}}{M_k} = \frac{\exp\left(\frac{\mu_f}{\sigma}\cos(\theta_{k+\frac{1}{2}} - \bar{\theta})\right)}{\exp\left(\frac{\mu_f}{\sigma}\cos(\theta_k - \bar{\theta})\right)} = \exp\left(\frac{\mu_f}{\sigma}[\cos(\theta_{k+\frac{1}{2}} - \bar{\theta}) - \cos(\theta_k - \bar{\theta})]\right)$$

$$= \exp\left(-2\frac{\mu_f}{\sigma}\sin\left(\theta_k - \bar{\theta} + \frac{\Delta\theta}{4}\right)\sin\left(\frac{\Delta\theta}{4}\right)\right),$$

where we have used the identity $\cos\alpha - \cos\beta = -2\sin\frac{\alpha+\beta}{2}\sin\frac{\alpha-\beta}{2}$. We deduce

$$\frac{M_{k+\frac{1}{2}}}{M_k} \leq \exp\left(2\frac{\mu_f}{\sigma}\sin\left(\frac{\Delta\theta}{4}\right)\right),$$

and find:

$$\max|b_k| \leq 2\exp\left(2\frac{\mu_f}{\sigma}\sin\left(\frac{\Delta\theta}{4}\right)\right) = 2 + \frac{\mu_f}{\sigma}\Delta\theta + \mathcal{O}(\Delta\theta^2).$$

This leads to the tractable (sufficient) CFL condition:

$$\frac{2\sigma\Delta t}{\Delta\theta^2} < \exp\left(-2\frac{\mu_f}{\sigma}\sin\left(\frac{\Delta\theta}{4}\right)\right). \tag{20}$$

Algorithm 1 Collision part Eq. (15)

1: **procedure** COLLISION($f(\theta_k), \Delta t$)
2:     $\mathbf{j} = \sum_k \omega_k f_k \Delta\theta$; $\bar{\theta} = angle(\mathbf{j})$
3:     $M_k = \exp(\frac{\mu_f}{\sigma}\cos(\theta_k - \bar{\theta}))$
4:     **for** k in $1 : N$ **do**
5:         $Q_N(f)_k = \frac{\sigma}{\Delta\theta^2} \cdot (a_k f_{k-1} - b_k f_k + c_k f_{k+1})$
6:     **end for**
7:     $f \mathrel{+}= \Delta t \cdot Q_N(f)$
8:     Return f
9: **end procedure**

4.1.3 Adaptive Time Step for the Collision

One of the difficulties in computing an approximate solution to (5) is coping with the associated CFL condition (19). Indeed, the existence of a locally high $|j(f)|$ greatly decreases the right-hand side of (19), which penalizes the whole algorithm. Hence using a global CFL condition for the transport part and the collision part can lead to extremely long computation time.

We propose to decouple the time steps for the transport part and the collision equation at each time step, by using an adaptive method for the latter. Technically, we use the maximal time step associated with the CFL condition to solve the transport part (14), $\Delta t = \Delta x$, which incidentally has the advantage of minimizing numerical diffusion. Then, for each (x_i, y_j), we consider (15) as a differential equation with final time Δt, which we solve by using the method described in Sect. 4.1 with a variable time step δt that needs to be recomputed at each time $0 \leq s \leq \Delta t$:

$$\delta t(s) := \min\left(\frac{\Delta\theta^2}{2\sigma} \exp\left(\mu_f \frac{-2\sin(\frac{\Delta\theta}{4})}{\sigma}\right), \Delta t - s\right).$$

This method also works for a constant relaxation $\mu(f) = \mu_0$, and can be preferred because it minimizes the numerical diffusion in the transport equation. Particularly when the constant $\sigma = \frac{\mu}{D}$ is large, in which case the collision CFL (20) is much smaller than the transport CFL (21).

A comparison between the errors done by the standard and adaptive schemes, respectively, is shown in Fig. 5.

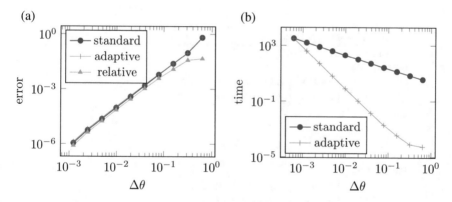

Fig. 5 Comparison of the standard and adaptive methods for the collision operator (Vicsek). Parameters are: $\mu = 1.0$, $\sigma = 0.2$, $\rho = 1.0$, $\Delta t = 8.458 \cdot 10^{-7}$ (standard), $\Delta t = 0.1$ (adaptive), $T = 1.0$. The initial condition is $f_0(\theta) = \rho\left(1 + \frac{1}{5}\sum_{k=1}^{5} \cos(p_k\theta)\right)$ where $p_1 = 1$ and p_k is the prime following p_{k-1}. (**a**) Error in the standard and adaptive models. The *standard* and *adaptive* curves show the maximum of the difference between the computed distributions and the one with lowest $\Delta\theta$. The "relative" curve shows the maximum of the difference between the standard and adaptive solutions. (**b**) Comparison of the time (in seconds) needed to compute the solution. Notice that the two curves converge to the leftmost point, which was taken as a reference. Understandably, the computational advantage of the adaptive method increases with $\Delta\theta$

4.2 Numerical Scheme for the Transport Operator

We use an upwind finite-difference method to solve the transport equation (14). More accurate schemes are available (such as WENO scheme [21]); however, stability and positivity preserving are more essential in our study, since we are more interested in the emergence of robust structures in the long-time dynamics, than in the precise long-time behavior of a particular initial condition. We fix $M > 0$ and consider a uniform discretization of the interval $[0, L)$ in M points with $x_i = i \Delta x$, $y_j = j \Delta y$, and $\Delta x = \Delta y = \frac{L}{M}$. To discretize the kinetic equation, we use

$$\cos\theta \, \partial_x f = \begin{cases} \cos\theta \, \frac{f(x_i) - f(x_{i-1})}{\Delta x} + \mathcal{O}(\Delta x), & \text{if } \cos\theta \geq 0 \\ \cos\theta \, \frac{f(x_{i+1}) - f(x_i)}{\Delta x} + \mathcal{O}(\Delta x), & \text{if } \cos\theta \leq 0, \end{cases}$$

and similarly for $\sin(\theta) \partial_y f$. Using this discretization, the standard Euler scheme gives as CFL condition:

$$c \frac{\Delta t}{\Delta x} < 1. \tag{21}$$

Algorithm 2 Transport part Eq. (14)

1: **procedure** TRANSPORT($f(x_i, y_j, \theta_k), \Delta t$)
2:    **for** i, j, k **do**

3:        $F_{i+\frac{1}{2}, j, k} = \begin{cases} c \cos\theta_{i+\frac{1}{2}} \, f_{i,j,k} & \text{if } \cos\theta_{i+\frac{1}{2}} \leq 0 \\ c \cos\theta_{i+\frac{1}{2}} \, f_{i+1,j,k} & \text{if } \cos\theta_{i+\frac{1}{2}} \geq 0 \end{cases}$

4:    **end for**
5:    **for** i, j, k **do**
6:        $f_{i,j,k} \mathrel{+}= -\frac{\Delta t}{\Delta x} \cdot (F_{i+\frac{1}{2}, j, k} - F_{i-\frac{1}{2}, j, k})$
7:    **end for**
8:    **for** i, j, k **do**

9:        $F_{i, j+\frac{1}{2}, k} = \begin{cases} c \sin\theta_{j+\frac{1}{2}} \, f_{i,j,k} & \text{if } \sin\theta_{i+\frac{1}{2}} \leq 0 \\ c \sin\theta_{j+\frac{1}{2}} \, f_{i,j+1,k} & \text{if } \sin\theta_{i+\frac{1}{2}} \geq 0 \end{cases}$

10:   **end for**
11:   **for** i, j, k **do**
12:       $f_{i,j,k} \mathrel{+}= -\frac{\Delta t}{\Delta y} \cdot (F_{i, j+\frac{1}{2}, k} - F_{i, j-\frac{1}{2}, k})$
13:   **end for**
14:   Return f
15: **end procedure**

4.3 Summary: Full Scheme

The full algorithm is finally a splitting between the transport and collision part. Notice that the time step Δt should satisfy both CFL conditions (19) and (21).

In general, the collisional CFL (19) is more restrictive. Therefore, the transport equation will be solved with a small CFL corresponding to large numerical viscosity. Since we aim at studying the large-time behavior of the dynamics, this numerical viscosity might drastically change the outcome. Thus, we propose to use an adaptive method for the collisional operator, described in Sect. 4.1.3.

Algorithm 3 Collision part Eq. (15)

1: **procedure** COLLISIONADAPT($f(\theta_k)$, Δt)
2:     Find K such that $\delta t = \Delta t / K$ satisfies (20)
3:     **for** s in $1 : K$ **do**
4:         $f = $ **Collision**($f, \delta t$)
5:     **end for**
6:     Return f
7: **end procedure**

Algorithm 4 Full kinetic Eq. (4)

1: Fix $\Delta t < \min(\Delta x, \Delta y)/c$
2: $t = 0$
3: **while** $t < T$ **do**
4:     $f^* = $ **Transport**($f^n, \Delta t$)
5:     **for** i, j **do**
6:         $f_{i,j,k}^{n+1} = $ **CollisionAdapt**($f_{i,j,k}^*, \Delta t$)
7:     **end for**
8:     $t += \Delta t$
9: **end while**
10: Return f

---

5 Numerical Experiments

5.1 Homogeneous Case

To first investigate our numerical scheme, we study the homogeneous equation, thus solving only the collision operator (15). We present our numerical experiments with the Vicsek model (4); however, results are similar with the DFL dynamics except that the time step Δt may have to be adapted (since the CFL condition depends on $|\mathbf{j}|$ which varies over time).

As a first sanity check, we estimate the accuracy of the scheme. With this aim, we fix a final time $T = 1$ and time step $\Delta t = 0.001$. Then, we vary the meshgrid in θ, taking $\Delta \theta \in \{\frac{2\pi}{8}, \frac{2\pi}{16}, \ldots, \frac{2\pi}{128}\}$ and estimate the L^2 error with the reference solution f_{ref} computing with $\Delta \theta = \frac{2\pi}{256}$. For the initial condition, we use a smooth initial condition:

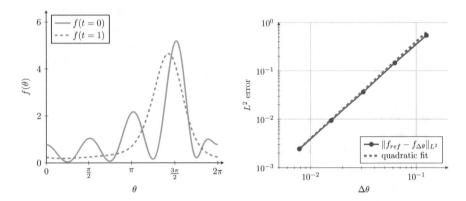

Fig. 6 **Left:** the initial condition f_0 (22) and the reference solution f_{ref} (dashed) computed after $t = 1$ (with $\Delta t = 10^{-3}$ and $\Delta\theta = \frac{2\pi}{256}$). **Right:** L^2 error in log scale between the solution $f_{\Delta\theta}$ with the reference solution f_* at $t = 1$. We observe a quadratic accuracy

$$f_0(\theta) = (1.1 + \cos 4\theta) \cdot \exp\left(-\cos\left(\pi(s + s^8)\right)\right), \qquad \text{with } s = \theta/2\pi. \quad (22)$$

We use a rather complicated expression to make sure that f_0 is non-symmetric. When f_0 is symmetric, the mean direction $\bar\theta$ is preserved over time, thus the Vicsek dynamics (6) becomes a linear evolution equation. Twisting the initial condition f_0 guarantees to have a fully non-linear equation.

In Fig. 6—left, we plot the initial condition f_0 along with the reference solution f_{ref} at $t = 1$. The L^2 error for various discretizations is given in log scale in Fig. 6—right. We observe that the error is decaying quadratically as expected.

Moreover, we also investigate the large-time behavior of the solution. First, we measure the evolution of the free entropy $\mathcal{F}$ over time and we observe that it is strictly decreasing (Fig. 7—left). Second, we estimate the rate of convergence of $f(t)$ toward an equilibrium distribution. Using semi-log scale in Fig. 7—right, we observe a linear decay indicating that the convergence is exponential. Note that this is expected, owing to the exponential convergence proved in [12, 14].

5.2   Band Formation

In the Vicsek model (4), we did not observe the formation of any bands. Rather, the dynamics always converge to a robust global alignment dynamics, where the spatial distribution (first moment of f) converges to a constant. The typical long-time behavior is represented in Fig. 8. We postulate that the long-time behavior of this equation is just to converge to a uniform distribution of von Mises equilibria to the homogeneous equation.

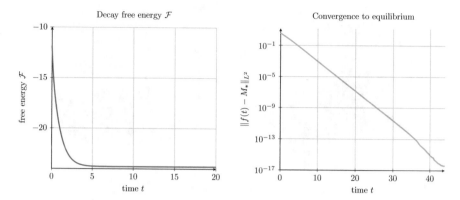

Fig. 7 Left: evolution of the free energy $\mathcal{F}$ (12) over time. The function is strictly decreasing. **Right:** L^2 error between the solution $f(t)$ and its equilibrium distribution M_*. Since it uses semilog scale, the convergence is actually exponential

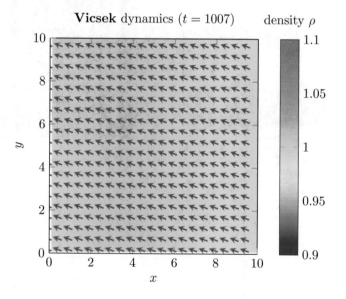

Fig. 8 Typical shape of the long-time solution observed in the case of the Vicsek interaction model (obtained starting from a random initial condition) at $t = 1000$. Parameters are: $\mu = 1.0$, $\sigma = 0.2$, $c = 1.0$, $L = 10.0$, $\Delta x = \Delta y = 0.1$, $\Delta \theta = \frac{2\pi}{30}$

On the other hand, the DFL model (5) of interaction leads to the observation of bands. Typically, for a fixed set of parameters, we observed two different scenarios regarding the behavior of the local density ρ and mean value $\bar{\rho}$

$$\rho(t, \mathbf{x}) = \int_{\mathbb{S}^1} f(t, \mathbf{x}, \omega) d\omega, \qquad \bar{\rho} = \frac{\int_{[0,L]^2 \times \mathbb{S}^1} f(\mathbf{x}, \omega) d\mathbf{x} d\omega}{2\pi L^2}. \tag{23}$$

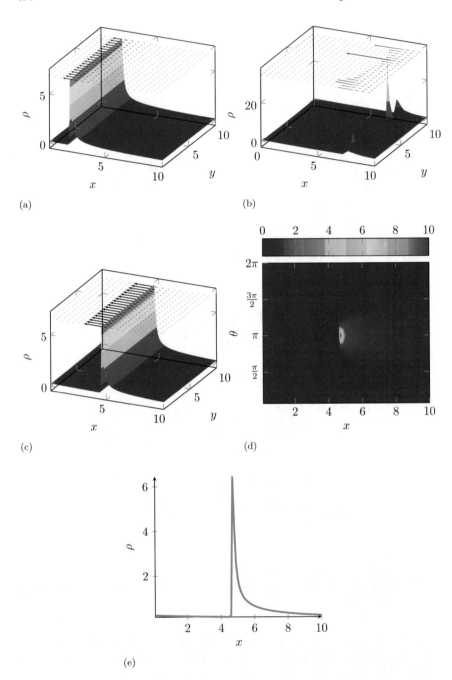

Fig. 9 Different observations of bands. The surface plot corresponds to the observed density $\rho = \int_{\mathbb{S}^1} f(t, \mathbf{x}, \omega) d\omega$. The arrows on the top correspond to the local flux $\mathbf{j}(\mathbf{x})$. Figure **a**, **b** was obtained starting from the same random initial condition. Figure **c** was obtained starting from a

For a fixed mean value ρ, when the strength of interaction μ is small compared to the diffusion parameter σ (i.e., $\mu \gg \sigma$), we observe that the solution converges to a uniform steady state. However, when $\mu \ll \sigma$, we observe the formation of bands as shown in Fig. 9 (in which the x-axis has been reversed to provide better aesthetics). Thus, we retrieve an equivalent of the phase transition dynamics noticed by Frouvelle and Liu in [14]. Those bands were first noticed starting from a random initial condition (Fig. 9a), though we do not expect the randomness of the initial condition to be an important factor in the band formation phenomenon. Even though they literally emerge from chaos, they appear to be only meta-stable, as their small inhomogeneity in the direction perpendicular to the propagation may amplify slowly by attracting the neighbor particles and finally lead to high and localized concentrations as shown in Fig. 9b. At this point the computation is difficult to continue due to the extremely high computation times required by the CFL condition (20). Starting from an initial condition which is homogeneous in one direction (e.g., in y), however, the observed bands are very stable in time and can be kept alive for apparently an arbitrarily long time (the homogeneity being preserved by our scheme). Such a band is represented in Fig. 9c. The initial condition we used is the following:

$$f_0(x, y, \theta) = \bar{\rho}\left(1 + \frac{1}{10}\sum_{k=1}^{5}\cos(p_k\theta) + \cos\left(2p_k\pi\frac{x}{L}\right)\right).$$

Bands were also observed in a modified one-dimensional model which we encoded to take advantage of the preservation of homogeneity in one direction. A resulting band is presented in Fig. 9d, e.

Numerical evidences show that the bands cannot be understood as traveling wave solutions to the kinetic equation (4), as one may believe at first sight. Indeed, there remains an inner motion inside the bands that we can reveal by monitoring the maximal value of ρ through time (see Fig. 10). This reveals an asymptotically periodic behavior that strongly resembles the notion of pulsating fronts, which has been extensively studied in the context of reaction–diffusion equations [24]. A deeper analytical understanding of this phenomenon is left for future work.

Finally, to strengthen the link between the phase transition and the formation of bands, we show in Fig. 11 two kinds of entropy computed for a range of values of the diffusion coefficient σ and the mean value of the initial condition $\bar{\rho}$. Figure 11a

---

Fig. 9 (continued) homogeneous in y initial condition. Figure **d** was obtained by a one-dimensional version of the code. In all cases, the same set of parameters was used: $\mu = 1.0$, $\sigma = 0.2$, $c = 1.0$, $L = 10.0$, $\Delta x = \Delta y = 0.1$, $\Delta\theta = \frac{2\pi}{30}$. (**a**) Starting from a random initial condition $t = 1000.0$ ($\bar{\rho} \approx 0.0766$). (**b**) Starting from a random initial condition $t = 1227.0$. (**c**) Starting from a homogeneous initial condition, $t = 1007.0$ ($\bar{\rho} = 0.0763$). (**d**) Pseudo-1D code, $t = 1005.0$ ($\bar{\rho} = 0.0763$). Here the real value of f is represented as a function of x and θ. (**e**) Pseudo-1D code, $t = 1005.0$ ($\bar{\rho} = 0.0763$). ρ is represented as a function of x

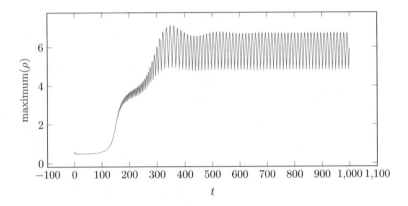

Fig. 10 Maximal value of $\rho(t, \mathbf{x})$ as a function of t

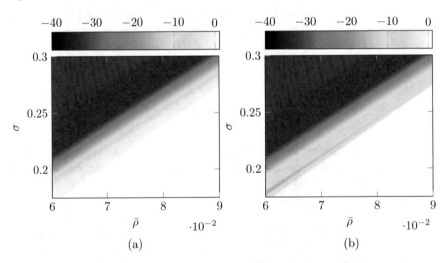

$$(a) \qquad\qquad\qquad (b)$$

Fig. 11 Entropy as a function of $\bar{\rho}$ and σ. The entropies were computed on the numerical solution of the DFL equation at time $t = 600.0$, with parameters $\mu = 1.0$, $\Delta t = 0.1$, $\Delta x = \Delta y = 0.1$, and $\Delta\theta = \frac{2\pi}{30}$. (**a**) Log-entropy against the uniform distribution $i \ln \mathcal{E}_u[f]$. (**b**) Log-entropy against the von Mises distribution $\ln \mathcal{E}_{VM}[f]$

represents the log of the entropy of f computed against the uniform distribution of the same mass:

$$\mathcal{E}_u[f] = \int_0^L \int_0^L \int_0^{2\pi} f(t, \mathbf{x}, \theta) \log\left(\frac{f(t, \mathbf{x}, \theta)}{\bar{\rho}}\right) d\theta d\mathbf{x}.$$

This entropy quantifies of much the distribution f differs from the uniform distribution. Similarly, Fig. 11b represents the log of the generalized entropy of f computed against the corresponding von Mises distribution:

$$\mathcal{E}_{VM}[f] = \int_0^L \int_0^L \int_0^{2\pi} f(t, \mathbf{x}, \theta) \log\left(\frac{f(t, \mathbf{x}, \theta)}{M[\bar{\rho}](\theta)}\right) d\theta d\mathbf{x},$$

where $M[\bar{\rho}](\theta) = 2\pi \bar{\rho} \frac{\exp(\frac{\mu\kappa}{d}\cos(\theta))}{\int_0^{2\pi} \exp(\frac{\mu\kappa}{d}\cos(\theta))d\theta}$ is the only candidate as stationary von Mises distribution, κ satisfying the compatibility condition

$$2\pi \bar{\rho} \frac{\int_0^{2\pi} \cos\theta \exp\left(\frac{\mu\kappa}{d}\cos(\theta)\right) d\theta}{\int_0^{2\pi} \exp\left(\frac{\mu\kappa}{d}\cos(\theta)\right) d\theta} = \kappa.$$

Let us recall that the latter has only one solution $\kappa = 0$ when $\sigma \geq \pi\mu\bar{\rho}$, and has exactly one positive solution when $\sigma < \pi\mu\bar{\rho}$ [14]. This entropy quantifies how much f differs from a distribution which would be composed of the same von Mises distribution at each point $\mathbf{x}$.

The match between the two plots suggests that the latter stationary state candidate is never stable except when $\kappa = 0$. Indeed for $\sigma \geq \pi\mu\bar{\rho}$, the uniform distribution is stable for the homogeneous problem and $\kappa = 0$; this corresponds to the top-left part of Fig. 11a, which suggests that this stability is transferred to the inhomogeneous problem (5). For $\sigma < \pi\mu\bar{\rho}$, however, we have $\kappa > 0$ and the stable state for the homogeneous problem is described by the corresponding von Mises distribution $M[\bar{\rho}](\theta)$; Fig. 11b suggests that the inhomogeneous problem behaves otherwise, neither the uniform nor the homogeneous von Mises distribution corresponding to the long-time behavior of the equation, except possibly in a very small area near $\sigma \approx \pi\mu\bar{\rho}$. Instead, the instability of both homogeneous stationary states could be at the origin of the formation of bands.

6   Conclusion

In this manuscript, we have introduced a numerical scheme to solve kinetic equations related to the Vicsek model. Despite the additional difficulties to treat non-conservative dynamics (e.g., no Rankine–Hugoniot conditions, no entropy), the numerical scheme is able to preserve positivity and entropy which allow to study the long-time behavior of the dynamics. Of particular interest is the formation of bands that were first observed at the microscopic model (i.e., particle simulations). The scheme is able to capture such band formation at the kinetic level.

It remains many open questions related to the kinetic equation of the Vicsek model. First, we still do not know if asymptotically there is an analytical description of the band. Having an *Ansatz* will help to study the convergence of the dynamics toward a band formation. However, the lack of entropy in the non-homogeneous case makes this task complicated. Another open problem is to investigate how the Vicsek model in a certain regime (i.e., low density, large coefficients) is able to create bands. We do not observe such band formation for the original Vicsek model at the kinetic level, bands only form with the DFL dynamics. This might indicate that the propagation of chaos, which links particle and kinetic formulations, is not valid in this regime.

References

1. M. Aldana and C. Huepe. Phase Transitions in Self-Driven Many-Particle Systems and Related Non-Equilibrium Models: A Network Approach. *Journal of Statistical Physics*, 112(1):135–153, 2003.
2. F. Bolley, J. Cañizo, and J. Carrillo. Mean-field limit for the stochastic Vicsek model. *Applied Mathematics Letters*, 25(3):339–343, 2012.
3. S. Camazine, J. L Deneubourg, N. R Franks, J. Sneyd, G. Theraulaz, and E. Bonabeau. Self-organization in biological systems. *Princeton University Press; Princeton, NJ: 2001*, 2001.
4. J. Carrillo, A. Chertock, and Y. Huang. A finite-volume method for nonlinear nonlocal equations with a gradient flow structure. *Communications in Computational Physics*, 17(01):233–258, 2015.
5. H. Chaté, F. Ginelli, and G. Grégoire. Comment on "phase transitions in systems of self-propelled agents and related network models". *Physical review letters*, 99(22):229601, 2007.
6. H. Chaté, F. Ginelli, G. Grégoire, F. Peruani, and F. Raynaud. Modeling collective motion: variations on the Vicsek model. *The European Physical Journal B*, 64(3-4):451–456, 2008.
7. P. Degond, A. Frouvelle, and J-G. Liu. Macroscopic limits and phase transition in a system of self-propelled particles. *Journal of nonlinear science*, 23(3):427–456, 2013.
8. P. Degond, A. Frouvelle, and J-G. Liu. Phase transitions, hysteresis, and hyperbolicity for self-organized alignment dynamics. *Archive for Rational Mechanics and Analysis*, 216(1):63–115, 2015.
9. P. Degond, J-G. Liu, S. Motsch, and V. Panferov. Hydrodynamic models of self-organized dynamics: derivation and existence theory. *Methods and Applications of Analysis*, 20(2):89–114, 2013.
10. P. Degond and S. Motsch. Continuum limit of self-driven particles with orientation interaction. *Mathematical Models and Methods in Applied Sciences*, 18(1):1193–1215, 2008.
11. G. Dimarco and S. Motsch. Self-alignment driven by jump processes: Macroscopic limit and numerical investigation. *Mathematical Models and Methods in Applied Sciences*, 26(07):1385–1410, 2016.
12. A. Figalli, M-J. Kang, and J. Morales. Global well-posedness of the spatially homogeneous Kolmogorov–Vicsek model as a gradient flow. *Archive for Rational Mechanics and Analysis*, 227(3):869–896, 2018.
13. F. Filbet and C-W. Shu. Discontinuous-Galerkin methods for a kinetic model of self-organized dynamics. *arXiv preprint arXiv:1705.08129*, 2017.
14. A. Frouvelle and J-G. Liu. Dynamics in a kinetic model of oriented particles with phase transition. *SIAM Journal on Mathematical Analysis*, 44(2):791–826, 2012.
15. I. Gamba, J. Haack, and S. Motsch. Spectral method for a kinetic swarming model. *Journal of Computational Physics*, 297:32–46, 2015.
16. I. Gamba and M-J. Kang. Global weak solutions for Kolmogorov–Vicsek type equations with orientational interactions. *Archive for Rational Mechanics and Analysis*, 222(1):317–342, 2016.
17. S. Gottlieb, C-W. Shu, and E. Tadmor. Strong stability-preserving high-order time discretization methods. *SIAM review*, 43(1):89–112, 2001.
18. G. Grégoire and H. Chaté. Onset of collective and cohesive motion. *Physical review letters*, 92(2):025702, 2004.

19. M. Nagy, I. Daruka, and T. Vicsek. New aspects of the continuous phase transition in the scalar noise model (SNM) of collective motion. *Physica A: Statistical Mechanics and its Applications*, 373:445–454, 2007.

20. K. Oelschläger. A law of large numbers for moderately interacting diffusion processes. *Zeitschrift für Wahrscheinlichkeitstheorie und verwandte Gebiete*, 69(2):279–322, 1985.

21. C-W. Shu. High order weighted essentially nonoscillatory schemes for convection dominated problems. *SIAM review*, 51(1):82–126, 2009.

22. T. Vicsek, A. Czirók, E. Ben-Jacob, I. Cohen, and O. Shochet. Novel type of phase transition in a system of self-driven particles. *Physical Review Letters*, 75(6):1226–1229, 1995.

23. T. Vicsek and A. Zafeiris. Collective motion. *Physics Reports*, 517(3):71–140, 2012.

24. J. Xin. Front propagation in heterogeneous media. *SIAM review*, 42(2):161–230, 2000.

Singular Cucker–Smale Dynamics

Piotr Minakowski, Piotr B. Mucha, Jan Peszek, and Ewelina Zatorska

Abstract This chapter is dedicated to the singular models of flocking. We give an overview of the existing literature starting from microscopic Cucker–Smale (CS) model with singular communication weight, through its mesoscopic mean-field limit, up to the corresponding macroscopic regime. For the microscopic CS model and its selected variants, the collision-avoidance phenomenon is discussed. For the kinetic mean-field model, we sketch the existence of global-in-time measure-valued solutions, paying special attention to weak-atomic uniqueness of solutions. Ultimately, for the macroscopic singular model, we provide a summary of existence results for the Euler-type alignment system. This includes the existence of strong solutions on a one-dimensional torus, and the extension of this result to higher dimensions by restricting the size of the initial data. Additionally, we present the pressureless Navier–Stokes-type system corresponding to particular choice of alignment kernel. This system is then compared—analytically and numerically—to the porous medium equation.

P. Minakowski
Institute of Analysis and Numerics, Otto von Guericke University Magdeburg, Magdeburg, Germany
e-mail: piotr.minakowski@ovgu.de

P. B. Mucha
Institute of Applied Mathematics and Mechanics, University of Warsaw, Warsaw, Poland
e-mail: p.mucha@mimuw.edu.pl

J. Peszek (✉)
Center for Scientific Computation and Mathematical Modeling (CSCAMM), University of Maryland, College Park, MD, USA

Institute of Mathematics of the Polish Academy of Sciences, Warsaw, Poland
e-mail: j.peszek@mimuw.edu.pl

E. Zatorska
Department of Mathematics, University College London, London, UK
e-mail: e.zatorska@ucl.ac.uk

© Springer Nature Switzerland AG 2019
N. Bellomo et al. (eds.), *Active Particles, Volume 2*, Modeling and Simulation in Science, Engineering and Technology,
https://doi.org/10.1007/978-3-030-20297-2_7

1 Introduction

The phenomenon of *flocking* can be understood as a general tendency of self-propelled particles (or agents) to organize their dynamics based on the behaviour of their neighbours. It is a process in which an ensemble of particles align their velocities, remaining in a close proximity to each other, typical for herds of mammals, schools of fish, or flocks of birds. Surprisingly enough, due to the inherent flexibility of mathematical modelling, this basic phenomenon can be used to describe a variety of seemingly unrelated processes. Indeed, achieving consensus, the synchronization of motion, or the emergence of complex structures and patterns are observable in much more diverse areas like: distribution of goods [71], spacecraft formation [58], sensor networks [16] and digital media arts [44], as well as emergence of languages in primitive societies [48]. For the survey on the multi-agent systems and their applications we refer the reader to [56].

The Cucker–Smale (CS) flocking model, introduced in the seminal work of Cucker and Smale [25], is a basic example of a flocking model that concentrates on the alignment of particles' velocities. In physical language it means that the velocity vector of each individual is determined in terms of positions and momenta of other members of the group. Perhaps the simplest example is a *flock of two birds* (see [25, Section IV]) with positions and velocities equal to $(x_i(t), v_i(t))$, for $i = 1, 2$. Their dynamics can be described as follows:

$$\frac{\mathrm{d}}{\mathrm{d}t} x = v, \qquad \frac{\mathrm{d}}{\mathrm{d}t} v = -\frac{v}{(1 + |x|)^\alpha} \quad \text{with } \alpha > 0, \tag{1}$$

where

$$x(t) = x_1(t) - x_2(t), \qquad v(t) = v_1(t) - v_2(t).$$

As a basic feature of the model one gets alignment of velocities, i.e.

$$v(t) \to 0 \quad \text{as} \quad t \to \infty.$$

The complete information of the dynamics of system (1) is encoded in the communication weight, which in this case is equal to $(1 + |x|)^{-\alpha}$. The communication weight represents the perception of the particles and, in general, tames the long-range interactions between the particles. The same principle as in (1) applied to multiple agents leads to the general CS flocking model

$$\begin{cases} \dfrac{\mathrm{d}}{\mathrm{d}t} x_i = v_i, & \text{(2a)} \\[3mm] \dfrac{\mathrm{d}}{\mathrm{d}t} v_i = \dfrac{1}{N} \displaystyle\sum_{j=1}^{N} (v_j - v_i) \psi(|x_i - x_j|), & \text{(2b)} \end{cases}$$

with the initial conditions

$$(x_1, \ldots, x_N)(0) = \mathbf{x}_0, \ (v_1, \ldots, v_N)(0) = \mathbf{v}_0.$$

Here, N denotes the number of the particles, $x_i(t)$ and $v_i(t)$ are the position and the velocity of ith particle at the time t, and ψ is the aforementioned communication weight, which is usually assumed to be positive, non-increasing and smooth.

After the introduction of the CS model in 2007, inspired to some degree by the work or Vicsek et al. [73] from 1995, extensive studies were carried out in various directions including: asymptotics [25, 32], pattern formation [57, 58], collision-avoidance [1, 23], variants of the model with preferences [24, 47], leadership [22, 62] and additional deterministic or stochastic forces [6, 7, 15, 29]. Further directions include the study of the kinetic [9, 32, 34] and hydrodynamic [30, 31, 42] limits of the CS particle system and their coupling with classical equations of hydrodynamics [2, 3, 17, 55]. Other interesting variants of the CS model revolve around changing the symmetric all-to-all character of the interactions. As an example, let us mention interactions with cone-shaped sensitivity regions, introduced in [53] and developed in [12], weighted normalization [53] (known as the Motsch–Tadmor model) or topological interactions [35, 67]. For an exhaustive overview of the research on the CS model with a regular communication weight we refer the reader to [19], and the references therein. In the present survey we omit certain directions that were covered in [19] focusing solely on the case of singular ψ. We explain this direction of research below.

Most of the mathematical analysis of system (2) has been done for the case of regular, bounded communication weight. The CS model with such weight can be treated as a nonlinear ODE system with a Lipschitz continuous nonlinearity. Such systems are relatively well understood, as far as their basic properties like well-posedness are concerned. Also from the application perspective, regular communication weight is often perfectly suitable. However, certain phenomena exhibiting strongly local interactions require the use of singular communication weights that blow up whenever any two particles collide. One of the most distinctive characteristics of the CS model with singular weight that opens further possibilities of application is that the particles avoid collisions regardless of the initial data. This chapter is dedicated to the description of broad dynamics arising from the singular CS model. We consider system (2) not only as a law governing the motion of a number of particles, but also discuss the possible dynamics in different scales: meso- and macroscopic. The first one is relevant for systems with very large number of particles, while the second (hydrodynamical) level becomes convenient when condensation of particles is too large to distinguish the evolution of single ones.

The rest of the chapter is organized as follows. In Sect. 2, we present the necessary preliminaries. Section 3 is dedicated to the singular CS particle system and, particularly, to the collision-avoidance and its applications. In Sect. 4, we present the results concerning the kinetic CS equation. Finally, in Sect. 5, we discuss a particular version of hydrodynamic CS model, known as the fractional Euler alignment system, which we also compare to the porous medium equation.

2 Preliminaries

Before presenting the results directly related to the singular CS model, let us introduce basic properties that can be derived from the structure of system (2). First, we define the four basic states of the dynamics.

Definition 2.1

i) We say that ith and jth particles *collide* at time t_0 iff

$$x_i(t_0) = x_j(t_0).$$

ii) We say that ith and jth particles *stick together* at time t_0 iff they collide at t_0 and

$$v_i(t_0) = v_j(t_0).$$

iii) We say that the system (2) *aligns asymptotically* iff

$$\sum_{i,j=1}^{N} |v_i(t) - v_j(t)|^2 \to 0, \qquad \text{as} \qquad t \to \infty.$$

iv) We say that the system (2) *flocks asymptotically* iff it aligns asymptotically and there exists a constant $D > 0$ such that

$$\limsup_{t \to \infty} \max_{i,j=1,\dots,N} |x_i(t) - x_j(t)| \le D.$$

Remark 2.1 Definition 2.1 is admissible only in the case of the Cucker–Smale particle system (2), while throughout the chapter we will also discuss the kinetic and hydrodynamic regimes. Note, unlike points (i) and (ii), points (iii) and (iv) can be easily generalized to the dynamics of continua.

Summing Eq. (2b) with respect to $i = 1, \dots, N$ we deduce that

$$\frac{\mathrm{d}}{\mathrm{d}t} \sum_{i=1}^{N} v_i = \frac{1}{N} \sum_{i,j=1}^{N} (v_j - v_i) \psi(|x_i - x_j|) = 0,$$

which means that the average velocity is constant, i.e. $\dfrac{1}{N} \displaystyle\sum_{i=1}^{N} v_i = const.$ Therefore, throughout the chapter, we assume without a loss of generality that

$$\frac{1}{N}\sum_{i=1}^{N} v_i = 0, \qquad \text{and} \qquad \frac{1}{N}\sum_{i=1}^{N} x_i = 0. \tag{3}$$

This assumption implies equivalence between alignment and dissipation of the kinetic energy:

$$E_k := \frac{1}{2}\sum_{i=1}^{N} v_i^2 = \frac{1}{4N}\sum_{i,j=1}^{N} |v_i - v_j|^2. \tag{4}$$

The following proposition combines two most important structural properties of the CS system: dissipation of the kinetic energy and the boundedness of the velocity.

Proposition 2.1 *Let (x, v) be a smooth solution to system (2). Then*

$$\frac{d}{dt}\sum_{i=1}^{N} v_i^2 = -\frac{1}{N}\sum_{i,j=1}^{N} (v_i - v_j)^2 \psi(|x_i - x_j|) \leq 0. \tag{5}$$

In particular, the kinetic energy $E_k(t)$ is bounded by the initial kinetic energy $E_k(0)$ and it holds that $|v_i(t)| \leq \sqrt{2E_k(0)}$ for all $t \geq 0$ and all $i = 1, \ldots, N$.

Proof By (2b), we have

$$\frac{d}{dt}\sum_{i=1}^{N} v_i^2 = \frac{2}{N}\sum_{i,j=1}^{N} v_i (v_j - v_i)\psi(|x_i - x_j|)$$

$$= \frac{1}{N}\sum_{i,j=1}^{N} v_i (v_j - v_i)\psi(|x_i - x_j|) + \frac{1}{N}\sum_{i,j=1}^{N} v_j (v_i - v_j)\psi(|x_i - x_j|)$$

$$= -\frac{1}{N}\sum_{i,j=1}^{N} (v_i - v_j)^2 \psi(|x_i - x_j|), \tag{6}$$

where equality (6) is obtained by swapping the indexes i and j in the second term. □

Remark 2.2 Proposition 2.1 is based only on the structure of the CS system, and holds for any nonnegative communication weight ψ. The symmetry of $x \mapsto \psi(|x|)$ allows to swap indexes in (6).

Throughout the chapter, we assume that the singular communication weight is of the form

$$\psi(s) = s^{-\alpha}, \qquad \alpha > 0. \tag{7}$$

This form is convenient for distinction between two cases: the *weakly singular* case corresponding to $\alpha \in (0, 1)$, and the *strongly singular* case corresponding to $\alpha \geq 1$. However, such exact form is not necessary, and the results can be generalized to an arbitrary weight that is positive, locally Lipschitz continuous on $(0, \infty)$, and singular at 0. In such case the weakly and strongly singular cases translate to integrability of the weight around 0, or lack thereof, respectively.

3   Singular Cucker–Smale Model: Particle System

3.1   Motivation

One of the most desirable qualitative features of the CS system (2), either for standard or for the singular weight, is collision-avoidance. Such property is required in the case where the agents naturally avoid collisions, such as behaviour of flocks of animals or control over autonomous sensors or robots. One of the approaches in the study of collision-avoidance comes from [1] where the authors establish a set of initial data such that the regular CS particle system admits no collisions. Generally speaking, the initial total kinetic energy has to be small compared to the initial minimal distance between the particles. Then the alignment force on the right-hand side of (5) dissipates the kinetic energy and the rate of the dissipation increases as the distance between particles becomes smaller.

This effect motivates consideration of the CS with the singular communication weight (7), for which one expects that the rate of dissipation of the kinetic energy becomes infinite as the particles collide. It proves to be a good approach as it leads to collision-avoidance that is *unconditional*, i.e. it does not rely on the initial configuration. The CS particle system with weight (7) provides an interesting mathematical challenge owing to the fact that its nonlinear right-hand side is not Lipschitz continuous. However, since the system looses its regularity only at times of collisions, careful qualitative analysis of the model provides information required for the quantitative analysis. The following subsections are dedicated to the summary of results in two main areas of analysis of this model: simultaneous qualitative and quantitative analysis, and asymptotics. For detailed account c.f. [13, 59, 60].

At the end of the section we also provide examples of the influence of the singular kernel when coupled with a pattern-inducing control.

3.2   Collision-Avoidance

The dynamics of the singular CS model is defined by the right-hand side of (2b) and particularly by the interplay between $(v_j - v_i)$ which tends to zero and

$\psi(|x_i - x_j|)$ which tends to infinity at the time of collision between ith and jth particle. Interestingly, the results of such interplay vary dramatically depending on the values of exponent α. Suppose N particles are governed by the CS model with a singular weight (7) on the time interval $[0, T]$. Then, equalities (4) and (5) together give

$$\frac{d}{dt} E_k = -\frac{2}{N} \sum_{i,j=1}^{N} (v_i - v_j)^2 \psi(|x_i - x_j|).$$

It implies that the kinetic energy dissipates at the integrable rate, i.e.

$$\frac{2}{N} \int_0^T \sum_{i,j=1}^{N} (v_i - v_j)^2 \psi(|x_i - x_j|) \leq E_k(0). \tag{8}$$

Therefore, if the function

$$\theta_{ij}(t) := \psi(|x_i(t) - x_j(t)|) \tag{9}$$

integrates to infinity in a neighbourhood of $t_0 \in [0, T]$ then, in order to ensure (8), we necessarily need

$$(v_i(t) - v_j(t))^2 \to 0, \tag{10}$$

for $t \to t_0$. This suggests that to analyse collisions one should first analyse maps θ_{ij}. In the case of strongly singular kernel ψ with $\alpha \geq 1$, assuming that ith and jth particles collide at t_0, for $s \nearrow t_0$ we have

$$|x_i(s) - x_j(s)| \leq M(t_0 - s),$$

where $M := \sqrt{2E_k(0)}$ is the uniform bound for the velocity, c.f. Proposition 2.1. Therefore,

$$\theta_{ij}(s) = \psi(|x_i(s) - x_j(s)|) \geq \psi(M(t_0 - s)) = M^{-\alpha}|t_0 - s|^{-\alpha},$$

which for $\alpha \geq 1$ is non-integrable in any neighbourhood of t_0, and thus indeed (10) is necessary. On the other hand, for $\alpha \in (0, 1)$ the function $|t_0 - s|^{-\alpha}$ is integrable and the above argumentation is inconclusive. It was shown in [59] that with $\alpha \in (0, 1)$ functions θ_{ij} can be either integrable or non-integrable.

We summarize the above consideration in the following remark.

Remark 3.1 Assuming that t_0 is a time of collision between ith and jth particles we learn the following:

1. For $\alpha \geq 1$, function θ_{ij} is non-integrable at t_0. Consequently $(v_i - v_j)^2 \to 0$. Thus if any particles collide, then they stick together at the same time. Later we show that collisions are actually impossible.
2. For $\alpha \in (0, 1)$, function θ_{ij} may be either integrable or non-integrable at t_0. If it is integrable, we have a collision between particles and if it is non-integrable, then $(v_i - v_j)^2 \to 0$ and particles stick together.

With this information we present main results concerning the qualitative and quantitative analysis of the singular CS model. First, let us focus on the case of strongly singular weight with $\alpha \geq 1$, which leads to collision-avoidance.

Theorem 3.1 ([13]) *Let $\alpha \geq 1$ and $T \in (0, \infty]$. Then the CS particle system with singular weight (7) admits a unique non-collisional smooth solution $(\mathbf{x}, \mathbf{v})$ provided that initial data are non-collisional (we recall Definition 2.1(i) for the notion of a collision).*

Idea of the Proof The local existence of solutions is clear as the system is singular only at times of collision, and we begin with non-collisional initial data. This solution can be prolonged until the first time of collision, which as we prove, never happens. As observed in Remark 3.1, if the particles collide, their relative velocity tends to zero. However, the rate of the alignment outweighs the speed with which the particles approach each other, and as a result, the collision never occurs. We present the proof of this fact for the simplest case of two-particle system in $\mathbb{R}$.

Assume that $t < t_0 < \infty$, where t_0 is a time of the first collision, and denote $x(t) := x_1(t) - x_2(t)$ and $v(t) := v_1(t) - v_2(t)$. Using (5), we get

$$\frac{d}{dt}v^2 = -2\psi(|x|)v^2,$$

equivalently, dividing by $|v|$, we have

$$\frac{d}{dt}|v| = -\psi(|x|)|v|,$$

and so, by Gronwall's lemma

$$|v(s)| \leq e^{-\int_0^s \psi(|x(\sigma)|)d\sigma}|v_0|,$$

for any $s < t_0$. Then the primitive function of ψ, denoted by Ψ, satisfies

$$|\Psi(|x(t)|)| \leq |\Psi(|x(0)|)| + |v_0| \int_0^t \psi(|x(s)|)e^{-\int_0^s \psi(|x(\sigma)|)d\sigma}\,ds \leq C.$$

Since Ψ is singular at 0 (singularity of the order $s^{1-\alpha}$ for $\alpha > 1$ and of the order $\ln s$ for $\alpha = 1$), we conclude that there exists $\delta > 0$ such that $|x(t)| \geq \delta$ as $t \nearrow t_0$ which contradicts the assumption that t_0 is a time of collision.

In [13] we expand on the above idea. We divide the particles into two groups: A—of all particles colliding with ith particle at the time t_0, and B—of all remaining particles. Then, in a neighbourhood of t_0 there exists a minimal distance between groups A and B. Therefore the dynamics of group A is influenced by the singular interaction within A and a negligible, bounded interaction between the groups. Thus half of the singular interaction within A outweighs the influence of B and then the remaining half is used to prove that the collision cannot happen similarly to the case of two particles.   □

Our next goal is to discuss the case of $\alpha \in (0, 1)$. With $\alpha \geq 1$ trajectories of particles cannot cross while, as we show later, the weakly singular kernel with $\alpha < 1$ admits not only collisions between the particles but also sticking. This leads to two hypothetical problems.

The first problem is related to the uniqueness. Since ψ is singular at 0 the loss of uniqueness can happen at any time at which any two particles are stuck together. For example, if ith and jth particle have the same position and velocity in the time interval (t_1, t_2), then their trajectories may separate at any time $t > t_2$ as one observers in the case of the basic example $\dot{x} = x^{\frac{1}{3}}$. The second problem arises from the fact that upon approaching the first (or any other) time of sticking of particles, we lose the absolute continuity of the solution, and its derivative cannot be defined in the weak sense.

We deal with these two delicate hypothetical problems using an admittedly heavy-handed approach. In a neighbourhood of any time at which no particles are stuck together the problems do not exist. As we approach a time of sticking t_0 we say that the solution exists in a classical sense up to any $t < t_0$ and its position and velocity components are continuous at t_0^-. Thus, even though the classical meaning of the solution is no longer available, we still can prolong it up to t_0. When t_0 is reached, we redefine the system in order to remove the possibility of the separation of the trajectories. Then, we re-initiate the solution starting from t_0 which, again, is classical up to the second time of sticking t_1. This procedure is repeated up to at most N times, since this is the maximum number of sticking between the particles (after we ensured that the trajectories cannot separate). To employ this strategy we denote

$$B_i(t) := \left\{ k \in \{1, \ldots, N\} : x_k(t) \neq x_i(t) \text{ or } v_k(t) \neq v_i(t) \right\},$$

which can be interpreted as the set of indexes of all the particles that are distinct from ith particle. Then we define the solution as follows.

Definition 3.1 (Piecewise Weak Solutions) Let $\{t_n : n = 0, 1, \ldots, K\}$ with $K \leq N$ be the set of all times when the particles stick together. For $n \geq -1$, on each interval $[t_n, t_{n+1}]$ (we assume that $t_{-1} = 0$) we consider the problem

$$
\begin{cases}
\dfrac{\mathrm{d}}{\mathrm{d}t} x_i = v_i, & \text{(11a)} \\[3mm]
\dfrac{\mathrm{d}}{\mathrm{d}t} v_i = \dfrac{1}{N} \displaystyle\sum_{j \in B_i(t_n)} (v_j - v_i)\psi(|x_i - x_j|), & \text{(11b)}
\end{cases}
$$

with the initial data $(\mathbf{x}(t_n), \mathbf{v}(t_n))$.

We say that $(\mathbf{x}, \mathbf{v})$ is a *piecewise weak* solution of (2) in the time interval $[0, T)$ with the initial data $(\mathbf{x}_0, \mathbf{v}_0)$ if and only if the function $(\mathbf{x}, \mathbf{v})(t)$ is continuous on $[0, T)$ and it solves (11) on each interval $[t_n, t]$ for all $t \in [t_n, t_{n+1}]$ and $(\mathbf{x}(t_{-1}), \mathbf{v}(t_{-1})) = (\mathbf{x}_0, \mathbf{v}_0)$.

Theorem 3.2 ([59, 60]) *Let* $\alpha \in (0, 1)$ *and* $T \in (0, \infty]$. *Then for any initial data* $(\mathbf{x}_0, \mathbf{v}_0) \in \mathbb{R}^{2d}$, *system (2) with the singular communication weight (7) admits a unique solution in the sense of Definition 3.1.*

Idea of the Proof The proof of existence can be found in [59], while the proof of uniqueness can be found in [60]. Existence outside of times of collision is straightforward. At any time of collision we prove that due to the relatively small singularity exponent α the function θ_{ij} (see (9)) is integrable and thus $\mathbf{v}$ is absolutely continuous, which grants existence also in points of collision. Existence in points of sticking is dealt with mostly by the definition of the solution itself. Continuity of the velocity at the times of sticking is the only remaining problem, which is resolved by a careful elementary analysis of the dynamics. □

It turns out, see [60], that after restricting the range of singularity to $\alpha \in (0, \frac{1}{2})$, one obtains existence of classical solutions which, by uniqueness, coincide with the piecewise-weak solutions.

Theorem 3.3 ([60]) *Let* $\alpha \in (0, \frac{1}{2})$ *and* $T \in (0, \infty]$. *Given initial data* $(\mathbf{x}_0, \mathbf{v}_0) \in \mathbb{R}^{2d}$, *system (2) with the singular communication weight (7) admits a unique classical solution with absolutely continuous velocity component.*

The advantage obtained by assuming $\alpha \in (0, \frac{1}{2})$ is that the communication weight becomes square-integrable. In the proof we estimate the right-hand side of (2b) using Young's inequality with exponent 2, which leads to doubling exponent α.

Introduction of the piecewise-weak solutions and the whole approach to the quantitative analysis of the CS model with singularity $\alpha \in (0, 1)$ is based on the problems appearing at the times of sticking between particles. Therefore, a natural question arises if such phenomenon can even occur. In [59] a detailed analysis of the two-particle in $\mathbb{R}^d$ case was performed which we sketch below.

Assuming without a loss of generality that $x_1 + x_2 \equiv 0$ and $v_1 + v_2 \equiv 0$ (see (3)) we end up with two particles that move either on two parallel lines or on the same line. We omit the first possibility since it naturally leads to no collisions and focus on the situation which is equivalent to two particles in $\mathbb{R}$. Assume that $x_2(0) > x_1(0)$ and to make the particles move in the direction of each other we are forced to assume $v_2(0) - v_1(0) < 0$. Denoting $x := x_2 - x_1 > 0$ we use (2) to obtain

$$\ddot{x} = -\dot{x}\psi(x) \tag{12}$$

in $[0, t_0]$, where t_0 is the first time of collision between the particles.

Proposition 3.1 ([59]) *Let $\alpha \in (0, 1)$ and let x be the unique solution[1] of (12) with the initial data $x(0) > 0$ and $\dot{x}(0) < 0$. Then the following are equivalent:*

1. *There exists a time $0 < t_0 < \infty$ such that $x(t_0) = \dot{x}(t_0) = 0$.*
2. *We have*

$$\dot{x}(0) = -\Psi(x(0)), \tag{13}$$

where $\Psi(s) := \frac{1}{1-\alpha}s^{1-\alpha}$ is a primitive of ψ.

The proof can be found in [59]. Here we shall only make the observation that integrating (12) in the time interval $[0, t]$ leads to

$$\dot{x}(t) + \Psi(x(t)) = \dot{x}(0) + \Psi(x(0)) \overset{(13)}{=} 0.$$

It means that the initial condition (13) places the solution on the trajectory described by $\dot{x} = -\Psi(x)$. For such trajectory if $x = 0$, then $\Psi(x) = 0$ and thus $\dot{x} = 0$, which implies that the particles stick together whenever they collide. Then the proof of the proposition revolves around showing that this is the only trajectory leading to a collision and that the collision happens in a finite time.

3.3 Asymptotics

The asymptotics of the singular CS model is mostly the same as the asymptotics of the regular one, since it is related to the integrability of ψ away from zero. This case was thoroughly studied on particle, kinetic and hydrodynamic levels by Ha and Liu in [32], Carrillo et al. in [10], Ha and Tadmor in [34] and others. We also recommend the survey [19], where the regular CS model was discussed. Although the asymptotic behaviour of solutions for singular kernels is not significantly different, we discuss it here for the sake of completeness using the results from [32] as an example. Since the integrability of $\psi(s)$ at $s = \infty$ is a major factor, we again distinguish based on α.

Proposition 3.2 (Unconditional Flocking, [32]) *Let $\alpha \in (0, 1]$ and let (x, v) be a solution to (2) with $|\mathbf{x_0}| \neq 0$. Then there exist positive constants x_m and x_M such that*

---

[1]Which exists by Theorem 3.2.

$$x_m \leq \sum_{i,j=1}^{N} (x_i - x_j)^2 \leq x_M, \qquad \|\mathbf{v}\| := \sum_{i,j=1}^{N} (v_i - v_j)^2 \leq \|\mathbf{v}_0\| e^{-\psi(x_M)t}.$$

Proposition 3.3 (Conditional Flocking, [32]) *Let $\alpha > 1$ and let $(\mathbf{x}, \mathbf{v})$ be a solution to (2) with $|\mathbf{x}_0| \neq 0$. Suppose the initial configuration $(\mathbf{x}_0, \mathbf{v}_0)$ satisfies*

$$\left(\sqrt{\sum_{i,j=1}^{N} (x_{0i} - x_{0j})^2} \right)^{1-\alpha} \geq (\alpha - 1) \sqrt{\sum_{i,j=1}^{N} (v_{0i} - v_{0j})^2}.$$

Then there exist positive constants x_m and x_M such that

$$x_m \leq \sum_{i,j=1}^{N} (x_i - x_j)^2 \leq x_M, \qquad \|\mathbf{v}\| := \sum_{i,j=1}^{N} (v_i - v_j)^2 \leq \|\mathbf{v}_0\| e^{-\psi(x_M)t}.$$

Remark 3.2 The class of admissible solutions required by Propositions 3.2 and 3.3 includes the classes provided by Theorems 3.2 and 3.3 for $\alpha < 1$ and by Theorem 3.1 for $\alpha \geq 1$. Observe that singularity with $\alpha = 1$ is the only value that satisfies the assumptions of Theorem 3.1 and Proposition 3.2 and thus, leads to, both, unconditional flocking and collision-avoidance. Of course, as explained below equation (7), the choice of the communication weight of the form (7) is quite arbitrary and, in practice, the only requirement for the lack of collisions is the nonintegrability of ψ near 0. On the other hand for the unconditional flocking nonintegrability of ψ at infinity is required.

3.4 Variants of the Model

From the perspective of applications, it is often useful to modify the Cucker–Smale model to adapt it to a particular phenomena. We recall the wide range of modifications, presented in the introduction, from models with leaders [22, 24, 62] and preferences [24, 47], models with time-delay [28], up to models with various additional external or internal forces (deterministic and stochastic) [7, 15, 29, 57]. We also refer to the survey in [19]. However, in the case of the CS model with a singular communication weight the well-posedness theory is relatively recent and thus not many additional directions were pursued as of yet. Moreover, as presented in the previous sections, the dynamics of the singular CS model either admits sticking of the trajectories of the particles or does not allow any collisions at all. In the first case, admittedly, not many perspectives of applications were discovered and in the second case, the dynamics is essentially equivalent to the regular CS model with an added bonus of initial-data-independent collision-avoidance. In particular, singular CS model with $\alpha \geq 1$ seems to be viable for most modifications that the regular CS model underwent with some additional mathematical challenge. That being said, in the remainder of this section we present results directly related to the singular CS model.

Nonlinear velocity coupling [50] combines the CS model from [11], with the discussion on collision-avoidance from [13]. The result in [50] provides the extension of results from [13] for the CS particle system with the right-hand side of the form

$$\sum_{j=1}^{N} a(v_j - v_i)\psi(|x_i - x_j|).$$

Here, e.g. $a(v) \approx x|x|^\beta$ for $\beta \geq 0$. This direction is motivated by real-life phenomena, since there is no definitive reason to narrow the admissible interactions only to those with linear dependence on the velocity.

Thermomechanical CS model [18] was proposed in [33]. The authors introduce an intrinsic property of the particles, interpreted as their temperature, governed by an additional set of differential equations that describe the heat-transfer between the particles. In [18], the authors consider the thermomechanical CS particle system with a strongly singular communication weight, providing conditions on the singularity α in terms of the heat-transfer law that result in the collision-avoidance or in the lack thereof.

Bonding-Force and Decentralized Control In [43] the authors provide a new insight into the CS model with the bonding force from [57] with a strongly singular communication weight (7), $\alpha \geq 1$

$$\begin{cases} \dfrac{dx_i}{dt} = v_i, \quad x_i, v_i \in \mathbb{R}^d, \quad i = 1, 2, \cdots, N, \quad t > 0 & \text{(14a)} \\[2ex] \dfrac{dv_i}{dt} = \dfrac{K_1}{N} \sum_{j=1}^{N} \psi(|x_j - x_i|)(v_j - v_i) + \dfrac{\tilde{K}}{N} \sum_{j=1}^{N} \dfrac{(v_i - v_j) \cdot (x_i - x_j)}{2|x_i - x_j|^2}(x_j - x_i) \\[2ex] \hphantom{\dfrac{dv_i}{dt} =} + \dfrac{K_2}{N} \sum_{j=1}^{N} \dfrac{|x_j - x_i| - 2R}{2|x_j - x_i|}(x_j - x_i). & \text{(14b)} \end{cases}$$

In (14b), the latter two terms constitute the aforementioned bonding force, whose main purpose is to force the particles to stay within the distance $2R$ from each other. As a result of this force, system (14) exhibits many desirable properties like asymptotic convergence to characteristic patterns seen in Fig. 1a, b, where the particles end up spread out uniformly within a given radius, predefined by the bonding force. On top of that, system (14) inherits the collision-avoidance of the classical strongly singular CS particle system.

Another modification of the singular CS model is through the addition of a decentralized control [4, 39]. Here we present the idea from [20] to assign a

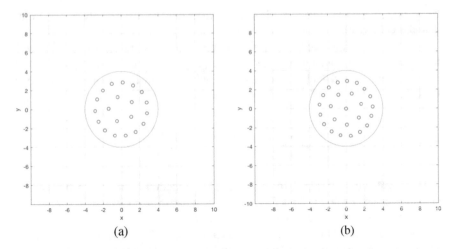

Fig. 1 The particles governed by system (14) coverage to a symmetric pattern that depends not only on the number of the particles (as seen on the pictures) but also on the initial configuration. Blue circles represent the particles, while the large red circle represents the $2R$ radius around the origin. (a) $N = 20$, (b) $N = 25$

provisional order to the agents and make each agent (except for the first one) synchronize its position with the position of the previous one. The system reads

$$\frac{dx_i}{dt} = v_i, \quad i = 1, \ldots, N, \quad t > 0,$$

$$\frac{dv_i}{dt} = \frac{K}{N} \sum_{j=1}^{N} \psi(|x_i - x_j|)(v_j - v_i) + u_i. \tag{15}$$

Given $z_i \in \mathbb{R}^d$ for $i = 1, \ldots, N - 1$, the control term $u := (u_1, \ldots, u_N)$ is given by

$$u_1 = -\phi(|x_1 - x_2 - z_1|^2)(x_1 - x_2 - z_1),$$

$$u_N = \phi(|x_{N-1} - x_N - z_{N-1}|^2)(x_{N-1} - x_N - z_{N-1}),$$

$$u_i = \phi(|x_{i-1} - x_i - z_{i-1}|^2)(x_{i-1} - x_i - z_{i-1})$$

$$\quad - \phi(|x_i - x_{i+1} - z_i|^2)(x_i - x_{i+1} - z_i),$$

for $i \in \{2, \ldots, N - 1\}$, where ϕ is a smooth weight of the form $\phi(s) = (1 + s)^{-\beta}$, $\beta > 0$. Thus, through this control, in theory, ith particle adjusts its position in a way that minimizes u_i, which is by having $x_{i-1} - x_i$ converge to z_{i-1}. Therefore, by prescribing proper coordinates z_i, one can force ith particle to attain any position respective to $(i - 1)$th particle.

The main area of application is in the control of unmanned aerial vehicles. The advantage of the decentralized control is that it allows for emergence of a wide variety of patterns.[2] Further advantage of the decentralized control is that each agent is required to "remember" only its relative position to a single other agent. The disadvantage is that it requires input of z_i, while the bonding force achieves symmetric patterns depicted in Fig. 1a, b automatically.

In [20] decentralized control was added to the singular CS model resulting in pattern formation with collision-avoidance presented in the following theorem.

Theorem 3.4 ([20])

(A) Consider system (15) with $\alpha \geq 1$ and $\beta > 0$ subjected to non-collisional initial data $(\mathbf{x}_0, \mathbf{v}_0)$ (see Definition 2.1). Then there exists a global smooth, non-collisional solution.

(B) Moreover if $\alpha \geq 2$ and one of the two following hypotheses holds:

  (i) $\beta \leq 1$;
  (ii) $\beta > 1$ and

$$\sum_{i=2}^{N} \int_{|x_{0i-1}-x_{0i}-z_{i-1}|^2}^{\infty} \phi(r)\, dr > \frac{4}{MN} \sqrt{\sum_{i,j=1}^{N} |v_{0i} - v_{0j}|^2}.$$

Then we have

$$\sup_{0 \leq t \leq \infty} \max_{1 \leq i,j \leq N} |x_i(t) - x_j(t)| < \infty \text{ and } \max_{1 \leq i,j \leq N} |v_i(t) - v_j(t)| \to 0 \text{ as } t \to \infty.$$

(C) Finally if

$$\liminf_{t \to \infty} |x_i(t) - x_j(t)| > 0 \tag{16}$$

for all $i, j \in \{1, \ldots, N\}$. Then there exists a limit $\lim_{t \to \infty} x(t) =: x^{\infty}$ satisfying

$$x_i^{\infty} = x_{i-1}^{\infty} - z_{i-1} \quad \text{for all} \quad i = 1, \ldots, N.$$

The proof of the existence follows from the collision-avoidance exactly as in Theorem 3.1. As in the proof of Theorem 3.1 collision-avoidance is shown by dividing the interactions between the particles into two groups: A—singular interactions, and B—bounded interactions. Here all the interactions originating

---

[2]See, for example, the Olympic Rings symbol simulated in the video available at https://youtu.be/C7UDGRudsyA, produced within the work [20] through the manipulation of z_i.

from the control are added to the group B (since they are non-singular). Large-time behaviour is shown by elementary but careful analysis in the spirit of [32].

Remark 3.3 It is worthwhile to have a closer look at part (C) of the above theorem and particularly assumption (16). The reason to exclude the asymptotic collisions is related to collision-avoidance. Take, for example, two particles in $\mathbb{R}$, with $z_1 = -1$ then the resulting pattern has to be of the form $x_2^\infty = x_1^\infty + 1 > x_1^\infty$. However if initially $x_1(0) > x_2(0)$, then the particles change order, which means that they collide. This is however impossible by Theorem 3.4(A). In other words we need to exclude the situations when the control leads to a finite-time collision. Of course in $d \geq 2$ such situation is very unlikely.

4 Singular Cucker–Smale Model: Kinetic Equation

4.1 Formal Derivation

The particle model provides the most precise description of the evolution of the particles, but in the case of large N, it quickly becomes impractical. With very large N, the microscopic models are too computationally intensive, and it is much more efficient to perform numerical simulations for what we call *mean-field limit system*, see, for example, [5]. This can be viewed as a model in the *mesoscopic* scale. Instead of tracing position and momentum of each particle, we look for the distribution (or probability) of the particles at time t, position x and with velocity v. Hence, the sough object of analysis is a distribution of the type

$$f(t, x, v) : [0, T) \times \mathbb{R}_x^d \times \mathbb{R}_v^d \to \mathbb{R}. \tag{17}$$

At the right-hand side of the above expression, $\mathbb{R}$ should be viewed only formally, since f might be barely a measure. The evolution of f is described by a Vlasov-type system, which, along with Boltzmann equation, is the backbone of kinetic theory [69]. The methodology developed to deal with Vlasov-type equations is robust but in the case of systems with singular interactions there is no general approach.

Before we deliberate further on the matter, let us explain the link between the particle and the kinetic models. One of the fundamental challenges of mathematical physics is to provide the description of continua originating from the agent-based models. In the case of the CS model, such a formal limiting passage was introduced in [34], followed by [10], and is used as a basis for rigorous derivation of variants of the kinetic CS models. Assume that we have an N particle system in the following form:

$$\begin{cases} \dfrac{\mathrm{d}}{\mathrm{d}t} x_i = v_i, \\[2mm] \dfrac{\mathrm{d}}{\mathrm{d}t} v_i = \sum_{j=1}^{N} m_j (v_j - v_i) \phi(|x_i - x_j|), \end{cases} \qquad (18)$$

where total mass of the particles reads $\sum_{j=1}^{N} m_j = 1$ and for simplicity equals one. Note that (18) becomes (2) with $m_i = \frac{1}{N}$ for all $i = 1, ..., N$. Here, for the sake of clarity of presentation, we skip the dependence of $x_i = x_i^N$, $v_i = v_i^N$ and $m_i = m_i^N$ on N but it should be noted that, naturally, the solution itself changes with N.

Next, having a solution to (18), whose existence was discussed in the previous section, we aim to let $N \to \infty$ and define (17) as a limit of solutions to the particle system (18) written as follows:

$$f_N(t, x, v) = \sum_{i=1}^{N} m_i \delta_{x_i(t)} \otimes \delta_{v_i(t)}, \qquad (19)$$

where x_i and v_i denote a position and velocity of the ith particle obtained by solving the CS particle system (2). We will refer to (19) as the *atomic solution*.

The so-called empirical measure f_N, defined in (19), satisfies

$$\frac{\mathrm{d}}{\mathrm{d}t} f_N = 0, \quad \text{with} \quad \frac{\mathrm{d}}{\mathrm{d}t} = \partial_t + \frac{\mathrm{d}x}{\mathrm{d}t} \nabla_x + \frac{\mathrm{d}v}{\mathrm{d}t} \nabla_v \qquad (20)$$

in the following distributional sense: for any smooth test function

$$\Phi : [0, T) \times \mathbb{R}_x^d \times \mathbb{R}_v^d \to \mathbb{R}$$

with a compact support in $\mathbb{R}_x^d \times \mathbb{R}_v^d$ and $\Phi|_{t=T} \equiv 0$, we have

$$\int_0^T \int_{\mathbb{R}_x^d} \int_{\mathbb{R}_v^d} f_N(t, x, v) \frac{\mathrm{d}}{\mathrm{d}t} \Phi(t, x, v) \mathrm{d}v \mathrm{d}x \, \mathrm{d}t = \int_{\mathbb{R}_x^d} \int_{\mathbb{R}_v^d} f_N|_{t=0} \Phi(0, x, v) \mathrm{d}x \mathrm{d}v.$$

Then by the definition of f_N (and particularly (2)) it is easy to show that

$$\int_0^T \int_{\mathbb{R}_x^d} \int_{\mathbb{R}_v^d} f_N[\partial_t + v \cdot \nabla_x + F(f_N) \cdot \nabla_v] \Phi(t, x, v) \mathrm{d}v \mathrm{d}x \, \mathrm{d}t$$

$$= \int_{\mathbb{R}_x^d} \int_{\mathbb{R}_v^d} f_N|_{t=0} \Phi(0, x, v) \mathrm{d}x \mathrm{d}v, \qquad (21)$$

where

$$F(f_N)(t, x, v) = \sum_{j=1}^{N} m_j(t)(v_j - v)\psi(|x - x_j|)$$

$$= \int_{\mathbb{R}_x^d} \int_{\mathbb{R}_v^d} (w - v)\psi(|x - y|) f_N(t, y, w) dy dw.$$

Then, taking the limit in (21) as $N \to \infty$, assuming that each term is well defined and smooth, and that

$$\lim_{N \to \infty} f_N = f, \tag{22}$$

we deduce that f satisfies the same equation as f^N. This is a distributional version of the Vlasov-type equation

$$f_t + v \cdot \nabla_x f + \text{div}_v(F(f)f) = 0 \tag{23}$$

with

$$F(f) = \int_{\mathbb{R}_y^d} \int_{\mathbb{R}_w^d} (w - v)\psi(|x - y|) f(t, y, w) dy dw. \tag{24}$$

Note that a very convenient property of Vlasov-type equations is that due to the *nonlocal* interactions (contrary to Boltzmann equation) the solution of the particle system already is a distributional solution. This is the reason why such a simple approximation is possible.

In general, the limit passage $N \to \infty$ requires some more information about uniform estimates for f_N. This is in fact the gist of the problem, and we postpone the discussion on this issue to the following sections. Here let us briefly mention that even defining $F(f)f$ is not straightforward. For the singular weight ψ, if f is a Radon measure, then $F(f)$ is an L^p-function, and product of an L^p-function with a measure might not be well defined.

As for the topology of convergence, for the method shown above, a suitable choice is the Wasserstein distance. For the sake of this survey we shall introduce a simple version of such metrics, i.e. the bounded-Lipschitz distance. Given two Radon measures f_1 and f_2 let

$$d_1(f_1, f_2) = \sup \left\{ \int_{\mathbb{R}^d} (f_1 - f_2)\phi dx : \phi \in Lip(\mathbb{R}^d), Lip(\phi) \le 1, |\phi|_\infty \le 1 \right\}, \tag{25}$$

where $Lip(\mathbb{R}^d)$ is the space of Lipschitz continuous functions and $Lip(\phi)$ is the Lipschitz constant of function ϕ. Then d_1 is the bounded-Lipschitz distance, sometimes referred to as Monge–Kantorovich–Rubinstein distance and it is equivalent to the Wasserstein-1 distance, we refer the reader to [69] or [74]. By $\mathcal{M}$ we denote the metric space of all nonnegative Radon measures with topology generated by d_1. Its subspace of probabilistic measures is denoted by $\mathscr{P}_1$.

4.2 Local-in-Time Well-Posedness

A possible approach to this subject, including singular communication weight, has been presented in [11]. To the best of our knowledge, it was the first existence result for the singular CS kinetic equation. Here we present only a special case of this result tailored to the singular CS model. Interested reader is refereed to [11], where a more general variant with nonlinear velocity coupling was analysed.

Theorem 4.1 ([11]) *Suppose* $\alpha \in (0, d-1)$ *and* $p > 1$ *fulfils* $(\alpha + 1)p' < d$ *with* $p' : \frac{1}{p} + \frac{1}{p'} = 1$. *If initial datum* f_0 *is nonnegative and has a compact support in the velocity space and*

$$f_0 \in (L^1 \cap L^p)(\mathbb{R}^d \times \mathbb{R}^d) \cap \mathscr{P}_1(\mathbb{R}^d \times \mathbb{R}^d),$$

then there exists $T > 0$ *such that there exists a unique weak solution*

$$0 \le f \in L^\infty(0, T; (L^1 \cap L^p)(\mathbb{R}^d \times \mathbb{R}^d)) \cap \mathscr{C}([0, T], \mathscr{P}_1(\mathbb{R}^d \times \mathbb{R}^d))$$

for system (23) on time interval $[0, T]$.

Furthermore, if f_i *with* $i = 1, 2$ *are two such solutions, then the following* d_1-*stability estimate holds:*

$$\frac{\mathrm{d}}{\mathrm{d}t} d_1(f_1(t), f_2(t)) \le C d_1(f_1(t), f_2(t)) \quad for \quad t \in [0, T]$$

for a positive constant C.

The key point of Theorem 4.1 is the uniqueness. It follows from the fact that the singularity allows to consider a functional setting such that the field $F(f)$ is indeed Lipschitz continuous. The proof is based on the theory of optimal transport to control two solutions considered in setting of the flow generated by v and $F(f)$. Since, as we mentioned, the regularity is sufficient to define characteristics. Hence comparison of two solutions in the Wasserstein metric is possible.

4.3 Global-in-Time Measure-Valued Solutions

The local existence result presented in the previous section requires the initial data in the $L^p(\mathbb{R}^d \times \mathbb{R}^d)$ space. In consequence, the solution itself is also an L^p-function, which rules out a very interesting class of solutions—atomic solutions—given by (19).

The following result from [54] embraces the rich dynamics of the CS model admitting solutions that for all $t > 0$ live in the space of Radon measures, which we denote here by $\mathcal{M}$. Clearly, such class includes the atomic solutions. The price that needs to be paid for such a wide class of solutions is reduction of the range of singularity to $\alpha \in (0, \frac{1}{2})$, so that we can operate within the framework of higher regularity for the particle system granted by Theorem 3.3.

Theorem 4.2 ([54]) *Let $0 < \alpha < \frac{1}{2}$. For any compactly supported initial data $0 \le f_0 \in \mathcal{M}$ and any $T > 0$, Cucker–Smale's flocking model (23) admits at least one weak solution $0 \le f \in \mathscr{C}_{weak}([0, T], \mathcal{M}(\mathbb{R}^d \times \mathbb{R}^d))$ with $\partial_t f \in L^p(0, T; (C^1(\mathbb{R}^d \times \mathbb{R}^d))^*)$ for some $p > 1$ (here $(C^1)^*$ is the dual space of C^1).*

Moreover if f_0 is atomic, then f is atomic too, hence, by Theorem 3.3, it is unique and it corresponds to the solution of the particle system (2).

Idea of the Proof The proof of existence is based on the idea of mean-field limit, presented at the beginning of this section. It involves

i) Given initial data $f_0 \in \mathcal{M}$, we approximate it by atomic measures of the form

$$f_{N,0}(t, x, v) = \sum_{i=1}^{N} m_{0i}^N \delta_{x_{0i}^N} \otimes \delta_{v_{0i}^N}.$$

ii) For each fixed N, thanks to the results from Sect. 3.2 we establish the existence of solutions to particle system (18) with initial data $(\mathbf{x}_0^N, \mathbf{v}_0^N)$. By (19), such solutions correspond to atomic solutions f_N of Vlasov-type equation (23) (satisfied at least in the distributional sense).

iii) We converge with $N \to \infty$ and define f—the candidate for the solution of (23) associated with initial data f_0, as a limit of f_N in the bounded-Lipschitz distance.

iv) Through sufficient information on the uniform regularity of the approximate solutions we prove that f satisfies (23).

The existence of a weak* limit f is straightforward by Banach–Alaoglu theorem. The difficult part is to prove that f is the sought solution to (23). The crucial element of the proof is to show the convergence of the nonlinear alignment force term (compare with (21))

$$\int_0^T \int_{\mathbb{R}^{2d}} F(f_N) f_N \nabla_v \Phi dx \, dv \, dt, \quad \text{as} \quad N \to \infty. \tag{26}$$

In particular if $f_N \overset{*}{\rightharpoonup} f$ then it is not clear whether $F(f_N) f_N \overset{*}{\rightharpoonup} F(f) f$. It is useful to look at (26) as

$$\int_0^T \int_{\mathbb{R}^{4d}} g d\mu_n dt \quad \text{for } d\mu_N := f_N(x, v, t) \otimes f_N(y, w, t) dx \, dv \, dy \, dw$$

$$\text{and } g(x, y, v, w, t) := \psi(|x - y|)(w - v)\nabla_v \Phi(t, x, v).$$

Note that thanks to the above representation the convergence of a product in (26) is reduced to a convergence of the measure μ_N tested by the function g. To overcome the difficulty that g is not Lipschitz continuous we approximate it by a Lipschitz continuous family $g_m \to g$, such that $|g_m| \nearrow |g|$. First we converge with $N \to \infty$ and then with $m \to \infty$. Then a detailed analysis, based on the uniform regularity of trajectories provided by Theorem 3.3, leads to the proof that $F(f)$ is an integrable function with respect to measure $df = f dx dv$. Hence, we are able to define $F(f)f$, even though we cannot do it for $F(f)h$ for general $h \in \mathcal{M}$.   □

Theorem 4.2 provides the existence of weak measure-valued solutions globally in time, but does not solve the problem of uniqueness. In fact, the regularity is insufficient to estimate the difference between the two supposedly distinct solutions. However, for a special case of atomic solutions we are able to obtain the so-called *weak-atomic uniqueness* result. It states that if initial data is atomic, then the solution is atomic as well, and thus, by Theorem (3.3), it is unique. The proof is based on the analysis of possible supports of constructed solutions. Its main steps are presented in the following section.

4.4   Weak-Atomic Uniqueness

We present the idea of weak-atomic uniqueness: we aim to prove that if the initial data is atomic, then the solution must be atomic too.

It is sufficient to concentrate our attention on small times near $t = 0$. Let f_0 be atomic (recall (19)). Our goal is to restrict f_0 to small balls with just one particle (say $\underline{i}$th particle). Then we use the local propagation of the support to prove that the measure that initially formed the $\underline{i}$th particle remains atomic for short time. Since

$$f_0 = \sum_{i=1}^{N} m_i \delta_{x_{0i}} \otimes \delta_{v_{0i}},$$

we have a finite number of initial positions and velocities of the particles (x_{0i}, v_{0i}) for $i = 1, ..., N$. It implies that there exists $R_1 > 0$ such that for all $R < R_1$, we have

$$f_0|_{B_i(R)} = m_i \delta_{x_{0i}} \otimes \delta_{v_{0i}} \tag{27}$$

where $B_i(R)$ is a ball centred at (x_{0i}, v_{0i}) with radius R. In order to finish the proof it suffices to show that there exists T^* such that

$$f^D(t) := f(t)|_{B_{\underline{i}}(\frac{R}{4})} = m_{\underline{i}} \delta_{x_{\underline{i}}(t)} \otimes \delta_{v_{\underline{i}}(t)}, \quad \text{for } t \in [0, T^*]. \tag{28}$$

In other words, restriction of f to a small ball centred initially in an atom (denoted by f^D) is precisely the said atom (it does not disperse). Denoting

$$f^C(t) := f(t) - f^D(t)$$

we observe that f^D and f^C satisfy the following equation on $[0, T^*]$:

$$\partial_t f^D + v \cdot \nabla_x f^D + \text{div}[(F(f^C) + F(f^D)) f^D] = 0. \tag{29}$$

Let us introduce

$$\begin{cases} \dfrac{d}{dt} x_a(t) = v_a(t) & \text{(30a)} \\[2mm] \dfrac{d}{dt} v_a(t) = \displaystyle\int_{\mathbb{R}^{2d}} \psi(|x_a(t) - y|)(w - v_a(t)) f^C(y, w, t) dy\, dw, & \text{(30b)} \end{cases}$$

with the initial data $(x_a(0), v_a(0)) = (x_{0i}, v_{0i})$. A critical property of the measure-valued solutions that can be derived from the construction is the local propagation of the support. Thus, since by definition f^D is just a single particle at $t = 0$, then even if it dissolves, for a short time it will still be contained within a cone of the form $x_{0i} + t B(v_{0i}, \epsilon)$ for a small $\epsilon > 0$. In particular f^D remains separated from f^C, which allows to control the singularity of ψ. This ensures that the right-hand side of (30b) is smooth and thus (30) has exactly one smooth solution in $[0, T^*]$. Our goal is to show that f^D is supported on the curve $(x_a(t), v_a(t))$ and that in fact (28) holds with $(x_i, v_i) \equiv (x_a, v_a)$.

We test (29) with $(v - v_a(t))^2$ and with $|x - x_a(t)|^2 t^{-1}$ getting, after some observations, that

$$\frac{d}{dt} \left(\int_{\mathbb{R}^{2d}} (t^{-1} f^D |x - x_a(t)| + f^D |v - v_a(t)|^2) dx dv \right)$$

$$+ \frac{1}{2} \int_{\mathbb{R}^{2d}} t^{-2} f^D |x - x_a(t)|^2 dx dv$$

$$\leq A(t) \int_{\mathbb{R}^{2d}} t^{-1} f^D |x - x_a(t)|^2 + f^D |v - v_a(t)|^2 dx dv,$$

with $A(t) \sim t^{-1/2-\alpha}$. In order to obtain the above inequality, we again control the singularity by ensuring the separation of f^D and f^C, thanks to the cone-shaped propagation of f^D.

By Gronwall's lemma we get

$$\int_{\mathbb{R}^{2d}} (t^{-1} f^D |x - x_a(t)|^2 + f^D |v - v_a(t)|^2) dx dv \equiv 0$$

on $[0, T^*]$. Thus on $[0, T^*]$ we have $x \equiv x_a$ and $v \equiv v_a$ on the support of f, which is exactly equivalent to (28) what concludes the proof.

We have proved that f^D is mono-atomic. Then, repeating the procedure for all atoms (the number is finite) we conclude that f is atomic on a time interval $[0, T^*]$ with possibly smaller, but still positive $T^* > 0$. This procedure works till the first moment of sticking of an ensemble of particles. To reach the moment of sticking of the particles we use regularity in time granted by Theorem 4.2: $\partial_t f \in L^p(0, T; (C^1)^*)$ with $T > T_1$, where T_1 is the time of first sticking. Then we continue our procedure from T_1 till next time of sticking, up to the final time of existence T.

5 Singular Cucker–Smale Model: Hydrodynamics

5.1 Formal Derivation

Kinetic theory provides a way to model interacting particles on the *mesoscopic* scale level, it reduces the computational complexity. However, while the reduction is significant, we still end up with evolutionary equation in twice the spatial dimension, as both x and v belong to $\mathbb{R}^d$. In case when the number of particles is significantly large, the *macroscopic* scale can be employed instead. In such a scale the evolution of local in the phase space averages of density and velocity is studied. This way a hydrodynamic limit is obtained and it further reduces the computational complexity of the model.

In case of the CS model, it can be understood by taking formally

$$f(t, x, v) = \rho(t, x) \otimes \delta_{u(t,x)}(v)$$

in (23) and testing it with functions $\phi = 1$ and $\phi = v$, respectively, obtaining the following continuity and momentum equations

$$\begin{cases} \partial_t \rho + \mathrm{div}(u\rho) = 0, & \text{(31a)} \\[2mm] \partial_t(\rho u) + \mathrm{div}(\rho u \otimes u) = \rho \int_\Omega (u(y) - u(\cdot))\psi(|y - \cdot|)\rho(y)\mathrm{d}y. & \text{(31b)} \end{cases}$$

Here $\Omega = \mathbb{R}^d$ or $\Omega = \mathbb{T}^d$. The system (31) is a hydrodynamical version of the CS model, and we refer to it as the *fractional Euler alignment system*.

We dedicate this section to the discussion of the recent contributions related to this system, focusing particularly on the singular case with weight (7). The past contributions, related mostly to the regular CS model, such as [14, 30, 31, 70] can be found in the survey article [19]. Here, we also mention a recent paper [38] that follows the ideas from [14, 70] related to the existence of a critical threshold.

Before we proceed, let us rewrite Eq. (31b) in a way that is often more convenient. Developing both terms on the right-hand side of (31b) and applying Eq. (31a) one obtains

$$\rho \partial_t u + \rho (u \cdot \nabla) u = \rho \int_\Omega (u(y) - u(\cdot)) \psi(|y - \cdot|) \rho(y) dy,$$

which then, divided by ρ, yields the velocity equation

$$\partial_t u + (u \cdot \nabla) u = \int_\Omega (u(y) - u(\cdot)) \psi(|y - \cdot|) \rho(y) dy. \qquad (32)$$

Note that for

$$\rho = \tilde\rho \equiv const.$$

the right-hand side of (32) is the fractional Laplacian of order $\frac{\gamma}{2} \in (0, 1)$ defined as

$$\Lambda^\gamma u(x) = (-\Delta)^{\frac{\gamma}{2}} u(x) := c \int_\Omega \frac{u(x) - u(y)}{|x - y|^{d+\gamma}} dy, \qquad \gamma := \alpha - d \in (0, 2), \qquad (33)$$

which transforms (32) into a variant of pressureless fractional Euler equation

$$\partial_t u + (u \cdot \nabla) u + c\tilde\rho(-\Delta)^{\frac{\gamma}{2}} u = 0. \qquad (34)$$

We emphasize the introduction of the exponent $\gamma = \alpha - d$ which helps to express the singularity in the hydrodynamic variant of the CS model. This of course puts α in the range $(d, d+2)$. In equations similar to (34) the regularity of solutions is granted by the fractional elliptic term. Therefore, to control the regularity of the fractional Euler alignment system (31) when $\rho \neq const.$ the boundedness of the density ρ from below is essential. It is a reflection of a ubiquitous, in hydrodynamics, problem of the control of the vacuum.

Rigorous derivation of the fractional Euler alignment system or other hydrodynamic limits of the Cucker–Smale kinetic equation (like in [61]) is still mostly open. We refer to papers [30, 31] and to the survey [19], noting that, for the most part, they deal with the model with the regular communication weight.

5.2 Strong Theory on $\mathbb{T}^1$

The most recent developments in the study of (31) are due to Do et al. [27] and independently due to Shvydkoy and Tadmor [63, 65, 66]. The focus of their research is system (31) on a one-dimensional torus $\mathbb{T}$. The crucial discovery is that the quantity

$$e(t, x) := u_x(t, x) - \Lambda^\gamma \rho(t, x)$$

$$= u_x(t, x) + \int_{\mathbb{T}} \psi(|x - y|)(\rho(t, x) - \rho(t, y)) dy \tag{35}$$

for $\alpha = 1 + \gamma$, satisfies the continuity equation

$$e_t + (ue)_x = 0. \tag{36}$$

Thus, $q := \frac{e}{\rho}$, defined for positive ρ is transported with u, i.e.

$$q_t + uq_x = 0,$$

and thus, in particular, its extrema are constant. Of course, such direct comparison between the regularity of ρ and u strongly relies on the one-dimensional domain and as of today, its multidimensional generalization is unknown.

The discovery of the conservation law (36) is the basis of multiple results for the fractional Euler alignment system on one-dimensional torus. Methodology used in [27] and in [63, 65, 66] varies strongly but the main difficulty boils down to the control of the minimal values of the density ρ in (31b). In [27, 65], quantity e was used to provide a bound on the decay of ρ of order $\frac{1}{1+t}$, which was sufficient for the proof of the existence of smooth solutions and for flocking. However, as an example of application of quantity e we shall present the uniform lower bound on the density obtained later in [63].

Lemma 5.1 ([63]) *Let (u, ρ) be a smooth solution to (31) on $\mathbb{T}$. Assume further that $\rho_0 > 0$ and that $|q_0|_\infty < \infty$. Then there exists a constant $C_0 > 0$ such that*

$$\rho(t, x) \geq C_0 \text{ for all } x \in \mathbb{T}, \ t \geq 0.$$

Proof Using the definition of q we find out that (31a) is equivalent to

$$\rho_t + u\rho_x = -q\rho^2 + \rho \int_{\mathbb{T}} \psi(|x - y|)(\rho(y) - \rho(\cdot)) dy,$$

which evaluated at $\rho_m(t) := \min_{x \in \mathbb{T}} \rho(t, x)$ leads to

$$\frac{d}{dt} \rho_m = -q\rho_m^2 + \underbrace{\rho_m \int_{\mathbb{T}} \psi(|x - y|)(\rho(y) - \rho_m) dy}_{\geq 0}$$

$$\geq -q\rho_m^2 + \rho_m \int_{\mathbb{T}} \psi_m (\rho(y) - \rho_m) dy$$

$$\geq -(|q_0|_\infty + \psi_m)\rho_m^2 + \mathcal{M} \psi_m \rho_m,$$

where $\psi_m := \psi(1) = \inf_{x,y\in\mathbb{T}} \psi(|x - y|)$ and $\mathcal{M} := \rho(\mathbb{T})$. In the above inequality we replaced q with $|q_0|_\infty$ by the virtue of the fact that q is transported so it retains the values of its extrema. Then we conclude that $\rho_m \geq \min\{\rho_m(0), \frac{\mathcal{M}\psi_m}{|q_0|_\infty + \psi_m}\}$.

The control over the lower bound of ρ provides the opportunity to prove further results. As explained at the beginning of this section the crucial application of the lower bound is in ensuring that the ellipticity granted by the right-hand side of (31b) or (32) does not disappear. This ellipticity is used to obtain a variety of results, which we summarize in the following theorems.

Theorem 5.1 ([65]) *For $\gamma \in [1, 2)$, the fractional Euler alignment system (31) with periodic initial data $(u_0, \rho_0) \in H^4 \times H^{3+\gamma}$, such that $\rho_0(x) > 0$ for all $x \in \mathbb{T}$, has a unique global solution in the same class of regularity.*

The above theorem is the first result on the fractional Euler alignment system (31) that establishes global and unconditional existence of strong solutions for singular communication weight ψ. The proof is based on the e quantity and on a Beale–Kato–Majda type criterion. It also relies on the maximum principles for dissipative non-local operators due to Constantin and Vicol [21], and on the work on equations with fractional diffusion by Silvestre [68]. The authors address also the Navier–Stokes case of $\gamma = 2$ obtaining solutions in the class $H^2 \times H^3$.

Theorem 5.2 ([27]) *For $\gamma \in (0, 1)$, the fractional Euler alignment system (31) with periodic smooth initial data (u_0, ρ_0), such that $\rho_0(x) > 0$ for all $x \in \mathbb{T}$, has a unique global smooth solution.*

The above result is based on the lower bound on the density granted by the application of the quantity e (however it should be noted that at the time of publication of [27], Lemma 5.1 was not known and the authors used a weaker, time dependent bound $\rho \gtrsim (1 + t)^{-1}$). This allows to obtain local well-posedness in the form of a Beale–Kato–Majda type existence criterion, which reduces the global well-posedness to the question of regularity. Global regularity is then shown by a rescaling argument together with a modulus of continuity breaking-point method. The proof was first performed in the simplified case of $q \equiv 0$ and then generalized to $q \not\equiv 0$. Methods used in [27] are based on previous works due to Kiselev such as [45].

Theorem 5.3 ([63, 66]) *For $\gamma \in (0, 2)$, the fractional Euler alignment system (31) with periodic initial data $(u_0, \rho_0) \in H^4 \times H^{3+\gamma}$, such that $\rho_0(x) > 0$ for all $x \in \mathbb{T}$, has a unique global smooth solution $(u, \rho) \in L^\infty([0, \infty); H^4 \times H^{3+\gamma})$. Moreover the solution converges exponentially fast to a flocking state*

$$\bar{\rho} = \rho_\infty(x - t\bar{u}) \in H^\gamma$$

travelling with a finite speed $\bar{u}$, so that for any $s < \gamma$ there exist $C = C_s$ and $\delta = \delta_s$ with

$$\|u(t) - \bar{u}\|_{H^3} + \|\rho(t) - \bar{\rho}(t)\|_{H^s} \le C e^{-\delta t}, \qquad t > 0,$$

where $\bar{u}$ is the initial average velocity of the system.

The proof of the above theorem can be found in [66] but in reality, it is a final, rectified version of an effort started in [65] and continued through [63]. The proof of global well-posedness is based again on a Beale–Kato–Majda type existence criterion, this time however the issue of global regularity is addressed by controlling higher order derivatives of the solution and particularly—their exponential decay. It results in a proof that is simpler than the proofs found in [27, 63, 65] (even though it relies on results due to Constantin and Vicol [21] and Silvestre [68]).

5.3   Comparison with the PM Equation: Theoretical Results

We will now summarize the results on a particular model closely related to system (31). The starting point of this section is a reformulation of system (31) with $\gamma = 2$ in [65]. Interestingly enough, the same reformulation was noticed and employed earlier in the context of compressible fluids equations with density-dependent viscosity coefficients, see, for example, [8]. Taking formally $\gamma = 2$ in (33), we get the following analogue of (36) with $e = v_x$:

$$(v_x)_t + (u v_x)_x = 0, \tag{37}$$

where

$$v_x = u_x - \Lambda^2 \varrho = u_x + \varrho_{xx}. \tag{38}$$

Therefore, for $v(t, x)$ vanishing at $x = \pm\infty$, Eq. (37) leads to a simple transport equation for v, and so, the system (31) takes the form

$$\begin{cases} \varrho_t + (\varrho u)_x = 0, \\ v_t + u v_x = 0. \end{cases} \tag{39}$$

A formal calculation shows that this system is equivalent to the pressureless compressible Navier–Stokes system with density dependent viscosity:

$$\begin{cases} \varrho_t + (\varrho u)_x = 0, \\ (\varrho u)_t + (\varrho u^2)_x - (\varrho^2 u_x)_x = 0. \end{cases} \tag{40}$$

As we shall see in Theorem 5.4 below, the solution to this system, at least for conveniently chosen initial data, converges in some sense to the solution of the porous medium (PM) equation. The correspondence between both systems can be explained using the definition of v (38) again. It allows to rewrite system (39) as

$$\begin{cases} \varrho_t - \left(\dfrac{\varrho^2}{2}\right)_{xx} = -(\varrho v)_x, \\ (\varrho v)_t + (\varrho u v)_x = 0. \end{cases} \qquad (41)$$

In particular, the continuity equation becomes the porous-medium (PM) equation with force, whose potential solves the continuity equation.

This observation was a core of paper [37], in which the authors studied a generalization of system (40) on $\mathbb{R}$, augmented by the pressure term $\varepsilon (\varrho^\gamma)_x$ with $\gamma > 1$, and the initial condition

$$(\varrho, \varrho u)(0, x) = (\varrho_0, m_0)(x). \qquad (42)$$

They showed that if $(\sqrt{\varrho} v)(0, x) = \frac{m_0}{\sqrt{\varrho_0}} + \sqrt{\varrho_0} \varrho_{0x} = 0$, then for $\varepsilon \to 0$ the density solving the Navier–Stokes system converges to the solution of the porous medium equation

$$\begin{cases} \tilde{\varrho}_t - \left(\dfrac{\tilde{\varrho}^2}{2}\right)_{xx} = 0, \\ \tilde{\varrho}(0, x) = \varrho_0(x). \end{cases} \qquad (43)$$

emanating from the same initial data. The same technique can be used to prove the existence of solutions to system (40), and to show that when $(\sqrt{\varrho} v)(0, x) \to 0$, the same porous medium equation (43) is recovered.

Theorem 5.4 *Let the initial data (ϱ_0, m_0) satisfy*

$$\varrho_0 \geq 0, \ \varrho_0 \in L^1(\mathbb{R}) \cap L^\infty(\mathbb{R}), \quad (\varrho_0)_x^{\frac{3}{2}} \in L^2(\mathbb{R}),$$
$$\frac{m_0^2}{\varrho_0} \in L^1(\mathbb{R}), \quad \frac{|m_0|^{2+\kappa}}{\varrho_0^{1+\kappa}} \in L^1(\mathbb{R}), \ for \ some \ \kappa > 0, \qquad (44)$$

and let

$$\left\| \frac{m_0}{\sqrt{\varrho_0}} + \sqrt{\varrho_0} \varrho_{0x} \right\|_{L^2(\mathbb{R})} \leq \eta. \qquad (45)$$

1. System (40) with initial data (42) admits a global in time weak solution $(\varrho_\eta, \sqrt{\varrho_\eta} u_\eta)$, that is:

   • *The density $\varrho_\eta \geq 0$ a.e., and the following regularity properties hold*

$$\varrho_\eta \in L^\infty(0, T; L^1(\mathbb{R})) \cap \mathscr{C}([0, +\infty), (W^{1,\infty}(\mathbb{R}))^*),$$

$$\left(\varrho_\eta^{\frac{3}{2}}\right)_x \in L^\infty(0, T; L^2(\mathbb{R})), \quad \sqrt{\varrho_\eta}u_\eta \in L^\infty(0, T; L^2(\mathbb{R})),$$

where $(W^{1,\infty}(\mathbb{R}))^*$ is the dual space of $W^{1,\infty}(\mathbb{R})$.

- For any $t_2 \geq t_1 \geq 0$ and any $\psi \in C^1([t_1, t_2] \times \mathbb{R})$, the continuity equation is satisfied in the following sense:

$$\int_{\mathbb{R}} \varrho_\eta \psi(t_2) \, dx - \int_{\mathbb{R}} \varrho_\eta \psi(t_1) \, dx = \int_{t_1}^{t_2} \int_{\mathbb{R}} (\varrho_\eta \psi_t + \varrho_\eta u_\eta \psi_x) \, dx \, dt. \quad (46)$$

Moreover, for $\varrho_\eta v_\eta = \varrho_\eta u_\eta + \varrho_\eta \varrho_{\eta x}$, the following equality is satisfied

$$\int_{\mathbb{R}} \varrho_\eta \psi(t_2) \, dx - \int_{\mathbb{R}} \varrho_\eta \psi(t_1) \, dx$$

$$= \int_{t_1}^{t_2} \int_{\mathbb{R}} \left(\varrho_\eta \psi_t + \varrho_\eta v_\eta \psi_x - \varrho_\eta \varrho_{\eta x} \psi_x\right) \, dx \, dt. \quad (47)$$

- For any $\psi \in C_c^\infty([0, T) \times \mathbb{R})$ the momentum equation is satisfied in the following sense:

$$\int_{\mathbb{R}} m_0 \psi(0) \, dx + \int_0^T \int_{\mathbb{R}} \left(\sqrt{\varrho_\eta}(\sqrt{\varrho_\eta}u_\eta)\psi_t + (\sqrt{\varrho_\eta}u_\eta)^2 \psi_x\right) \, dx \, dt$$

$$-\langle \varrho_\eta^2 u_{\eta x}, \psi_x \rangle = 0, \quad (48)$$

where the diffusion term is defined as follows:

$$\langle \varrho_\eta^2 u_{\eta x}, \psi_x \rangle = -\int_0^T \int_{\mathbb{R}} \varrho_\eta^{\frac{3}{2}} \sqrt{\varrho_\eta} u_\eta \psi_{xx} \, dx \, dt$$

$$-\frac{4}{3} \int_0^T \int_{\mathbb{R}} \left(\varrho_\eta^{\frac{3}{2}}\right)_x \sqrt{\varrho_\eta} u_\eta \psi_x \, dx \, dt. \quad (49)$$

2. For $\eta \to 0$, ϱ_η converges strongly to $\tilde{\varrho}$—the strong solution to the porous medium equation (43) in the following sense: there exists a constant $C > 0$ depending on ϱ_0 such that

$$\|(\tilde{\varrho} - \varrho_\eta)(t)\|_{H^{-1}(\mathbb{R})} \leq C\eta^{\frac{1}{2}}t^{\frac{1}{2}}. \quad (50)$$

Note that in assumptions of this theorem we only require that $\varrho_0 \geq 0$, in particular, the initial density can be compactly supported. Then, the natural question is how does the support of the density propagate in time. For the degenerate Navier–Stokes equations the answer to this question is only partial, see, for example, [40, 75]. Estimate (50) allows us to compare weak solutions to the Navier–Stokes system with strong solutions of the porous medium equation. From the classical theory we know that the interface between fluid and the vacuum for the latter moves with the finite speed. For the special class of self-similar solutions, the so-called Barenblatt solutions, one can even give exact formula for the velocity of this motion, see [72].

Idea of the Proof The proof of existence of weak solutions starts from the approximate Navier-Stokes system augmented by the artificial viscosity term:

$$\begin{cases} \varrho_{\varepsilon t} + (\varrho_\varepsilon u_\varepsilon)_x = 0, \\ (\varrho_\varepsilon u_\varepsilon)_t + (\varrho_\varepsilon u_\varepsilon^2)_x - (\varrho_\varepsilon^2 + \varepsilon \varrho_\varepsilon^\theta u_{\varepsilon x})_x = 0, \\ \varrho_\varepsilon(0, x) = \varrho_{\varepsilon,0}(x), \quad \varrho_\varepsilon(0, x) u_\varepsilon(0, x) = m_{\varepsilon,0}(x), \end{cases} \tag{51}$$

where $\varepsilon > 0$, $\theta \in (0, 1/2)$ are constant. Furthermore, we cut the domain into a bounded interval $\Omega = [-M, M]$ for M large and we supplement system (51) with the boundary conditions $u_\varepsilon|_{\partial\Omega} = 0$. We assume that $\varrho_{\varepsilon,0} > C(\varepsilon) > 0$, and that $\varrho_{\varepsilon,0}, m_{\varepsilon,0}$ converge to ϱ_0, m_0 in the following sense:

$$\varrho_{\varepsilon,0} \to \varrho_0, \quad \text{strongly in } L^1(\Omega),$$

$$\left(\varrho_{\varepsilon,0}^{\frac{3}{2}}\right)_x \to \left(\varrho_0^{\frac{3}{2}}\right)_x \quad \text{strongly in } L^2(\Omega),$$

$$\frac{m_{\varepsilon,0}^2}{\varrho_{\varepsilon,0}} \to \frac{m_0^2}{\varrho_0} \quad \text{strongly in } L^1(\Omega), \tag{52}$$

$$\frac{|m_{\varepsilon,0}|^{2+\kappa}}{\varrho_{\varepsilon,0}^{1+\kappa}} \to \frac{|m_0|^{2+\kappa}}{\varrho_0^{1+\kappa}} \quad \text{strongly in } L^1(\Omega).$$

The existence of the classical solutions for ε, M being fixed can be obtained as in the works of Jiu and Xin [41], and of Li et al. [46] that treat the full Navier–Stokes system including the pressure term. The most important here are the lower and upper estimates of the density that however do not depend on the presence of the pressure. The way to obtain them is to rewrite system (51) in the mass Lagrangian coordinates

$$y = \int_{-M}^x \varrho(\tau, s)ds, \qquad \tau = t. \tag{53}$$

Since the total mass is given and bounded we have

$$\int_{-M}^M \varrho(t, s)ds = L < \infty, \tag{54}$$

and so $y \in [0, L] = \Omega_L$. Using (53), system (40) may be transformed into the following one:

$$\begin{cases} \varrho_\tau + \varrho^2 u_y = 0, \\ u_\tau - ((\varrho^3 + \varepsilon\varrho^{1+\theta})u_y)_y = 0, \\ \varrho_\varepsilon(0, y) = \varrho_{\varepsilon,0}(y), \quad \varrho_\varepsilon(0, y)u_\varepsilon(0, y) = m_{\varepsilon,0}(y), \end{cases} \tag{55}$$

where the boundary conditions are now equal to $u|_{\partial\Omega_L} = 0$. Then, for regular solutions $\varrho_\varepsilon, u_\varepsilon$ of (55) s.t. $\varrho_\varepsilon > 0$, we have the following a-priori estimates:

i) the energy estimate in the Lagrangian coordinates

$$\int_{\Omega_L} \frac{u_\varepsilon^2}{2}(T)\, dy + \int_0^T\!\!\int_{\Omega_L} \varrho_\varepsilon(\varrho_\varepsilon^2 + \varepsilon\varrho_\varepsilon^\theta)(u_{\varepsilon\eta})^2\, dy\, d\tau \le \int_{\Omega_L} \frac{u_\varepsilon^2}{2}(0)\, dy; \tag{56}$$

ii) the entropy estimate in the Lagrangian coordinates

$$\int_{\Omega_L} \left(u_\varepsilon^2 + (\varrho_\varepsilon^2)_\eta^2 + \varepsilon^2(\varrho_\varepsilon^\theta)_\eta^2 \right)(T)\, dy + \int_0^T\!\!\int_{\Omega_L} \varrho_\varepsilon(\varrho_\varepsilon^2 + \varepsilon\varrho_\varepsilon^\theta)(u_{\varepsilon\eta})^2\, dy\, d\tau$$

$$\le \int_{\Omega_L} \left(u_\varepsilon^2 + (\varrho_\varepsilon^2)_\eta^2 + \varepsilon^2(\varrho_\varepsilon^\theta)_\eta^2 \right)(0)\, dy. \tag{57}$$

As a consequence of these estimates we can show the following lemma.

Lemma 5.2 *Let $\varrho_\varepsilon, u_\varepsilon$ be a regular solution of (55), satisfying (56) and (57). Then there exists constants C and $C(\varepsilon)$ such that*

$$0 < C(\varepsilon) \le \varrho_\varepsilon \le C. \tag{58}$$

For the proof of this lemma we refer, to [41], Lemma 3.3. Having these estimates at hand, proving the existence of regular solutions to (55) is a classical result, see, e.g., [49, Chapter 7].

The existence of smooth solutions to (55) allows to come back to the system written in the Eulerian coordinates (51) for ε, M being fixed, and to translate the estimates from above to the following ones:

i') the energy estimate in the Eulerian coordinates

$$\int_\Omega \frac{\varrho_\varepsilon u_\varepsilon^2}{2}(T)\, dx + \int_0^T\!\!\int_\Omega (\varrho_\varepsilon^2 + \varepsilon\varrho_\varepsilon^\theta)u_{\varepsilon x}^2\, dx\, dt$$

$$\le \int_\Omega \left(\varrho_\varepsilon \frac{u_\varepsilon^2}{2} + \frac{\varepsilon}{\gamma - 1}\varrho_\varepsilon^\gamma \right)(0)\, dx; \tag{59}$$

ii') the entropy estimate in the Eulerian coordinates

$$\int_{\Omega} \varrho_\varepsilon \frac{\left(u_\varepsilon + (1+\varepsilon \varrho_\varepsilon^{\theta-2})\varrho_{\varepsilon x}\right)^2}{2}(T) \, dx \leq \int_{\Omega} \varrho_\varepsilon \frac{\left(u_\varepsilon + (1+\varepsilon \varrho_\varepsilon^{\theta-2})\varrho_{\varepsilon x}\right)^2}{2}(0) \, dx. \tag{60}$$

In addition to that, as in the work of Mellet and Vasseur in [51], one can improve the uniform estimates of the velocity vector field

$$\int_{\Omega} \frac{\varrho_\varepsilon |u_\varepsilon|^{2+\kappa}}{2+\kappa}(T) \, dx + (\kappa + 1) \int_0^T \int_{\Omega} (\varrho_\varepsilon^2 + \varepsilon \varrho_\varepsilon^\theta)|u_\varepsilon|^\kappa |u_{\varepsilon x}|^2 \, dx \, dt$$

$$\leq \int_{\Omega} \frac{\varrho_\varepsilon |u_\varepsilon|^{2+\kappa}}{2+\kappa}(0) \, dx,$$

for some $\kappa > 0$. These estimates lead to the following bounds:

$$\|\sqrt{\varrho_\varepsilon} u_\varepsilon\|_{L^\infty(0,T;L^2(\Omega))} + \|(\varrho_\varepsilon^{\frac{3}{2}})_x\|_{L^\infty(0,T;L^2(\Omega))} + \varepsilon\|(\varrho_\varepsilon^{\theta-\frac{1}{2}})_x\|_{L^\infty(0,T;L^2(\Omega))}$$

$$+ \|\varrho_\varepsilon |u_\varepsilon|^{2+\kappa}\|_{L^\infty(0,T;L^1(\Omega))} + +\|(\varrho_\varepsilon + \sqrt{\varepsilon}\varrho_\varepsilon^{\theta/2})u_{\varepsilon x}\|_{L^2(0,T;L^2(\Omega))} \leq C. \tag{61}$$

Moreover, translating (58) into the Eulerian coordinates, uniformly in ε we have

$$\|\varrho_\varepsilon\|_{L^\infty(0,T;L^\infty(\Omega))} \leq C. \tag{62}$$

With these estimates compactness arguments from [51], [46] yield convergence as $\varepsilon \to 0$ of the approximate solution $(\varrho_\varepsilon, u_\varepsilon)$ to the weak solution (ϱ, u) specified in Theorem 5.4, on the domain $\Omega = [-M, M]$ with the no-slip boundary condition for u. In order to let $M \to \infty$, one can combine the diagonal procedure with the convergence of the initial data $(\varrho_{\varepsilon,0}, m_{\varepsilon,0}) \to (\varrho_0, m_0)$ as it was done in [41].

The proof of the second part of Theorem 5.4 corresponds to the pressureless limit studied in [37]. We recall the main steps. Recalling the notation $v_\eta = u_\eta + \varrho_{\eta x}$, the first part of Theorem 5.4 provides that the continuity equation

$$\varrho_{\eta t} - \frac{1}{2}\left(\varrho_\eta^2\right)_{xx} + (\varrho_\eta v_\eta)_x = 0, \tag{63}$$

it is satisfied in the sense of distributions. From (45) and (60) with $\varepsilon = 0$ it also follows that

$$\sup_{t \in [0,T]} \|\sqrt{\varrho_\eta} v_\eta(t)\|_{L^2(\mathbb{R})} \leq \eta. \tag{64}$$

Therefore, when $\eta \to 0$, we expect to show that the last term in equation (63) converges in the "H^{-1}" sense to the strong solution of the corresponding porous-medium equation (43) with the same initial data ϱ_0. Rigorous proof of this fact is based on the duality technique in the spirit of Vázquez (see [72] Section 6.2.1) that was used in [37] for the pressureless limit.

The convergence $\varrho_\eta \to \tilde{\varrho}$ as stated in (50) is relatively weak. Note that in [37] the convergence of weak solutions to Navier–Stokes system with viscosity coefficient of the form ϱ^α could have been significantly improved for $1 < \alpha \leq \frac{3}{2}$. This was possible, thanks to uniform boundedness of $\varrho_{\eta x}$ in $L^\infty(0, T; L^2(\mathbb{R}))$. In this case one can estimate the mass corresponding to ϱ_η inside the support of certain Barenblatt profile. This provides a partial information about evolution of the interface between the medium and vacuum. For further discussion on that matter we refer the reader to [37], Section 5, and to [36] for relevant results in the multidimensional case.

5.4 Comparison with the PM Equation: Numerical Illustration

In this section we present the results of numerical simulations for the pressureless Navier–Stokes system with density dependent viscosity (40) and the PM equation (43). We aim to illustrate analytical developments of Sect. 5.3 and demonstrate the evolution of ϱ, u and $\tilde{\varrho}$ with respect to various initial conditions. For the sake of numerical simulations we will always assume that $\varrho_0 > 0$, then the initial velocity u_0 may be extracted from m_0, moreover

$$v_0 = v(0, x) = \frac{m_0}{\varrho_0} + \varrho_{0x}.$$

We define parameter c as a ratio

$$\frac{m_0}{\varrho_0} = c \, \varrho_{0x}. \tag{65}$$

Note that for $c = -1$, $v_0 = 0$ and thus we expect that both approximate solutions to (41) and (43) coincide.

Instead of the whole $\mathbb{R}$, the computational domain is approximated by sufficiently large interval $\Omega = [-20, 20]$, and we employ zero Neumann boundary condition on $\partial\Omega$, namely $u_x \cdot n = 0$. As an initial density profile we choose smooth functions with compact support; two initial conditions are taken into account:

$$\varrho_0^1 = \frac{0.2}{1 + x^2}, \quad \varrho_0^2 = \frac{0.2}{1 + (x - 10)^2} + \frac{0.2}{1 + (x + 10)^2},$$

that are refereed to as Case 1 and Case 2, respectively. As a consequence of (65) the initial velocity reads

$$u_0^1 = -c \frac{0.4x}{(1 + x^2)^2}, \quad u_0^2 = -c \frac{0.4(x - 10)}{(1 + (x - 10)^2)^2} - c \frac{0.4(x + 10)}{(1 + (x + 10)^2)^2}.$$

For the spatial discretization, standard finite element method is employed where the discrete space for the density is one order higher than for the velocity, namely $(\rho_h, u_h) \in \mathbb{P}^3(\Omega) \times \mathbb{P}^2(\Omega)$, where $\mathbb{P}^k(\Omega)$ denotes continuous Lagrange element of order k. In time we use implicit time-stepping, with time step $\Delta t = \Delta x = 0.01$. The nonlinear problem is solved by means of the Newton method with the Jacobian computed by automatic differentiation. Implementation is based on the finite element library FEniCS.

To illustrate the dependence of the solutions to (40) and (43) on the initial value of v_0, the constant c has been chosen from a set

$$c \in \{-0.1, -0.5. -0.9, -1.0, -1.1, -1.5, -1.9\}. \tag{66}$$

This corresponds to

$$v_0 = 0.9\varrho_{0x}, 0.5\varrho_{0x}, 0.1\varrho_{0x}, 0, -0.1\varrho_{0x}, -0.5\varrho_{0x}, -0.9\varrho_{0x}.$$

Figures 2 and 3 demonstrate time development of the initial profiles, ϱ_0^1 and ϱ_0^2, respectively. For each of the values of parameter c we depict the profile of the density ϱ, the velocity u and the solution to the porous medium equation $\tilde{\varrho}$ at times $t = 0, 200, 400$. Note that at time $t = 0$, the graphs of ϱ and $\tilde{\varrho}$ coincide. In addition to that, in Figs. 4 and 5 we compare density profiles at time $t = 200$ for various values of the parameter c. As predicted by the theoretical considerations from the previous section, for $c \rightarrow -1$, which corresponds to $v_0 \rightarrow 0$, the profile of the density ϱ resembles more and more the profile of $\tilde{\varrho}$, at least up to some time (50). For $c = -1$, i.e. for $v_0 = 0$, the profiles of ϱ and $\tilde{\varrho}$ coincide. Interesting feature, which is not captured by the current analytical theory, is the behaviour of the solution for $c \nrightarrow -1$. On the one hand, we have creation of high concentration around the origin for $c > -1$ (see the top rows of Figs. 2 and 3, and the left parts of Figs. 4 and 5). On the other, for $c < -1$, there is a separation of the initial mass into two groups travelling in the opposite directions (see the bottom rows of Figs. 2 and 3, and the right-hand parts of Figs. 4 and 5). In the first scenario ($c > -1$), it seems that depending on the initial data, the final state may consist of more than one concentration picks (see the top row of Fig. 3 and the left part of Fig. 5). While for the second scenario ($c < -1$), after initial division, the parts of mass colliding around the origin accumulate and create steep profile (see bottom row of Fig. 3 and the right part of Fig. 5).

5.5 Strong Theory in $\mathbb{T}^d$ with Small Data

Generalization of the results by Do et al. and Shvydkoy and Tadmor, presented in Sect. 5.2, to higher dimensions remains open. It is unclear what object should replace quantity e from (36) and the intuitively natural candidate

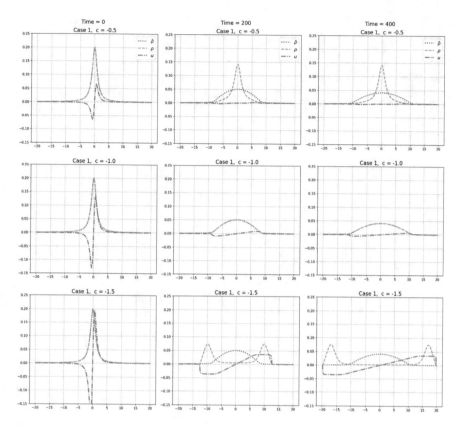

Fig. 2 Case:1 Evolution of ϱ, u solving (40) and of $\tilde{\varrho}$ solving (43), for $\varrho_0 = \varrho_0^1$ and $u_0 = u_0^1$. Comparison of the profiles for $t = 0$, $t = 200$, and $t = 400$, for different values of parameter c as specified by (66)

$$e = \mathrm{div}_x\, u - \Lambda^\gamma \rho(t, x)$$

does not satisfy the continuity equation like in the case of $d = 1$. However it still satisfies equation

$$e_t + \mathrm{div}_x(ue) = (\mathrm{div}_x\, u)^2 - \mathrm{Tr}(\nabla u)^2, \tag{67}$$

which can be used to derive an estimate on e. Such approach was applied by Shvydkoy in [64] leading to the following result.

Theorem 5.5 ([64]) *Let $\gamma \in (0, 2)$. There exists an $N \in \mathbb{N}$ such that for any sufficiently large $R > 0$ any initial condition $(u_0, \rho_0) \in H^m(\mathbb{T}^d) \times H^{m-1+\gamma}(\mathbb{T}^d)$, $m \geq d + 4$, satisfying*

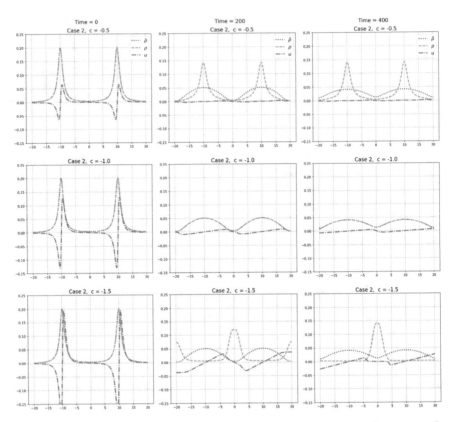

Fig. 3 Case 2: Evolution of ϱ, u solving (40) and of $\tilde{\varrho}$ solving (43), for $\varrho_0 = \varrho_0^2$ and $u_0 = u_0^2$. Comparison of the profiles for $t = 0$, $t = 200$, and $t = 400$, for different values of parameter c as specified by (66)

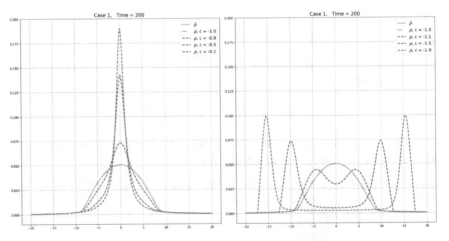

Fig. 4 Comparison of the density profiles for Case 1: $\varrho_0 = \varrho_0^1$, $u_0 = u_0^1$, at $t = 200$

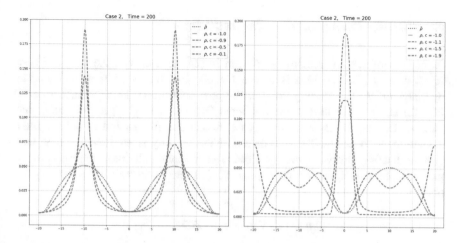

Fig. 5 Comparison of the density profiles for Case 2: $\varrho_0 = \varrho_0^2$, $u_0 = u_0^2$, at $t = 200$

$$|\rho_0|_\infty, |\rho_0^{-1}|_\infty, [u_0]_{\dot{W}^{3,\infty}}, [\rho_0]_{\dot{W}^{3,\infty}} \le R, \qquad \sup_{x,y\in\mathbb{T}^d} |u_0(x) - u_0(y)| \le \frac{1}{R^N},$$

$$(68)$$

gives rise to a unique global solution in class $C([0,\infty) : H^m(\mathbb{T}^d) \times H^{m-1+\gamma})$. Moreover, the solution converges to a flocking state exponentially fast. Here $\dot{W}^{3,\infty}$ is the homogeneous Sobolev space of functions with third weak derivative belonging to L^∞ and $[\cdot]_{\dot{W}^{3,\infty}}$ is its seminorm.

Conditions (68) should be viewed as assumptions on the smallness of initial data in terms of the initial deviation of the velocity u from the average. On top of that it is assumed that ρ_0 does not have singularities and is separated from 0.

Idea of the Proof The proof is performed in the spirit of [66]. Local existence, yet again, is reduced to the control over $|\nabla u|_\infty$ by a Beale–Kato–Majda type criterion. To prolong the existence a global estimate is required and to get it, similarly to the $d = 1$ case, the author makes sure that ρ is separated from 0. It is achieved using quantity e and thus the proof is based on its estimation. Such estimation is performed using Eq. (67), which enables to bound e in terms of $|\nabla u|_\infty$ and itself. To close the estimate smallness condition (68) is needed. Once existence in a sufficiently high class of regularity is established, flocking follows using elementary arguments based on the structure of the CS model (the argumentation is in essence the same as in the simplest case of the CS particle system with regular communication weight). $\qquad \square$

5.6 Strong Theory in $\mathbb{R}^d$ with Small Data

A mostly different approach is employed in [26] to deal with the case of $\Omega = \mathbb{R}^d$ with $d \geq 2$. The idea is to use the classical theory of fractional elliptic hydrodynamics. The starting point is the observation that the right-hand side of Eq. (32) is, up to a constant, the fractional Laplace operator of u. To pursue this idea we assume, similarly to the $\mathbb{T}^d$ case in Sect. 5.5, that ρ_0 and u_0 only slightly deviate from constants. Then smallness of the velocity is preserved in time and ensures that the density ρ also stays close to a constant for all times. To give a proper mathematical description of this approach let us rewrite the velocity equation (32) obtaining

$$\partial_t u + (u \cdot \nabla)u + c_\alpha (-\Delta)^{\frac{\gamma}{2}} u = \mathscr{B}, \tag{69}$$

where the reminder is defined as

$$\mathscr{B} := (1 - \rho)(-\Delta)^{\frac{\gamma}{2}} u + \mathscr{I},$$
$$\mathscr{I} := \int_{\mathbb{R}^d} \frac{u(x + h) - u(x)}{|x|^{d+\gamma}} (\rho(x + h) - \rho(x)) dh.$$

Then the system (31a) + (69) can be viewed as a compressible fractional Burgers system with a (hopefully manageable) right-hand side. Here also the effect of taking ρ close to a constant is apparent since then both terms that constitute $\mathscr{B}$ become small. Based on the works on fractional Burgers equation, such as [52] by Miao and Wu, it is reasonable to expect that the range of admissible γ would be $(0, 2)$. However for the sake of simplicity and accessibility we restrict the range of admissible singularity γ to $(1, 2)$ which allows us to simplify the system even further by transferring the convection to the right-hand side:

$$\partial_t u + c_\alpha (-\Delta)^{\frac{\gamma}{2}} u = \mathscr{R}, \tag{70}$$

with

$$\mathscr{R} := (1 - \rho)(-\Delta)^{\frac{\gamma}{2}} u + \mathscr{I} - (u \cdot \nabla)u.$$

Such presentation of the velocity equation provides an opportunity to consider the fractional Euler alignment system as a compressible fractional heat equation, whose theory boast a wide range of tools to chose from. We chose to use the Besov framework. Thus the fractional Euler alignment system is reduced to a well understood problem of compressible heat equation in the language of Besov spaces, with most of the difficulty moved to the external force $\mathscr{R}$ on the right-hand side. Such approach produced the following theorem.

Theorem 5.6 ([26]) *Assume that* $\gamma \in (1, 2)$ *and consider initial data* (ρ_0, u_0) *so that* u_0 *and* ∇u_0 *are in* $\dot{B}^{2-\gamma}_{d,1}$, *and* $\rho_0 - 1$ *and* $\nabla \rho_0$ *are in* $\dot{B}^1_{d,1}$. *There exists* $\varepsilon > 0$ *such that if in addition*

$$\|u_0\|_{\dot{B}^{2-\gamma}_{d,1}} + \|\rho_0 - 1\|_{\dot{B}^1_{d,1}} < \varepsilon, \tag{71}$$

then the fractional Euler system (31a) + (31b) *has a unique global solution* (ρ, u) *such that*

$$u, \nabla u \in \mathscr{C}_b(\mathbb{R}_+; \dot{B}^{2-\gamma}_{d,1}) \cap L^1(\mathbb{R}_+; \dot{B}^2_{d,1}) \quad and \quad (\rho - 1), \nabla \rho \in \mathscr{C}_b(\mathbb{R}_+; \dot{B}^1_{d,1}).$$

In the case where the smallness condition is fulfilled only by ρ_0, *there exists a unique solution* (ρ, u) *on some time interval* $[0, T]$ *with* $T > 0$ *so that*

$$u, \nabla u \in \mathscr{C}_b([0, T]; \dot{B}^{2-\gamma}_{d,1}) \cap L^1([0, T]; \dot{B}^2_{d,1}) \quad and \quad (\rho-1), \nabla \rho \in \mathscr{C}_b([0, T]; \dot{B}^1_{d,1}).$$

The proof of the above theorem follows by a standard iterative scheme with application of Besov techniques in fractional heat equation. The fractional Laplacian in $\mathscr{R}$ is dealt with by the fractional Laplacian on the left-hand side, thanks to the smallness of $\rho - 1$. The main difficulty lies in the control of $\mathscr{I}$, which is shown to satisfy the inequality

$$\|\mathscr{I}\|_{\dot{B}^{2-\gamma}_{d,1}} \leq C \|\nabla u\|_{\dot{B}^1_{d,1}} \|\rho - 1\|_{\dot{B}^1_{d,1}}.$$

Finally information provided by Theorem 5.6 leads to the following corollary regarding the asymptotic behaviour of the solutions.

Corollary 5.1 *Let* (ρ, u) *be a global in time solution given by Theorem 5.6. Then*

$$\|u(t)\|_{L^\infty} \to 0 \quad as \quad t \to \infty.$$

Here, let us note that, while the information provided by Theorem 5.6 enables the conclusion of the asymptotic decay of velocity, it is insufficient to ensure the exponential rate of the decay, since we are on the whole space $\mathbb{R}^d$ (unlike the $\mathbb{T}^d$ case presented in Sect. 5.5).

Acknowledgements PM was supported by the Deutsche Forschungsgemeinschaft (DFG, German Research Foundation)—314838170, GRK 2297 MathCoRe. JP was supported by the Polish MNiSW grant Mobilność Plus no. 1617/MOB/V/2017/0 and by the NSF grant RNMS11-07444 (KI-Net). EZ was supported by the UCL Department of Mathematics Grant, grant Iuventus Plus no. 0888/IP3/2016/74 of Ministry of Sciences and Higher Education RP, and by the Simons—Foundation grant 346300 and the Polish Government MNiSW 2015–2019 matching fund.

References

1. S. M. Ahn, H. Choi, S.-Y. Ha, and H. Lee. On collision-avoiding initial configurations to Cucker-Smale type flocking models. *Commun. Math. Sci.*, 10(2):625–643, 2012.

2. H.-O. Bae, Y.-P. Choi, S.-Y. Ha, and M.-J. Kang. Time-asymptotic interaction of flocking particles and an incompressible viscous fluid. *Nonlinearity*, 25(4):1155–1177, 2012.

3. H.-O. Bae, Y.-P. Choi, S.-Y. Ha, and M.-J. Kang. Global existence of strong solution for the Cucker-Smale-Navier-Stokes system. *J. Differential Equations*, 257(6):2225–2255, 2014.

4. L. Bakule. Decentralized control: An overview. *Annual Reviews in Control*, 32(1):87–98, 2008.

5. J. Barré, J. Carrillo, P. Degond, D. Peurichard, and E. Zatorska. Particle interactions mediated by dynamical networks: assessment of macroscopic descriptions. *J. Nonlinear Sci.*, 28(1):235–268, 2018.

6. N. Bellomo and S.-Y. Ha. A quest toward a mathematical theory of the dynamics of swarms. *Math. Models Methods Appl. Sci.*, 27(4):745–770, 2017.

7. F. Bolley, J. A. Cañizo, and J. A. Carrillo. Stochastic mean-field limit: non-Lipschitz forces and swarming. *Math. Models Methods Appl. Sci.*, 21(11):2179–2210, 2011.

8. D. Bresch and B. Desjardins. Existence of global weak solutions for a 2d viscous shallow water equations and convergence to the quasi-geostrophic model. *Communications in Mathematical Physics*, 238(1):211–223, Jul 2003.

9. J. A. Cañizo, J. A. Carrillo, and J. Rosado. A well-posedness theory in measures for some kinetic models of collective motion. *Math. Models Methods Appl. Sci.*, 21(3):515–539, 2011.

10. J. A. Carrillo, M. Fornasier, J. Rosado, and G. Toscani. Asymptotic flocking dynamics for the kinetic Cucker-Smale model. *SIAM J. Math. Anal.*, 42(1):218–236, 2010.

11. J. A. Carrillo, Y.-P. Choi, and M. Hauray. Local well-posedness of the generalized Cucker-Smale model with singular kernels. In *MMCS, Mathematical modelling of complex systems*, volume 47 of *ESAIM Proc. Surveys*, pages 17–35. EDP Sci., Les Ulis, 2014.

12. J. Carrillo, Y.-P. Choi, M. Hauray, and S. Salem. Mean-field limit for collective behavior models with sharp sensitivity regions. *Journal of the European Mathematical Society*, 21(1):121–161, 2018.

13. J. A. Carrillo, Y.-P. Choi, P. B. Mucha, and J. Peszek. Sharp conditions to avoid collisions in singular Cucker-Smale interactions. *Nonlinear Anal. Real World Appl.*, 37:317–328, 2017.

14. J. A. Carrillo, Y.-P. Choi, E. Tadmor, and C. Tan. Critical thresholds in 1D Euler equations with non-local forces. *Math. Models Methods Appl. Sci.*, 26(1):185–206, 2016.

15. J. A. Carrillo, A. Klar, S. Martin, and S. Tiwari. Self-propelled interacting particle systems with roosting force. *Math. Models Methods Appl. Sci.*, 20(suppl. 1):1533–1552, 2010.

16. L. Cheng, C. Wu, Y. Zhang, H. Wu, M. Li, and C. Maple. A survey of localization in wireless sensor network. *International Journal of Distributed Sensor Networks*, 8(12):962523, 2012.

17. Y.-P. Choi. Large-time behavior for the Vlasov/compressible Navier-Stokes equations. *J. Math. Phys.*, 57(7):071501, 13, 2016.

18. Y.-P. Choi, S.-Y. Ha, and J. Kim. Propagation of regularity and finite-time collisions for the thermomechanical Cucker-Smale model with a singular communication. *Netw. Heterog. Media*, 13(3):379–407, 2018.

19. Y.-P. Choi, S.-Y. Ha, and Z. Li. Emergent dynamics of the Cucker-Smale flocking model and its variants. In *Active particles. Vol. 1. Advances in theory, models, and applications*, Model. Simul. Sci. Eng. Technol., pages 299–331. Birkhäuser/Springer, Cham, 2017.

20. Y.-P. Choi, D. Kalise, J. Peszek, and A. A. Peters. A collisionless singular Cucker-Smale model with decentralized formation control. *arXiv:1807.05177*, 2018.

21. P. Constantin and V. Vicol. Nonlinear maximum principles for dissipative linear nonlocal operators and applications. *Geom. Funct. Anal.*, 22(5):1289–1321, 2012.

22. F. Cucker and J.-G. Dong. On the critical exponent for flocks under hierarchical leadership. *Math. Models Methods Appl. Sci.*, 19(suppl.):1391–1404, 2009.

23. F. Cucker and J.-G. Dong. Avoiding collisions in flocks. *IEEE Trans. Automat. Control*, 55(5):1238–1243, 2010.

24. F. Cucker and C. Huepe. Flocking with informed agents. *MathS in Action*, 1(1):1–25, 2008.

25. F. Cucker and S. Smale. Emergent behavior in flocks. *IEEE Trans. Automat. Control*, 52(5):852–862, 2007.

26. R. Danchin, P. B. Mucha, J. Peszek, and B. Wróblewski. Regular solutions to the fractional Euler alignment system in the Besov spaces framework. *Math. Models Methods Appl. Sci.*, 29(1):89–119, 2019.

27. T. Do, A. Kiselev, L. Ryzhik, and C. Tan. Global regularity for the fractional Euler alignment system. *Arch. Ration. Mech. Anal.*, 228(1):1–37, 2018.

28. R. Erban, J. Haškovec, and Y. Sun. A Cucker-Smale model with noise and delay. *SIAM J. Appl. Math.*, 76(4):1535–1557, 2016.

29. S.-Y. Ha, T. Ha, and J.-H. Kim. Asymptotic dynamics for the Cucker-Smale-type model with the Rayleigh friction. *J. Phys. A*, 43(31):315201, 19, 2010.

30. S.-Y. Ha, M.-J. Kang, and B. Kwon. A hydrodynamic model for the interaction of Cucker-Smale particles and incompressible fluid. *Math. Models Methods Appl. Sci.*, 24(11):2311–2359, 2014.

31. S.-Y. Ha, M.-J. Kang, and B. Kwon. Emergent dynamics for the hydrodynamic Cucker-Smale system in a moving domain. *SIAM J. Math. Anal.*, 47(5):3813–3831, 2015.

32. S.-Y. Ha and J.-G. Liu. A simple proof of the Cucker-Smale flocking dynamics and mean-field limit. *Commun. Math. Sci.*, 7(2):297–325, 2009.

33. S.-Y. Ha and T. Ruggeri. Emergent dynamics of a thermodynamically consistent particle model. *Arch. Ration. Mech. Anal.*, 223(3):1397–1425, 2017.

34. S.-Y. Ha and E. Tadmor. From particle to kinetic and hydrodynamic descriptions of flocking. *Kinet. Relat. Models*, 1(3):415–435, 2008.

35. J. Haskovec. Flocking dynamics and mean-field limit in the Cucker-Smale-type model with topological interactions. *Phys. D*, 261:42–51, 2013.

36. B. Haspot. From the highly compressible Navier–Stokes equations to fast diffusion and porous media equations, existence of global weak solution for the quasi-solutions. *Journal of Mathematical Fluid Mechanics*, 18(2):243–291, Jun 2016.

37. B. Haspot and E. Zatorska. From the highly compressible Navier-Stokes equations to the porous medium equation – rate of convergence. *Discrete & Continuous Dynamical Systems - A*, 36(6):3107–3123, 2016.

38. S. He and E. Tadmor. Global regularity of two-dimensional flocking hydrodynamics. *C. R. Math. Acad. Sci. Paris*, 355(7):795–805, 2017.

39. A. Jadbabaie, J. Lin, and A. S. Morse. Correction to: "Coordination of groups of mobile autonomous agents using nearest neighbor rules" [IEEE Trans. Automat. Control **48** (2003), no. 6, 988–1001; MR 1986266]. *IEEE Trans. Automat. Control*, 48(9):1675, 2003.

40. S. Jiang, Z. Xin, and P. Zhang. Global weak solutions to 1d compressible isentropic Navier-Stokes equations with density-dependent viscosity. *Methods Appl. Anal.*, 12(3):239–251, 2005.

41. Q. Jiu and Z. Xin. The Cauchy problem for 1d compressible flows with density-dependent viscosity coefficients, 2008.

42. T. K. Karper, A. Mellet, and K. Trivisa. Hydrodynamic limit of the kinetic Cucker-Smale flocking model. *Math. Models Methods Appl. Sci.*, 25(1):131–163, 2015.

43. J. Kim and J. Peszek. Cucker-Smale model with a bonding force and a singular interaction kernel. *arXiv:1805.01994*, 2018.

44. S. J. Kim, Y. Jeong, S. Park, K. Ryu, and G. Oh. *A Survey of Drone use for Entertainment and AVR (Augmented and Virtual Reality)*, pages 339–352. Springer International Publishing, Cham, 2018.

45. A. Kiselev. Nonlocal maximum principles for active scalars. *Adv. Math.*, 227(5):1806–1826, 2011.

46. H.-L. Li, J. Li, and Z. Xin. Vanishing of vacuum states and blow-up phenomena of the compressible Navier-Stokes equations. *Communications in Mathematical Physics*, 281(2):401, May 2008.

47. Z. Li. Effectual leadership in flocks with hierarchy and individual preference. *Discrete Contin. Dyn. Syst.*, 34(9):3683–3702, 2014.

48. V. Loreto and L. Steels. Social dynamics: Emergence of language. *Nature Physics*, 3:758–760, 2007.

49. A. Lunardi. *Analytic semigroups and optimal regularity in parabolic problems.* Modern Birkhäuser Classics. Birkhäuser/Springer Basel AG, Basel, 1995. [2013 reprint of the 1995 original] [MR1329547].

50. I. Markou. Collision-avoiding in the singular Cucker-Smale model with nonlinear velocity couplings. *Discrete Contin. Dyn. Syst.*, 38(10):5245–5260, 2018.

51. A. Mellet and A. Vasseur. On the barotropic compressible Navier–Stokes equations. *Communications in Partial Differential Equations*, 32(3):431–452, 2007.

52. C. Miao and G. Wu. Global well-posedness of the critical Burgers equation in critical Besov spaces. *J. Differential Equations*, 247(6):1673–1693, 2009.

53. S. Motsch and E. Tadmor. A new model for self-organized dynamics and its flocking behavior. *J. Stat. Phys.*, 144(5):923–947, 2011.

54. P. B. Mucha and J. Peszek. The Cucker-Smale equation: singular communication weight, measure-valued solutions and weak-atomic uniqueness. *Arch. Ration. Mech. Anal.*, 227(1):273–308, 2018.

55. P. B. Mucha, J. Peszek, and M. Pokorný. Flocking particles in a non-Newtonian shear thickening fluid. *Nonlinearity*, 31(6):2703–2725, 2018.

56. K.-K. Oh, M.-C. Park, and H.-S. Ahn. A survey of multi-agent formation control. *Automatica*, 53:424–440, 2015.

57. J. Park, H. J. Kim, and S.-Y. Ha. Cucker-Smale flocking with inter-particle bonding forces. *IEEE Trans. Automat. Control*, 55(11):2617–2623, 2010.

58. L. Perea, P. Elosegui, and G. Gomez. Extension of the Cucker-Smale control law to space flight formations. *Journal of Guidance, Control, and Dynamics*, 32(2):527–537, 2009.

59. J. Peszek. Existence of piecewise weak solutions of a discrete Cucker–Smale's flocking model with a singular communication weight. *J. Differential Equations*, 257(8):2900–2925, 2014.

60. J. Peszek. Discrete Cucker-Smale flocking model with a weakly singular weight. *SIAM J. Math. Anal.*, 47(5):3671–3686, 2015.

61. D. Poyato and J. Soler. Euler-type equations and commutators in singular and hyperbolic limits of kinetic Cucker-Smale models. *Math. Models Methods Appl. Sci.*, 27(6):1089–1152, 2017.

62. J. Shen. Cucker-Smale flocking under hierarchical leadership. *SIAM J. Appl. Math.*, 68(3):694–719, 2007/08.

63. R. Shvydkoy and E. Tadmor. Eulerian dynamics with a commutator forcing II: flocking. *Disc. and Cont. Dyn. Sys.*, 37(11):5503–5520, 2017.

64. R. Shvydkoy. Global existence and stability of nearly aligned flocks. *arXiv:1802.08926*, 2018.

65. R. Shvydkoy and E. Tadmor. Eulerian dynamics with a commutator forcing. *Transactions of Mathematics and Its Applications*, 1(1):tnx001, 2017.

66. R. Shvydkoy and E. Tadmor. Eulerian dynamics with a commutator forcing III. Fractional diffusion of order $0 < \alpha < 1$. *Physica D: Nonlinear Phenomena*, 376–377:131–137, 2018. Special Issue: Nonlinear Partial Differential Equations in Mathematical Fluid Dynamics.

67. R. Shvydkoy and E. Tadmor. Topological models for emergent dynamics with short-range interactions. *arXiv:1806.01371*, 2018.

68. L. Silvestre. Hölder estimates for advection fractional-diffusion equations. *Ann. Sc. Norm. Super. Pisa Cl. Sci. (5)*, 11(4):843–855, 2012.

69. H. Spohn. Large scale dynamics of interacting particles. *Springer-Verlag, Berlin and Heidelberg*, 1991.

70. E. Tadmor and C. Tan. Critical thresholds in flocking hydrodynamics with non-local alignment. *Philos. Trans. R. Soc. Lond. Ser. A Math. Phys. Eng. Sci.*, 372(2028):20130401, 22, 2014.

71. G. Toscani, C. Brugna, and S. Demichelis. Kinetic models for the trading of goods. *J. Stat. Phys.*, 151(3-4):549–566, 2013.

72. J. Vazquez. *The Porous Medium Equation: Mathematical Theory.* Oxford Mathematical Monographs. Oxford University Press, 2007.

73. T. Vicsek, A. Czirók, E. Ben-Jacob, I. Cohen, and O. Schochet. Novel type of phase transition in a system of self-driven particles. *Phys. Rev. Lett.*, 75:1226–9, 1995.
74. C. Villani. *Optimal transport*, volume 338 of *Grundlehren der Mathematischen Wissenschaften [Fundamental Principles of Mathematical Sciences]*. Springer-Verlag, Berlin, 2009. Old and new.
75. T. Yang and H. Zhao. A vacuum problem for the one-dimensional compressible Navier-Stokes equations with density-dependent viscosity. *J. Differential Equations*, 184(1):163–184, 2002.

A Stochastic-Statistical Residential Burglary Model with Finite Size Effects

Chuntian Wang, Yuan Zhang, Andrea L. Bertozzi, and Martin B. Short

Abstract Transience of spatio-temporal clusters of residential burglary is well documented in empirical observations, and could be due to *finite size effects* anecdotally. However a theoretical understanding has been lacking. The existing agent-based statistical models of criminal behavior for residential burglary assume deterministic-time steps for arrivals of events. To incorporate random arrivals, this article introduces a Poisson clock into the model of residential burglaries, which could set time increments as independently exponentially distributed random variables. We apply the Poisson clock into the seminal deterministic-time-step model in Short et al. (Math Models Methods Appl Sci 18:1249–1267, 2008). Introduction of the Poisson clock not only produces similar simulation output, but also brings in theoretically the mathematical framework of the Markov pure jump processes, e.g., a martingale approach. The martingale formula leads to a continuum equation that coincides with a well-known mean-field continuum limit. Moreover, the martingale formulation together with statistics quantifying the relevant pattern formation leads to a theoretical explanation of the *finite size effects*. Our conjecture is supported by numerical simulations.

The original version of the chapter has been revised. A correction to this chapter can be found at
https://doi.org/10.1007/978-3-030-20297-2_9.

C. Wang (✉)
The University of Alabama, Tuscaloosa, AL, USA
e-mail: cwang27@ua.edu

Y. Zhang
Peking University, Beijing, People's Republic of China
e-mail: zhangyuan@math.pku.edu.cn

A. L. Bertozzi
University of California, Los Angeles, Los Angeles, CA, USA
e-mail: bertozzi@ucla.edu

M. B. Short
Georgia Institute of Technology, Atlanta, GA, USA
e-mail: mbshort@math.gatech.edu

© Springer Nature Switzerland AG 2019, corrected publication 2023
N. Bellomo et al. (eds.), *Active Particles, Volume 2*, Modeling
and Simulation in Science, Engineering and Technology,
https://doi.org/10.1007/978-3-030-20297-2_8

1 Introduction

Crime is an unfortunate aspect of modern life that takes place in every major urban area. In the past 10 years quantitative scientists have been working in the burgeoning area of crime modeling and prediction (e.g., [3, 4, 28, 29, 46, 48, 52, 55, 57, 61–63, 67, 71–77, 81, 83, 84]). Crime is not uniformly distributed in space and time but rather exhibits spatio-temporal aggregates of criminal occurrences that are referred to as crime "hotspots." The first model to quantify such patterns is an agent-based human environment interaction model assuming deterministic-time steps (DTS Model [76]). Follow-up works show that such simple models can exhibit both crime displacement and crime suppression in the presence of police activity [73, 74]. In all these works *finite size effects* were observed: discrete simulations show both transient and stationary hotspots, while continuum simulations show only stationary hotspots. As criminal number decreases, discrete simulations exhibit more transience, and crime population is finite size actually in real life. Therefore, a deeper understanding of these effects is relevant to real crime statistics [28, 61]. Moreover, the DTS Model [76] assumes deterministic-time arrivals of events; however, criminals act randomly in reality. Models setting random arrivals are called for.

In this work, with the application of a Poisson clock to the DTS Model [76], the time increments are made into independently exponentially distributed random variables. Introduction of the Poisson clock not only exhibits similar simulation output, but also brings in theoretically the mathematical framework of Markov pure jump processes and interacting particle systems [14, 47, 49–51].

The Poisson clock is the basic element to build a Poisson point process, which is particularly suitable to model random arrivals. Poisson point process is one of the most studied and used point processes in probability and in more applied disciplines such as biology, economics, and physics (see, e.g., [2, 5, 13, 18, 27, 65]). Normally independent Poisson clocks are assumed for each agent. Nevertheless assuming a uniform Poisson clock is helpful here, as it is more computationally efficient. Moreover, a martingale approach for the Markov pure jump processes is applicable. The martingale formulation is a useful tool to analyze both statistical and stochastic aspects of the model, and expresses the model as the summation of two components: a predictable or deterministic component and a stochastic or unpredictable component. As far as we know this is the first time that the stochastic component of the model is mathematically demonstrated.

The deterministic component serves as a continuum model that is parallel to the discrete one. The continuum Poisson-clock model turns out to be the same as the continuum DTS Model [76]. More importantly, the martingale formulation leads to a theory for the *finite size effects*, thanks to the demonstration of the stochastic component. To quantify pattern formation for the *finite size effects*, we construct statistics to measure the degree of hotspot transience. These statics could be applied to general pattern formation problems. We find that a scaling property of the stochastic component with varying criminal population is the key to the theory of *finite size effects*. As the total criminal population decreases, the stochastic component increases while the deterministic component remains fixed. We conjecture that this is the mechanism behind the *finite size effects*. Our conjecture is supported by numerical simulations. The study is also in line with the scaling of initial data of hydrodynamic limit approximations for interacting particle systems [24]. In general the initial total particle number is assumed to increase as the lattice grid decreases so that the initial density remains fixed. This has been treated so far as merely a technicality and is rarely explored analytically nor computationally. However in the applications the details are important because the real world is of finite size.

The article is organized as follows. In Sect. 2, the discrete Poisson-clock model is introduced (Sect. 2.1) and the martingale formulation is established (Sect. 2.2). Based on the formulation a continuum model is derived (Sect. 2.3). The finite size effects are analyzed in Sect. 3. Statistics are constructed to quantitatively measure the degree of hotspot transience (Sect. 3.1). A theory of the *finite size effects* is explored (Sect. 3.2). Section 4 is about future works. The Appendix shows the derivation of the martingale formulation.

2 Poisson-Clock Model

2.1 Discrete Model

The discrete Poisson-clock model is the same as the DTS Model [76], with the only exception of introduction of the Poisson clock.

2.1.1 Introduction of the Poisson Clock

The discrete model consists of two components—the stationary burglary sites and a collection of burglar agents jumping from site to site. We assume the domain to be $\mathscr{D} := [0, L] \times [0, L]$ with the periodic boundary conditions.[1] The lattice grid over $\mathscr{D}$ has spacing $\ell = 1/N$, $N \in \mathbb{N}$. The grid points are denoted as $\mathbf{s} = (s_1, s_2)$, $s_1 = \ell, 2\ell, \cdots, L, s_2 = \ell, 2\ell, \cdots, L$. The collection of all the grid points is denoted as $\mathscr{S}^\ell$. Attached to each $\mathbf{s} \in \mathscr{S}^\ell$ is a pair $(n_\mathbf{s}^\ell(t), A_\mathbf{s}^\ell(t))$ representing the number of criminal agents and attractiveness at site $\mathbf{s}$ at time t. The attractiveness stands for the burglar's beliefs about the vulnerability and value of the target site. We also assume that $A_\mathbf{s}^\ell(t)$ consists of two parts, a dynamic term and a static background term

$$A_\mathbf{s}^\ell(t) = B_\mathbf{s}^\ell(t) + A0_\mathbf{s}^\ell. \tag{1}$$

---

[1]With minor changes we can also consider, e.g., the Dirichlet boundary conditions, which is more realistic.

Here $A0_s^\ell$ is not necessarily uniform over the lattice grids. The dynamic term $B_s^\ell(t)$ represents the component associated with repeat and near-repeat victimization, which will be discussed shortly. The initial data are given as $(n_s^\ell(0), B_s^\ell(0)) = (n0_s^\ell, B0_s^\ell)$.

We assume that the Poisson clock advances according to a Poisson process with rate $D\ell^{-2}$, D an absolute constant independent of ℓ. On average, the time increment δt is the inverse of the rate

$$\delta t \cong \frac{\ell^2}{D}. \tag{2}$$

This is analogous to the Brownian scaling used in the DTS Model [76]. Suppose that the clock advances at time t^-. At time t, the system gets updated as follows:

Step 1 Every agent chooses to burglarize with the probability

$$p_s^\ell(t) = 1 - e^{-\frac{A_s^\ell(t^-)\ell^2}{D}} \cong A_s^\ell(t^-)\delta t, \tag{3}$$

where **s** is his current location. This implies that burglary events occur roughly according to a Poisson process with the rate $A_s^\ell(t^-)$.

Step 2 If an agent chooses to burglarize, he will be immediately removed from the system (representing the criminal fleeing with his trophy). If he chooses not to, he will jump from site **s** to one of the neighboring sites, say **k**, with the probability proportional to the attractiveness of the target site

$$q_{s \to k}^\ell(t) = \frac{A_k^\ell(t^-)}{T_s^\ell(t^-)}, \tag{4}$$

where $T_s^\ell(t) := \sum_{s' \sim s} A_{s'}^\ell(t)$, and $s' \sim s$ indicates all of the neighboring sites of **s**.

Step 3 The attractiveness field gets updated according to the repeat and near-repeat victimization and the broken-windows effect. These concepts in criminology and sociology have all been empirically observed [8, 25, 33]. The "broken windows" theory argues that the visible signs of past crimes are likely to create an environment that encourages further illegal activities [83]. The so-called repeat and near-repeat events refer to the phenomenon that residential burglars prefer to return to a previously burglarized house and its neighbors [23, 39–41, 75]. The repeat victimization is modeled by letting B_s^ℓ depend upon previous burglary events at site **s**. The attractiveness increases if a burglary event occurred on that site, and this increase has a finite lifetime. Let $E_s^\ell(t)$ be the total number of burglary events that occurred at site **s** within this time step, then the repeat victimization can be modeled as

$$B_{\mathbf{s}}^{\ell}(t) = B_{\mathbf{s}}^{\ell}(t^{-})\left(1 - \frac{\omega\ell^2}{D}\right) + \theta E_{\mathbf{s}}^{\ell}(t), \qquad (5)$$

where ω and θ are absolute constants setting the speed of the decay and measuring the strength of the repeat victimization effect. The near-repeat victimization and broken windows effects are modeled by allowing $B_{\mathbf{s}}^{\ell}$ to spread in space from each house to its neighbors. To accomplish this, we modify (5) to read

$$B_{\mathbf{s}}^{\ell}(t) = \left[B_{\mathbf{s}}^{\ell}(t^{-}) + \frac{\eta}{4}\ell^2 \Delta^{\ell} B_{\mathbf{s}'}^{\ell}(t^{-})\right]\left(1 - \frac{\omega\ell^2}{D}\right) + \theta E_{\mathbf{s}}^{\ell}(t), \qquad (6)$$

where $\eta \in (0, 1)$ is an absolute constant that measures the significance of neighborhood effects, and Δ^{ℓ} is the discrete spatial Laplace operator associated with the lattice grid, namely

$$\Delta^{\ell} B_{\mathbf{s}}^{\ell}(t) = \ell^{-2}\left(\sum_{\substack{\mathbf{s}'\\ \mathbf{s}'\sim\mathbf{s}}} B_{\mathbf{s}'}^{\ell}(t) - 4B_{\mathbf{s}}^{\ell}(t)\right).$$

Step 4 Let Γ be an absolute constant indicating the growth rate of criminal population. At each site with probability $\Gamma\ell^2/D$, a new agent will be replaced. We assume that ℓ is small enough such that $\Gamma\ell^2/D < 1$.

Figure 1 presents a visual summary of the above steps in the form of a flowchart.

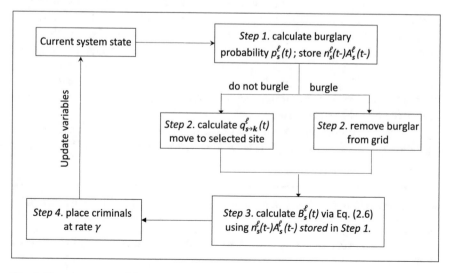

Fig. 1 Flowchart summarizing the discrete model

The spatially homogeneous equilibrium solutions are the same as in the DTS Model [76]. For simplicity from now on we always assume that $A0_s^\ell \equiv A0$. Then the homogeneous equilibrium values can be deduced as

$$\bar{B} = \frac{\theta \Gamma}{\omega}, \quad \bar{n}^\ell = \frac{\Gamma \ell^2}{D \left(1 - e^{-\frac{\ell^2 \bar{A}}{D}} \right)} \simeq \frac{\Gamma}{\bar{A}}. \tag{7}$$

where $\bar{A} = \bar{B} + A0$.

2.1.2 Numerical Simulations

To compare the Poisson-clock model and the DTS Model [76], we perform simulations of the attractiveness field $A_s^\ell(t)$ in Figs. 2, 3, and 4. The parameters are mostly the equivalent of those used to create the plots for the DTS Model in Fig. 3 of [76], and the same behavioral regimes are observed:

(1) Spatial homogeneity. In this regime, $A_s^\ell(t)$ does not vary essentially in time or in space. Very few visible hotspots, that is, the accumulation of $A_s^\ell(t)$ in time and in space, appear in the process.
(2) Dynamic hotspots. In this regime, hotspots form and are transient.
(3) Stationary hotspots. In this regime, hotspots will form and stay more or less stationary over time.

For all the simulations, the spatially homogeneous equilibrium value of the dynamic attractiveness $\bar{B}$ in (7) serves as a midpoint, and is shaded in green. A color key is given in the figures to document the false color map for the attractiveness. All the simulations were run with $L = 128$, $\ell = 1$, $\omega = 1/15$, $A0 = 1/30$, and the initial criminal number at each site $n0_s^\ell$ is set to be $\bar{n}^\ell$ on average.[2] In Figs. 2c, d and 4a, b, we set $\Gamma = 0.0019q$, $\theta = 5.6/q$, $q = 1, 10, 100$, and 1000, respectively. The same is true of Figs. 3c, d and 4c, d. As q increases, (7) implies that the initial criminal population and the criminal replacement rate both increase while the initial attractiveness field remains fixed.

Specific to the Poisson-clock model (Figs. 2a, b, 3a, b, and 4), we set $D = 100$. Specific to the DTS Model [76] (the remaining figures), we set $\delta t = 1/100$. Specific

---

Fig. 2 (continued) $B0_s^\ell \equiv \bar{B}$, $n0_s^\ell \cong \bar{n}^\ell$, $L = 128$, $\ell = 1$, $\omega = 1/15$, and $\eta = 0.2$. Specific to the Poisson-clock model (**a**) and (**b**), we set $D = 100$, $\Gamma = 0.0019q$, $\theta = 5.6/q$, $q = 1$ in (**a**), and $q = 10$ in (**b**). Specific to the DTS Model [76] (**c**) and (**d**), we set $\delta t = 1/100$, $\Gamma = 0.0019q$, $\theta = 5.6/q$, $q = 1$ in (**c**), and $q = 10$ in (**d**). (**b**, **d**) Spatially homogeneous regimes. (**a**, **c**) Dynamic hotspot regimes

---

[2]More precisely, the criminal agents are assumed to be uniformly (randomly) distributed over the 128×128 grids, while $\sum_{s \in \mathscr{L}^\ell} n0_s^\ell = 128^2 \bar{n}^\ell$.

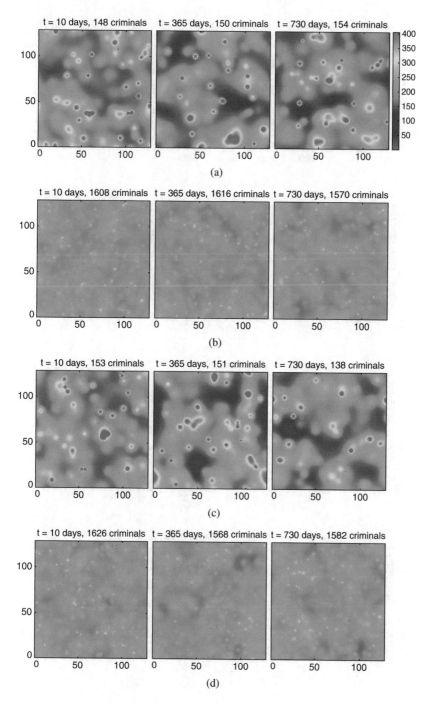

Fig. 2 Plot of the attractiveness $A_s^\ell(t)$ for the discrete Poisson-clock model and the DTS Models [76]. For both models, the initial conditions (at $t = 0$) and parameters are taken to be $A0 = 1/30$,

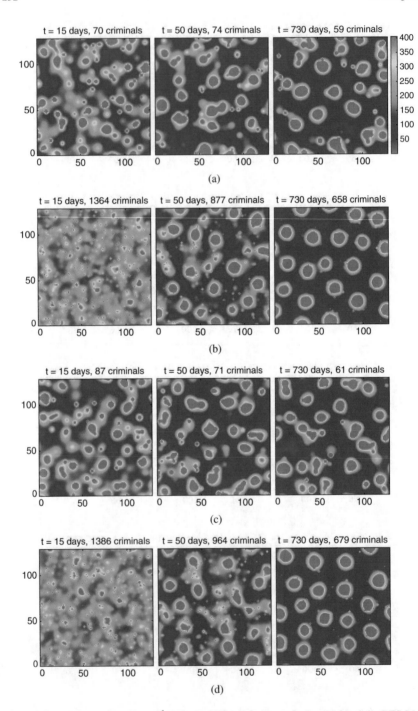

Fig. 3 Plot of the attractiveness $A_s^\ell(t)$ for the discrete Poisson-clock model and the DTS Model [76]. For both models, the parameters and initial conditions (at $t = 0$) are taken to be $L = 128$,

to the cases with zero hotspot formation (Figs. 2 and 4a, b), we set $\eta = 0.2$ and $BO_s^\ell \equiv \bar{B}$ for every $s \in \mathcal{S}$. Specific to the cases with hotspot formation (the remaining figures), η is set to be 0.03, and BO_s^ℓ is set to be $\bar{B}$ on every site except for 30 grid points randomly chosen a priori each gets increased by 0.002.

The same *finite size effects* as in the DTS Model [76] are observed, that is, the degree of hotspot "transience" seems to depend on the total criminal population. The regimes of transient hotspots seem to appear associated with low or vanishing criminal numbers and low numbers of events, while the regimes of stationarity, including the stationary hotspots or homogeneity regimes, occur more likely with large numbers of criminals and burglary events.

2.2 Martingale Formulation

For every t, we define $\left(B^\ell(t), n^\ell(t)\right) := \{\left(B_s^\ell(t), n_s^\ell(t)\right) : s \in \mathcal{S}^\ell\}$, and similarly we can define the stochastic processes $n^\ell(t)$, $A^\ell(t)$, $p^\ell(t)$, and $E^\ell(t)$, associated with the Poisson-clock model. For $f^\ell := \{f_s^\ell : s \in \mathcal{S}^\ell\}$ and $g^\ell := \{g_s^\ell : s \in \mathcal{S}^\ell\}$, we define the discrete inner product and L^p norm over the lattice $\mathcal{S}^\ell$:

$$\left\langle f^\ell, g^\ell \right\rangle := \ell^2 \sum_{s \in \mathcal{S}^\ell} f_s^\ell g_s^\ell, \quad \left| f^\ell \right|_p := \left(\ell^2 \sum_{s \in \mathcal{S}^\ell} \left| f_s^\ell \right|^p \right)^{1/p}, \quad p \geq 1.$$

Let $\phi^\ell = \{\phi_s^\ell : s \in \mathcal{S}^\ell\}$ be an arbitrary stationary scalar field, we define

$$\left\langle \left(B^\ell(t), n^\ell(t)\right), \phi^\ell \right\rangle := \left(\left\langle B^\ell(t), \phi^\ell \right\rangle, \left\langle n^\ell(t), \phi^\ell \right\rangle \right). \tag{8}$$

$\left\langle \left(B^\ell(t), n^\ell(t)\right), \phi^\ell \right\rangle$ is a Markov pure jump process with state space $\mathbb{R}^2$. Hence a martingale approach is applicable (e.g. [14, 15, 44, 51, 79]). The martingale formulation of the process can be derived based on the infinitesimal parameters. The components of the infinitesimal mean vector and diagonal components of the infinitesimal covariance matrix of $\left\langle \left(B^\ell(t), n^\ell(t)\right), \phi^\ell \right\rangle$ are denoted as follows:

$$\begin{cases} \text{infinitesimal mean for attractiveness } \mathscr{G}_1^\ell \left(\left\langle \left(B^\ell(t), n^\ell(t)\right), \phi^\ell \right\rangle \right), \\ \text{infinitesimal mean for criminal distribution } \mathscr{G}_2^\ell \left(\left\langle \left(B^\ell(t), n^\ell(t)\right), \phi^\ell \right\rangle \right), \\ \text{infinitesimal variance for attractiveness } \mathscr{V}_1^\ell \left(\left\langle \left(B^\ell(t), n^\ell(t)\right), \phi^\ell \right\rangle \right), \\ \text{infinitesimal variance for criminal distribution } \mathscr{V}_2^\ell \left(\left\langle \left(B^\ell(t), n^\ell(t)\right), \phi^\ell \right\rangle \right). \end{cases}$$

Fig. 3 (continued) $\ell = 1$, $\omega = 1/15$, $\eta = 0.03$, $A0 = 1/30$, $n0_s^\ell \cong \bar{n}^\ell$, and BO^ℓ is set to be $\bar{B}$ except for sites with slight perturbations. Specific to the Poisson-clock model **(a)** and **(b)**, we set $D = 100$, $\Gamma = 0.0019q$, $\theta = 5.6/q$, $q = 1$ in **(a)**, and $q = 10$ in **(b)**. Specific to the DTS Model [76] **(c)** and **(d)**, we set $\delta t = 1/100$, $\Gamma = 0.0019q$, $\theta = 5.6/q$, $q = 1$ in **(c)**, and $q = 10$ in **(d)**. **(b, d)** Spatially homogeneous regimes. **(a, c)** Dynamic hotspot regimes

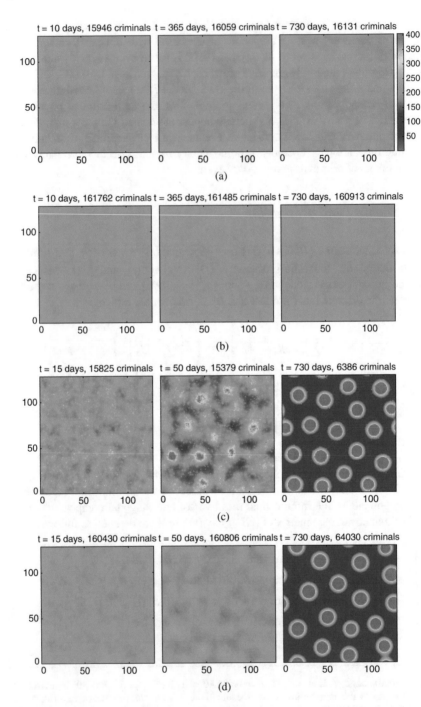

Fig. 4 Plot of the attractiveness $A_s^\ell(t)$ for the discrete Poisson-clock model. For both regimes, the initial conditions (at $t = 0$) and parameters are taken to be $A0 = 1/30$, $n0_s^\ell \cong \bar{n}^\ell$, $L = 128$,

Thus we have the following martingale formulation for $\left\langle \left(B^\ell(t), n^\ell(t) \right), \phi^\ell \right\rangle$:

Theorem 2.1 *For each fixed ℓ, before the possible blow-up time, the stochastic process $\left\langle \left(B^\ell(t), n^\ell(t) \right), \phi^\ell \right\rangle$ can be written as*

$$
\begin{cases}
\left\langle B^\ell(t), \phi^\ell \right\rangle = & \left\langle B0^\ell, \phi^\ell \right\rangle + \int_0^t \mathscr{G}_1^\ell \left(\left\langle \left(B^\ell(s), n^\ell(s) \right), \phi^\ell \right\rangle \right) ds \\
& + \mathscr{M}_1^\ell \left(\left\langle \left(B^\ell(t), n^\ell(t) \right), \phi^\ell \right\rangle \right), \\
\left\langle n^\ell(t), \phi^\ell \right\rangle = & \left\langle n0^\ell, \phi^\ell \right\rangle + \int_0^t \mathscr{G}_2^\ell \left(\left\langle \left(B^\ell(s), n^\ell(s) \right), \phi^\ell \right\rangle \right) ds \\
& + \mathscr{M}_2^\ell \left(\left\langle \left(B^\ell(t), n^\ell(t) \right), \phi^\ell \right\rangle \right).
\end{cases}
\tag{9}
$$

where $\mathscr{M}_i^\ell \left(\left\langle \left(B^\ell(t), n^\ell(t) \right), \phi^\ell \right\rangle \right)$, $i = 1, 2$, are martingales starting at $t = 0$ as zeros, whose variances can be characterized by the infinitesimal variances:

$$
Var \left(\mathscr{M}_i^\ell \left(\left\langle \left(B^\ell(t), n^\ell(t) \right), \phi^\ell \right\rangle \right) \right) = \int_0^t \mathbb{E} \left[\mathscr{V}_i^\ell \left(\left\langle \left(B^\ell(s), n^\ell(s) \right), \phi^\ell \right\rangle \right) \right] ds,
\tag{10}
$$

where $i = 1, 2$. Moreover for the infinitesimal means and variances we have

$$
\mathscr{G}_1^\ell \left(\left\langle \left(B^\ell(t), n^\ell(t) \right), \phi^\ell \right\rangle \right)
$$

$$
= \left\langle \left(1 - \frac{\omega \ell^2}{D} \right) \frac{\eta D}{4} \Delta^\ell B^\ell(t) - \omega B^\ell(t) + D\theta \ell^{-2} p^\ell(t) n^\ell(t), \phi^\ell \right\rangle,
\tag{11}
$$

$$
\mathscr{G}_2^\ell \left(\left\langle \left(B^\ell(t), n^\ell(t) \right), \phi^\ell \right\rangle \right)
$$

$$
= \ell^2 \sum_{s \in \mathscr{S}^\ell} \left[D\ell^{-2} \left(A_s^\ell(t) \sum_{s' \sim s} \frac{n_{s'}^\ell(t) \left(1 - p_{s'}^\ell(t) \right)}{T_{s'}^\ell(t)} - n_s^\ell(t) \right) + \Gamma \right] \phi_s^\ell,
\tag{12}
$$

$$
\mathscr{V}_1^\ell \left(\left\langle \left(B^\ell(t), n^\ell(t) \right), \phi^\ell \right\rangle \right)
$$

$$
= \frac{\ell^2}{D} \left(\mathscr{G}_1^\ell \left(\left\langle \left(B^\ell(t), n^\ell(t) \right), \phi^\ell \right\rangle \right) \right)^2
$$

Fig. 4 (continued) $\ell = 1$, $\omega = 1/15$, and $D = 100$. Specific to the spatially homogeneous regime (a) and (b), we set $B0^\ell \equiv \bar{B}$, $\eta = 0.2$, $\Gamma = 0.0019q$, $\theta = 5.6/q$, $q = 100$ in (a), and $q = 1000$ in (b). Specific to the dynamic hotspot regime (c) and (d), we set $\eta = 0.03$, $B0^\ell$ to be $\bar{B}$ except for sites with slight perturbations, $\Gamma = 0.0019q$, $\theta = 5.6/q$, $q = 100$ in (c), and $q = 1000$ in (d)

$$+ D\theta^2 \left\langle n^\ell(t) p^\ell(t)(1 - p^\ell(t)), \left(\phi^\ell\right)^2 \right\rangle, \tag{13}$$

$$\mathscr{V}_2^\ell \left(\left\langle \left(B^\ell(t), n^\ell(t)\right), \phi^\ell \right\rangle\right)$$

$$= \frac{\ell^2}{D} \left(\mathscr{G}_2^\ell \left(\left\langle \left(B^\ell(t), n^\ell(t)\right), \phi^\ell \right\rangle\right)\right)^2 + \ell^2 \Gamma \left(1 - \frac{\Gamma \ell^2}{D}\right) \left|\phi^\ell\right|_2^2$$

$$+ D \left\langle n^\ell(t) \left(1 - p^\ell(t)\right), \, p^\ell(t) f^\ell(t) + g^\ell(t) \right\rangle, \tag{14}$$

where

$$f_{\mathbf{s}}^\ell(t) = \left(\sum_{\substack{s' \\ s' \sim s}} \phi_{s'} \frac{A_{s'}^\ell(t)}{T_{\mathbf{s}}^\ell(t)}\right)^2, \quad g_{\mathbf{s}}^\ell(t) = \ell^2 \sum_{\substack{s' \\ s' \sim s}} \frac{A_{s'}^\ell(t)}{T_{\mathbf{s}}^\ell(t)} \left(\sum_{\substack{s'' \\ s'' \neq s' \\ s'' \sim s}} \nabla_{s' \to s''}^\ell \phi_{s'}^\ell \frac{A_{s''}^\ell(t)}{T_{\mathbf{s}}^\ell(t)}\right)^2. \tag{15}$$

Here $\nabla_{s' \to s''}^\ell \phi_{s'}^\ell$ denotes the discrete directional derivative from s' pointing towards s'', that is, $\nabla_{s' \to s''}^\ell \phi_{s'}^\ell = \left(\phi_{s'}^\ell - \phi_{s''}^\ell\right)/\ell$.

The proof of Theorem 2.1 is in the Appendix.

2.3 Continuum Model

The martingale formulation (9) implies that the process $\left\langle \left(B^\ell(t), n^\ell(t)\right), \phi^\ell \right\rangle$ is characterized by infinitesimal means and variances. The former gives the expected behavior and the latter the variance of the trajectories of the evolution. Thus the process can be viewed as the sum of differential equations and random fluctuations. From (13) and (14) we infer that the infinitesimal variances have a lower order of magnitude than ℓ, that is:

$$\mathscr{V}_1^\ell \left(\left\langle \left(B^\ell(t), n^\ell(t)\right), \phi^\ell \right\rangle\right)$$

$$\cong \frac{\ell^2}{D} \left(\mathscr{G}_1^\ell \left(\left\langle \left(B^\ell(t), n^\ell(t)\right), \phi^\ell \right\rangle\right)\right)^2 + \ell^2 \theta^2 \left\langle n^\ell(t) A^\ell(t), \left(\phi^\ell\right)^2 \right\rangle$$

$$\cong O(\ell^2), \tag{16}$$

$$\mathscr{V}_2^\ell \left(\left\langle \left(B^\ell(t), n^\ell(t)\right), \phi^\ell \right\rangle\right)$$

$$\cong \frac{\ell^2}{D} \left(\mathscr{G}_2^\ell \left(\left\langle \left(B^\ell(t), n^\ell(t) \right), \phi^\ell \right\rangle \right) \right)^2 + \ell^2 \Gamma \left| \phi^\ell \right|_2^2$$

$$+ D\ell^2 \left\langle n^\ell(t), \frac{1}{D} A^\ell(t) f^\ell(t) + g^\ell(t)\ell^{-2} \right\rangle$$

$$\cong O(\ell^2). \tag{17}$$

Thus for ℓ small, it is reasonable to set the continuum version of the differential equations as a continuum model. Let $n(\mathbf{x}, t)$, $A(\mathbf{x}, t)$, and $B(\mathbf{x}, t)$, $\mathbf{x} \in \mathscr{D}$ be the continuum versions of $n_s^\ell(t)$, $A_s^\ell(t)$, and $B_s^\ell(t)$, we have

$$\begin{cases} \dfrac{\partial B}{\partial t} = \dfrac{\eta D}{4} \Delta B - \omega B + \theta n A, \\[3mm] \dfrac{\partial n}{\partial t} = \dfrac{D}{4} \nabla \cdot \left(\nabla n - \dfrac{2n}{A} \nabla A \right) - nA + \Gamma, \\[3mm] n(0) = n0, \; B(0) = B0. \end{cases} \tag{18}$$

We note that (18) and the continuum DTS Model [76] ((3.2) and (3.5) in [76]) are the same. The derivation of (18) is similar to that of the hydrodynamic limit of interacting particle systems [19, 30, 44, 45, 69, 70, 79, 80].

We verify the validity of the continuum model through simulations. We use the same algorithm as the continuum DTS Model [76] ((3.11)–(3.13) in [76]). Figure 5 shows example output of the attractiveness $A(\mathbf{x}, t)$ in the cases of hotspot formation. The same color key is used as in Fig. 2. From (18) we infer that the parameters and data used to create Figs. 3a, b and 4c, d give rise to the same continuum attractiveness field. Hence we only display the output once here in Fig. 5. As for the case of zero hotspot formation, the parameters and data used to create

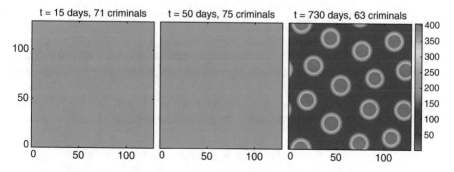

Fig. 5 Output of the attractiveness $A(\mathbf{x}, t)$ for the continuum Poisson-clock model with hotspot formation. The parameters and data are the equivalent of the discrete parameters used in Figs. 3 and 4c, d

Figs. 2a, b and 4a, b give rise to the same continuum attractiveness field, which is the equilibrium as the system stays at the equilibrium with equilibrium initial data.

As was observed in the DTS Model [76], the regimes of dynamic hotspots are absent in the continuum simulations. As the total criminal population decreases, distinctions between the behavioral regimes of the discrete and continuum simulations increase.

3 Mathematical Analysis of the *Finite Size Effects*

Hotspot transience is well documented in real crime statistics. It would be useful if we could guess the parameters of the model from observed pattern formation. The *finite size effects* are closely related to the degree of hotspot transience. Therefore, through the observation of the hotspot transience, we could estimate the number of criminals. This number is normally difficult to predict. To analyze the *finite size effects* quantitatively, we build a mathematical framework to quantify the relevant pattern formation. We propose a theory for the *finite size effects* based on a scaling property of the martingale formulation. Our conjecture is supported by qualitative and quantitative simulations.

3.1 *Quantification of the Pattern Formation*

3.1.1 Statistics of Degree of the Hotspot Transience

We construct the following statistics to measure the degree of hotspot transience.

(i) Relative Fisher information relative to the uniform measure over $\mathscr{S}^\ell$, logarithm mean type (see Appendix [11]):

$$\mathscr{I}^\ell(t) := \ell^{-2} \sum_{s \in \mathscr{S}} \left(A_s^\ell(t) - A_{s'}^\ell(t) \right) \left(\log A_s^\ell(t) - \log A_s^\ell(t) \right). \qquad (19)$$

As suggested in [20–22], the Fisher information is particularly suitable to measure the entropy of diffusion processes, which fits with the dynamics of the attractiveness.

(ii) Rate of change of $A^\ell(t)$ over time in the discrete L^p norm, $p \geq 1$. For a fixed time increment $\Delta t > 0$, we define the L^p-area rate of change as:

$$\delta_p^\ell(t) := \left| A^\ell(t + \Delta t) - A^\ell(t) \right|_p. \qquad (20)$$

(iii) Rate of change of the total area of certain types of regions. We focus on the regions with attractiveness higher than $2\bar{A}$ (red regions). For these regions, we define the relative overlapping area $\mathscr{O}^\ell(t)$ and non-overlapping area $\mathscr{N}^\ell(t)$ as follows:

$$\mathscr{O}^\ell(t) := \frac{1}{A_R^\ell} \sum_{\substack{s \in \mathscr{S}^\ell \\ A_s^\ell(t) \geq 2\bar{A} \\ A_s^\ell(t+\Delta t) \geq 2\bar{A}}} \mathbf{1}(s) \tag{21}$$

$$\mathscr{N}^\ell(t) := \frac{1}{A_R^\ell} \left[\sum_{\substack{s \in \mathscr{S}^\ell \\ A_s^\ell(t+\Delta t) \geq 2\bar{A}}} \mathbf{1}(s) + \sum_{\substack{s \in \mathscr{S}^\ell \\ A_s^\ell(t) \geq 2\bar{A}}} \mathbf{1}(s) \right] - 2\mathscr{O}^\ell(t), \tag{22}$$

where $\mathbf{1}(s)$ is an indicator function, and A_R^ℓ is the renormalization:

$$A_R^\ell = \sum_{\substack{s \in \mathscr{S}^\ell \\ A_s^\ell(t) \geq 2\bar{A}}} \mathbf{1}(s) \tag{23}$$

As for the continuum model, in a very similar way we can define all the above quantities, and we denote them as $\mathscr{I}(t)$, $\delta_p(t)$, $\mathscr{O}(t)$, and $\mathscr{N}(t)$.

3.1.2 Numerical Simulations

Example output of direct simulations for (19)–(22) can be seen in Figs. 6, 7, and 8. All the simulations are run with $\Delta t = 10$, $t \in [0, 730]$, and $p = 1$. Moreover, the blue, magenta, black, and red lines show results with increasing values of q and thus increasing criminal population. Figures 6a and 7a show results with zero hotspot formation, and the blue, magenta, black, and red lines represent results with the simulations in Figs. 2a, b and 4a, b, respectively. Figures 6b, 7b, and 8 show results with hotspot formation, and the blue, magenta, black, red, and green lines represent results with the simulations in Figs. 3a, b, 4c, d, and 5, respectively.

The simulation output shows that a larger degree of hotspot transience appears with a smaller criminal population. Also the continuum simulations have the lowest degree of hotspot transience. The peaks in Figs. 7b and 8a, b correspond to the initial emergence of the hotspots in the discrete and continuum models. During the emergence period the statistics increase as the hotspots form, and decrease and stabilize (or directly stabilize) as the hotspots stabilize. The same simulation results are also observed over other random paths. To conclude, the output matches well with the qualitative simulations in Figs. 2, 3, 4, and 5, which suggests that the above statistics are suitably chosen.

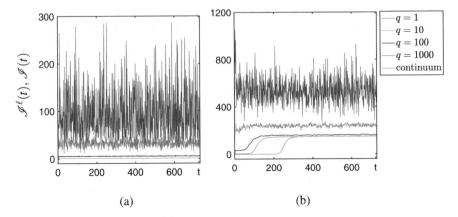

Fig. 6 Examples of the relative Fisher information $\mathscr{I}^{\ell}(t)$ and $\mathscr{I}(t)$ in the cases of zero hotspot formation and hotspot formation. (**a**) Results with the cases of zero hotspot formation, and the blue, magenta, black, and red lines represent the statistics of the discrete models plotted in Figs. 2a, b and 4a, b, respectively. (**b**) Results with the cases of hotspot formation, and the blue, magenta, black, red, and green lines show results with the simulations in Figs. 3a, b, 4c, d, and 5, respectively

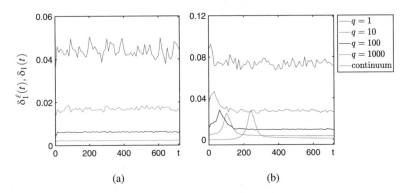

Fig. 7 Examples of the L^1-area rate of change $\delta_1^{\ell}(t)$ and $\delta_1(t)$ in the cases of zero hotspot formation and hotspot formation. (**a**) Results with zero hotspot formation, and the blue, magenta, black, and red lines represent the statistics of the discrete models plotted in Figs. 2a, b and 4a, b, respectively. (**b**) Results with hotspot formation, and the blue, magenta, black, red, and green lines show results with the simulations in Figs. 3a, b, 4c, d, and 5, respectively

3.2   Theory of the Finite Size Effects

With quantification of the degree of hotspot transience, we analyze the *finite size effects* mathematically.

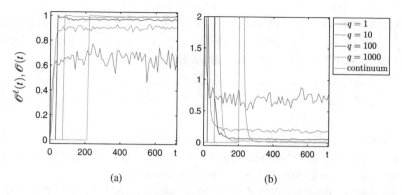

Fig. 8 Examples of the relative overlapping and non-overlapping area, $\mathscr{O}^\ell(t)$, $\mathscr{O}(t)$, $\mathscr{N}^\ell(t)$, and $\mathscr{N}(t)$ with hotspot formation. The blue, magenta, black, red, and green lines show results with the simulations in Figs. 3a, b, 4c, d, and 5, respectively

3.2.1 Scaling Property of the Stochastic Component

We study a scaling property related to the total criminal population. Specifically we vary the initial criminal number, the repeat victimization strength θ, and the replacement rate Γ. Fixing $\Theta > 0$, $\mathfrak{b} > 0$, for $q \in (0, D/\ell^2\mathfrak{b})$, we consider the discrete model with parameters and initial data scaled in a certain way:

$$\left(B^{\ell,(q)}(t), n^{\ell,(q)}(t)\right) := \left(B^\ell(t), n^\ell(t)\right)\Big|_{B0^\ell\cong\bar{\mathfrak{B}},\ n0^\ell\cong q\bar{n}^\ell,\ \theta=\frac{\Theta}{q},\ \Gamma=q\mathfrak{b}}, \qquad (24)$$

where $\bar{\mathfrak{B}}$, $\bar{n}^\ell$ are the homogeneous equilibrium values as in (7) with $\theta = \Theta$ and $\Gamma = \mathfrak{b}$:

$$\bar{\mathfrak{B}} = \frac{\Theta\mathfrak{b}}{\omega}, \quad \bar{n}^\ell = \frac{\mathfrak{b}\ell^2}{D\left(1 - e^{-\ell^2(\bar{\mathfrak{B}}+A0)D^{-1}}\right)} \cong \frac{\mathfrak{b}}{\bar{\mathfrak{B}} + A0}. \qquad (25)$$

As q increases, the initial criminal population and the criminal replacement rate both increase while the initial attractiveness field remains fixed. Applying (9) and (10) to $\left(B^{\ell,(q)}(t), n^{\ell,(q)}(t)\right)$ over a small time step δt we obtain

$$\begin{cases} \left\langle B^{\ell,(q)}(t+\delta t), \phi^\ell\right\rangle = \left\langle B^{\ell,(q)}(t), \phi^\ell\right\rangle + \mathscr{G}_1^{\ell,(q)}\left(t, \phi^\ell\right)\delta t + \mathscr{M}_1^{\ell,(q)}\left(t, \phi^\ell\right), \\[2mm] \left\langle n^{\ell,(q)}(t+\delta t), \phi^\ell\right\rangle = \left\langle n^{\ell,(q)}(t), \phi^\ell\right\rangle + \mathscr{G}_2^{\ell,(q)}\left(t, \phi^\ell\right)\delta t + \mathscr{M}_2^{\ell,(q)}\left(t, \phi^\ell\right), \end{cases}$$

$$(26)$$

where $\mathscr{G}_i^{\ell,(q)}(t,\phi^\ell)$ are the short notations for $\mathscr{G}_i^\ell\left(\left(B^{\ell,(q)}(t),n^{\ell,(q)}(t)\right),\phi^\ell\right)$, and $\mathscr{M}_i^{\ell,(q)}\left(t,\phi^\ell\right)$ for $\mathscr{M}_i^\ell\left(\left(B^{\ell,(q)}(t),n^{\ell,(q)}(t)\right),\phi^\ell\right)$, $i=1,2$. By (10) we obtain the standard deviation of the stochastic component

$$\sqrt{\mathrm{Var}\left(\mathscr{M}_i^{\ell,(q)}\left(t,\phi^\ell\right)\right)} \cong \sqrt{\mathbb{E}\left[\mathscr{V}_i^{\ell,(q)}\left(t,\phi^\ell\right)\right]\delta t}, \quad i=1,2, \tag{27}$$

where $\mathscr{V}_i^{\ell,(q)}(t,\phi^\ell)$'s are the short nations for $\mathscr{V}_i^\ell\left(\left(B^{\ell,(q)}(t),n^{\ell,(q)}(t)\right),\phi^\ell\right)$, $i=1,2$. This implies that the infinitesimal variances are the key to estimate deviation of the discrete model from the deterministic component.

We analyze the infinitesimal variance for attractiveness. By (11) and (13) we rewrite the infinitesimal variance as the summation of two components

$$\mathscr{V}_1^{\ell,(q)}\left(t,\phi^\ell\right) = \mathscr{V}_{1,1}^{\ell,(q)}\left(t,\phi^\ell\right) + \frac{1}{q}\mathscr{V}_{1,2}^{\ell,(q)}\left(t,\phi^\ell\right), \tag{28}$$

where

$$\mathscr{V}_{1,1}^{\ell,(q)}\left(t,\phi^\ell\right)$$
$$\cong \frac{\ell^2}{D}\left(\left\langle\left|\frac{\eta D}{4}\Delta^\ell B^{\ell,(q)}(t) - \omega B^{\ell,(q)}(t) + \frac{\Theta}{q}\left(B^{\ell,(q)}(t) + A0\right)n^{\ell,(q)}(t),\phi^\ell\right\rangle\right)^2, \tag{29}$$

and

$$\mathscr{V}_{1,2}^{\ell,(q)}\left(t,\phi^\ell\right) \cong q\ell^2\Theta^2\left\langle n^{\ell,(q)}(t)\left(B^{\ell,(q)}(t) + A0\right),\left(\phi^\ell\right)^2\right\rangle. \tag{30}$$

We perform estimates at the first time step. At time zero, from (29), (30), and (25) we infer that the two components are actually independent of q, namely

$$\mathscr{V}_{1,1}^{\ell,(q)}\left(0,\phi^\ell\right) = \frac{\ell^2}{D}\left(\left\langle -\omega\bar{\mathfrak{B}} + \Theta\left(\bar{\mathfrak{B}} + A0\right)\bar{\mathfrak{n}}^\ell,\phi^\ell\right\rangle\right)^2 = 0, \tag{31}$$

and

$$\mathscr{V}_{1,2}^{\ell,(q)}\left(0,\phi^\ell\right) = \ell^2\Theta^2\left\langle\bar{\mathfrak{n}}^\ell\left(\bar{\mathfrak{B}} + A0\right),\left(\phi^\ell\right)^2\right\rangle = \ell^2\Theta^2\mathfrak{b}\left|\phi^\ell\right|_2^2. \tag{32}$$

This together with (28) implies that at time zero we have

$$\mathscr{V}_1^{\ell,(q)}\left(0,\phi^\ell\right) \equiv q^{-1}\ell^2\Theta^2\mathfrak{b}\left|\phi^\ell\right|_2^2. \tag{33}$$

Thus we infer that the infinitesimal variance for the attractiveness is inversely proportional to q:

$$\mathscr{V}_1^{\ell,(q)}\left(0,\phi^\ell\right) \propto q^{-1}. \tag{34}$$

This implies that for $0 < q < \tilde{q} < D\mathfrak{b}/\ell^2$, we have

$$\mathscr{V}_1^{\ell,(q)}\left(0,\phi^\ell\right) > \mathscr{V}_1^{\ell,(\tilde{q})}\left(0,\phi^\ell\right). \tag{35}$$

This together with (27) implies that at the first time step we have

$$\text{Var}\left(\mathscr{M}_1^{\ell,(q)}\left(\delta t,\phi^\ell\right)\right) > \text{Var}\left(\mathscr{M}_1^{\ell,(\tilde{q})}\left(\delta t,\phi^\ell\right)\right). \tag{36}$$

To conclude at the first time step a smaller value of q leads to a larger deviation of $B^{\ell,(q)}(t)$ from the deterministic component, and hence the hotspots develop temporal transience in the discrete attractiveness simulations. We conjecture that (35) remains to be true in later times, namely

$$\mathscr{V}_1^{\ell,(q)}\left(t,\phi^\ell\right) > \mathscr{V}_1^{\ell,(\tilde{q})}\left(t,\phi^\ell\right), \text{ for } 0 < q < \tilde{q} < \frac{D\mathfrak{b}}{\ell^2} \text{ and } t > 0. \tag{37}$$

To conclude, the scaling property of the martingale formulation related to the criminal population possibly leads to the *finite size effects*.

3.2.2 Numerical Simulations

We perform direct simulations of the infinitesimal standard deviation for the attractiveness, $\sqrt{\mathscr{V}_1^{\ell,(q)}(t)}$, to check the validity of (37). Example output can be seen in Fig. 9. The test function is chosen to be

$$\phi^\ell(\mathbf{x}) = 1 + sin(\mathbf{x}_1)sin(\mathbf{x}_2)/20. \tag{38}$$

Figure 9a shows results in the cases with no hotspot. The blue, magenta, black, and red lines show results with the simulations in Figs. 2a, b and 4a, b, respectively. Figure 9b shows results with hotspot formation. The blue, magenta, black, and red lines represent results with the simulations in Figs. 3a, b and 4c, d, respectively. The same simulation results are also observed over other random paths. The output of the simulations supports our previous conjecture as in (37).

Furthermore, in Fig. 10 we check whether (34) is valid for later times:

$$\mathscr{V}_1^{\ell,(q)}\left(t,\phi^\ell\right) \cong \frac{1}{q}\ell^2\Theta^2\mathfrak{b}\left|\phi^\ell\right|_2^2, \quad t > 0. \tag{39}$$

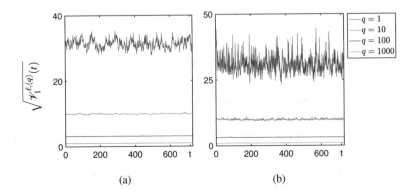

Fig. 9 Examples of the infinitesimal standard deviation for attractiveness, $\sqrt{\mathcal{V}_1^{\ell,(q)}(t)}$, for both zero hotspot formation and hotspot formation. (**a**) Results with no hotspot, and the blue, magenta, black, and red lines show results with the simulations in Figs. 2a, b and 4a, b, respectively. (**b**) Results with hotspot formation, and the blue, magenta, black, and red lines show results with the simulations in Figs. 3a, b and 4c, d, respectively

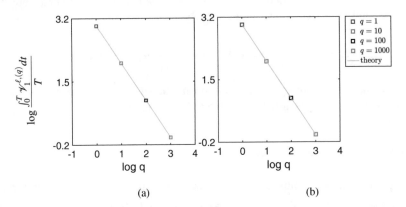

Fig. 10 Comparison of log–log plot of the theoretical and true scaling for both zero hotspot formation and hotspot formation. The straight lines show the theoretical scaling with slope -1 and the x-intercept $\ell^2\Theta^2\mathfrak{b}\,|\phi^\ell|_2^2 \cong 2.9892$. The points show the true scaling. (**a**) Results with no hotspot, and the points with x-axis as 0, 1, 2, and 3 show results with the simulations of the blue, magenta, black, and red lines in Fig. 9a, respectively. (**b**) Results with hotspot formation, and the points with x-axis as 0, 1, 2, and 3 show results with the simulations of the blue, magenta, black, and red lines in Fig. 9b, respectively

Taking the average of both sides of (39) over the time period $[0, T]$, we obtain

$$\frac{1}{T}\int_0^T \mathcal{V}_1^{\ell,(q)}\left(t, \phi^\ell\right) dt \cong \frac{1}{q}\ell^2\Theta^2\mathfrak{b}\,\left|\phi^\ell\right|_2^2. \tag{40}$$

Figure 10 shows the log–log plot of (40). The lines show the theoretical scaling with slope -1 and the x-intercept $\ell^2\Theta^2\mathfrak{b}\,|\phi^\ell|_2^2$. The points show the true scaling with the coordinates as:

$$x = \log q, \quad y = \log\left(\frac{1}{T}\int_0^T \mathscr{V}_1^{\ell,(q)}\left(t, \phi^\ell\right) dt\right), \quad q = 1, 10, 100, 1000.$$
(41)

The output shows that all the points fall onto the straight lines. This indicates stability of the Poisson-clock model with equilibrium initial value, which further indicates validity of the model.

4 Conclusion

In this paper, we apply the Poisson clock to the agent-based residential burglary DTS Model [76]. The time increments are exponentially distributed random variables, which are more suitable to model random arrivals. This is a Markov pure jump process, and a martingale approach is applicable. A martingale formulation is derived which consists of a deterministic and a stochastic component. It provides us with a tool to study both the statistical and stochastic features of the process. The model thus obtained is a stochastic-statistical residential burglary model (SSRB Model).

The deterministic part yields a continuum model. It is the same as the continuum version of the DTS Model [76]. We find that the continuum SSRB Model is a good approximation of the discrete SSRB Model under the circumstance of a large number of criminals. Moreover, for the *finite size effects*, we build statistics to quantify the relevant pattern formation, and find a theoretical explanation using a scaling property of the stochastic component. That is, as the criminal population decreases, the stochastic component increases in size, which leads to a larger deviation of the discrete model from the deterministic component. This scaling property can be proven at time zero with equilibrium initial data. Numerical simulations support our conjecture that the scaling property remains to be true in later times. This explains theoretically the reason why dynamic hotspots have been observed associated with small criminal population in the discrete simulations of the SSRB Model and the DTS Model [76]. And more, for general human behavior with similar aggregation pattern formation, our finding suggests that hotspot transience indicates a significant stochastic fluctuation in the martingale formulation, which could predict a small size of agents taking part in the activity.

There are two possible directions for the future work. On the one hand, we can assume independent Poisson clocks for each agent, which is more realistic to model typical criminal activities. Independent Poisson clocks will make the computations more complex though. It would be interesting to explore whether the *finite size effects* also occur. On the other hand, we can study parameterization of hotspot transience. This will give us a deeper understanding of the pattern formation in a complex system, e.g., pattern formation in fluid turbulence. This topic has been drawing a lot of interests recently [31, 32, 35, 38]. But quantitative studies are lacking as far as we know. The quantitative framework that we develop here could be applicable.

Acknowledgements We would like to thank the helpful discussions with Prof. A. Debussche, Prof. Andrea Montanari, Prof. Thomas Liggett, Prof. Carl Mueller, Prof. Wotao Yin, Mac Jugal Nankep Nguepedja, Da Kuang, Yifan Chen, Fangbo Zhang, Yu Gu, Jingyu Huang, Yatin Chow, Wuchen Li, and Kenneth Van. A. Bertozzi is supported by NSF grant DMS-1737770 and M. Short is supported by NSF grant DMS-1737925.

Appendix

To prove Theorem 2.1, we compute the infinitesimal means and variances of $\langle (B^\ell(t), n^\ell(t)), \phi^\ell \rangle$ for fixed ℓ [1, 12, 16, 34, 36, 43, 47, 58, 59, 66, 68].

We first study related random variables. We assume that the Poisson-clock advances at t^-, and analyze the transition at time t. Conditioned on $(B^\ell(\mathbf{s}, t^-), n^\ell(\mathbf{s}, t^-))$, we observe that $E_\mathbf{s}^\ell(t)$, $\mathbf{s} \in \mathscr{S}$, is a family of independently identically distributed Binomial random variables with the parameters $n^\ell(\mathbf{s}, t^-)$ and $p^\ell(\mathbf{s}, t^-)$, that is

$$\mathbb{P}\left(E_\mathbf{s}^\ell(t) = i \middle| B_\mathbf{s}^\ell(t^-), n_\mathbf{s}^\ell(t^-)\right) = \binom{n_\mathbf{s}^\ell(t^-)}{i} \left(p_\mathbf{s}^\ell(t^-)\right)^i (1 - p_\mathbf{s}^\ell(t^-))^{n_\mathbf{s}^\ell(t^-)-i}, \tag{42}$$

where $i = 0, 1, 2, \ldots, n_\mathbf{s}^\ell(t^-)$. Hence we have

$$\mathbb{E}\left[E_\mathbf{s}^\ell(t) \middle| B_\mathbf{s}^\ell(t^-), n_\mathbf{s}^\ell(t^-)\right] = p_\mathbf{s}^\ell(t^-)n_\mathbf{s}^\ell(t^-), \tag{43}$$

$$\mathrm{Var}\left(E_\mathbf{s}^\ell(t) \middle| \left(B_\mathbf{s}^\ell(t^-), n_\mathbf{s}^\ell(t^-)\right), \forall \mathbf{s} \in \mathscr{S}\right) = n_\mathbf{s}^\ell(t^-)p_\mathbf{s}^\ell(t^-)\left[1 - p_\mathbf{s}^\ell(t^-)\right]. \tag{44}$$

Let $J_{\mathbf{s},j}^\ell(t)$, $j = 1, 2, \ldots, n_\mathbf{s}^\ell(t^-)$ be a family of independently distributed Bernoulli random variables assuming 1 with probability $1 - p_\mathbf{s}^\ell(t^-)$. If the j-th agent at site $\mathbf{s}$ chooses not to burglarize, then $J_{\mathbf{s},j}^\ell(t)$ assumes 1. This implies that

$$\mathbb{E}\left[\left(J_{\mathbf{s},j}^\ell(t)\right)^2 \middle| \left(B_\mathbf{s}^\ell(t^-), n_\mathbf{s}^\ell(t^-)\right)\right] = \mathbb{E}\left[J_{\mathbf{s},j}^\ell(t) \middle| \left(B_\mathbf{s}^\ell(t^-), n_\mathbf{s}^\ell(t^-)\right)\right] = 1 - p_\mathbf{s}^\ell(t^-). \tag{45}$$

We also note that by our construction $E_\mathbf{s}^\ell(t) = \sum_{j=1}^{j=n_\mathbf{s}^\ell(t^-)}\left(1 - J_{\mathbf{s},j}^\ell(t)\right)$.

Let $\Phi_{\mathbf{s},j}^\ell(t)$, $j = 1, \ldots, n_\mathbf{s}^\ell(t^-)$ be a family of independently identically distributed random variables assuming $\phi_{\mathbf{s}'}$ with probability $A_{\mathbf{s}'}^\ell / T_\mathbf{s}^\ell$, for every $\mathbf{s}'$, $\mathbf{s}' \sim \mathbf{s}$. Then we have

$$\mathbb{E}\left[\Phi_{\mathbf{s},j}^\ell(t) \middle| \left(B_\mathbf{s}^\ell(t^-), n_\mathbf{s}^\ell(t^-)\right), \forall \mathbf{s} \in \mathscr{S}^\ell\right] = \sum_{\substack{\mathbf{s}' \\ \mathbf{s}' \sim \mathbf{s}}} \phi_{\mathbf{s}'}^\ell \frac{A_{\mathbf{s}'}^\ell(t^-)}{T_\mathbf{s}^\ell(t^-)}, \tag{46}$$

$$\mathbb{E}\left[\left(\Phi_{\mathbf{s},j}^\ell(t)\right)^2 \middle| \left(B_\mathbf{s}^\ell(t^-), n_\mathbf{s}^\ell(t^-)\right), \forall \mathbf{s} \in \mathscr{S}^\ell\right] = \sum_{\substack{\mathbf{s}' \\ \mathbf{s}' \sim \mathbf{s}}} \left(\phi_{\mathbf{s}'}^\ell\right)^2 \frac{A_{\mathbf{s}'}^\ell(t^-)}{T_\mathbf{s}^\ell(t^-)}. \tag{47}$$

Let the number of replaced criminals on site $\mathbf{s}$ at time t be $\xi_\mathbf{s}^\ell(t)$. Then $\xi_\mathbf{s}^\ell(t)$ is a family of Bernoulli random variables assuming 1 with probability $\Gamma\ell^2/D$. Hence we have

$$\mathbb{E}\left[\xi_\mathbf{s}^\ell(t)\Big|\left(B_\mathbf{s}^\ell(t^-), n_\mathbf{s}^\ell(t^-)\right)\right] = \frac{\Gamma\ell^2}{D}, \tag{48}$$

$$\text{Var}\left(\left(\xi_\mathbf{s}^\ell(t)\right)^2\Big|\left(B_\mathbf{s}^\ell(t^-), n_\mathbf{s}^\ell(t^-)\right)\right) = \frac{\Gamma\ell^2}{D}\left(1 - \frac{\Gamma\ell^2}{D}\right). \tag{49}$$

Because the decision to burglarize is (conditionally) independent from the decision to move for each of the burglars, $J_{\mathbf{s},j}^\ell(t)$ and $\Phi_{\mathbf{k},h}^\ell(t)$ are (conditionally) independent for any choices of $\mathbf{s}$, $\mathbf{k}$, j, and h. This implies that

$$\mathbb{E}\left[J_{\mathbf{s},j}^\ell(t)\Phi_{\mathbf{s},j}^\ell(t)\Big|\left(B_\mathbf{s}^\ell(t^-), n_\mathbf{s}^\ell(t^-)\right), \forall \mathbf{s} \in \mathscr{S}^\ell\right]$$

$$= \mathbb{E}\left[J_{\mathbf{s},j}^\ell(t)\Big|\left(B_\mathbf{s}^\ell(t^-), n_\mathbf{s}^\ell(t^-)\right), \forall \mathbf{s} \in \mathscr{S}^\ell\right]\mathbb{E}\left[\Phi_{\mathbf{s},j}^\ell(t)\Big|\left(B_\mathbf{s}^\ell(t^-), n_\mathbf{s}^\ell(t^-)\right), \forall \mathbf{s} \in \mathscr{S}^\ell\right]$$

$$= \text{(by (45) and (46))}$$

$$= \left[1 - p_\mathbf{s}^\ell(t^-)\right]\sum_{\substack{\mathbf{s}' \\ \mathbf{s}' \sim \mathbf{s}}} \phi_{\mathbf{s}'}^\ell \frac{A_{\mathbf{s}'}^\ell(t^-)}{T_\mathbf{s}^\ell(t^-)}, \tag{50}$$

$$\text{Var}\left(J_{\mathbf{s},j}^\ell(t)\Phi_{\mathbf{s},j}^\ell(t)\Big|\left(B_\mathbf{s}^\ell(t^-), n_\mathbf{s}^\ell(t^-)\right), \forall \mathbf{s} \in \mathscr{S}^\ell\right)$$

$$= \mathbb{E}\left[\left(J_{\mathbf{s},j}^\ell(t)\Phi_{\mathbf{s},j}^\ell(t)\right)^2\Big|\left(B_\mathbf{s}^\ell(t^-), n_\mathbf{s}^\ell(t^-)\right), \forall \mathbf{s} \in \mathscr{S}^\ell\right]$$

$$- \left[\mathbb{E}\left[J_{\mathbf{s},j}^\ell(t)\Phi_{\mathbf{s},j}^\ell(t)\Big|\left(B_\mathbf{s}^\ell(t^-), n_\mathbf{s}^\ell(t^-)\right), \forall \mathbf{s} \in \mathscr{S}^\ell\right]\right]^2$$

$$= \text{(by (45), (46), (47), and (45))}$$

$$= \left[1 - p_\mathbf{s}^\ell(t^-)\right]\sum_{\substack{\mathbf{s}' \\ \mathbf{s}' \sim \mathbf{s}}} \left(\phi_{\mathbf{s}'}^\ell\right)^2 \frac{A_{\mathbf{s}'}^\ell(t^-)}{T_\mathbf{s}^\ell(t^-)} - \left[1 - p_\mathbf{s}^\ell(t^-)\right]^2 \left[\sum_{\substack{\mathbf{s}' \\ \mathbf{s}' \sim \mathbf{s}}} \phi_{\mathbf{s}'}^\ell \frac{A_{\mathbf{s}'}^\ell(t^-)}{T_\mathbf{s}^\ell(t^-)}\right]^2, \tag{51}$$

Right after the Poisson clock advances we have the following transition:

$$\sum_{\mathbf{s} \in \mathscr{S}^\ell} n_\mathbf{s}^\ell(t)\phi_\mathbf{s}^\ell = \sum_{\mathbf{s} \in \mathscr{S}^\ell} \sum_{j=1}^{n_\mathbf{s}^\ell(t^-)} \left[1 - J_{\mathbf{s},j}^\ell(t)\right]\Phi_{\mathbf{s},j}^\ell(t) + \sum_{\mathbf{s} \in \mathscr{S}^\ell} \xi_\mathbf{s}^\ell(t)\phi_\mathbf{s}^\ell. \tag{52}$$

With the above random variables we compute the infinitesimal means and variances. In the computational steps we will drop the superscript ℓ for simplicity.

We compute the infinitesimal mean for $\langle B^\ell(t^-), \phi^\ell \rangle$. From (6) we have

$$
\mathcal{G}_1 \left(\langle B^\ell(t^-), \phi^\ell \rangle, \langle n^\ell(t^-), \phi^\ell \rangle \right)
$$

$$
= \frac{D}{\ell^2} \mathbb{E} \left[\ell^2 \sum_{s \in \mathscr{S}} \left[B_s(t) - B_s(t^-) \right] \phi_s \middle| (B(t^-), n(t^-)) \right]
$$

$$
= \text{by (43)}
$$

$$
= D \sum_{s \in \mathscr{S}} \left[\left(1 - \frac{\omega \ell^2}{D} \right) \frac{\eta \ell^2}{4} \Delta B_s(t^-) - \frac{\omega \ell^2}{D} B_s(t^-) + \theta p_s(t^-) n_s(t^-) \right] \phi_s,
$$

(53)

which implies (11).

We compute the infinitesimal variance of $\langle B^\ell(t^-), \phi^\ell \rangle$:

$$
\mathscr{V}_1^\ell \left(\left(\left(B^\ell(t), n^\ell(t) \right), \phi^\ell \right) \right)
$$

$$
= \lim_{\delta t \to 0} \frac{1}{\delta t} \mathbb{E} \left[\left(\langle B(\delta t + t^-), \phi \rangle - \langle B(t^-), \phi \rangle \right)^2 \middle| (B(t^-), n(t^-)) \right]
$$

$$
= \frac{D}{\ell^2} \mathbb{E} \left[\ell^4 \left(\sum_{s \in \mathscr{S}} B_s(t^-) \phi_s - \sum_{s \in \mathscr{S}} B_s(t) \phi_s \right)^2 \middle| (B(t^-), n(t^-)) \right]
$$

$$
= D\ell^2 \mathbb{E} \left[\sum_{s \in \mathscr{S}} B_s(t^-) \phi_s - \sum_{s \in \mathscr{S}} B_s(t) \phi_s \middle| (B(t^-), n(t^-)) \right]^2
$$

$$
+ D\ell^2 \text{Var} \left[\sum_{s \in \mathscr{S}} B_s(t^-) \phi_s - \sum_{s \in \mathscr{S}} B_s(t) \phi_s \middle| (B(t^-), n(t^-)) \right]
$$

$$
:= J_1 + J_2.
$$

(54)

For J_1, from (53) we have

$$
J_1 = \frac{\ell^2}{D} \mathcal{G}_1^2 \left(\langle B(t^-), \phi \rangle, \langle n(t^-), \phi \rangle \right).
$$

(55)

Then for J_2, we apply the independence of $E_s(t)$ for distinct $s \in \mathscr{S}$ and with (44) we obtain

$$J_2 = D\ell^2 \text{Var} \left(\sum_{s \in \mathscr{S}} \left[\left(1 - \frac{\omega \ell^2}{D} \right) \frac{\eta \ell^2}{4} \Delta B_s(t^-) - \frac{\omega \ell^2}{D} B_s(t^-) + \theta E_s(t) \right] \phi_s \middle| (B(t^-), n(t^-)) \right)$$

$$= D\ell^2 \text{Var} \left(\sum_{s \in \mathscr{S}} \theta \phi_s E_s(t) \middle| (B(t^-), n(t^-)) \right)$$

$$= D\ell^2 \sum_{s \in \mathscr{S}} \theta^2 \phi_s^2 \text{Var} \left(E_s(t) \middle| (B(t^-), n(t^-)) \right)$$

$$= D\ell^2 \sum_{s \in \mathscr{S}} \theta^2 \phi_s^2 n_s(t^-) p_s(t^-) \left[1 - p_s(t^-) \right]. \tag{56}$$

This together with (55) and (54) implies (13).

We compute the infinitesimal mean for $\langle n^\ell(t^-), \phi^\ell \rangle$. From (52) we have

$$\mathscr{G}_2 \left(\left\langle \left(B^\ell(t^-), n^\ell(t^-) \right), \phi^\ell \right\rangle \right)$$

$$= \frac{D}{\ell^2} \mathbb{E} \left[\ell^2 \sum_{s \in \mathscr{S}} n_s(t) \phi_s - \ell^2 \sum_{s \in \mathscr{S}} n_s(t) \phi_s \middle| (B(t^-), n(t^-)) \right]$$

$$= D \sum_{s \in \mathscr{S}} \mathbb{E} \left[\sum_{j=1}^{n_s(t^-)} \left[1 - J_{s,j}(t) \right] \Phi_{s,j}(t) + \xi_s(t) \phi_s - n_s(t^-) \phi_s \middle| (B(t^-), n(t^-)) \right]. \tag{57}$$

This together with (45), (46), and (48) implies

$$\mathscr{G}_2 \left(\left\langle \left(B^\ell(t^-), n^\ell(t^-) \right), \phi^\ell \right\rangle \right)$$

$$= D \sum_{s \in \mathscr{S}} \left[n_s(t^-) \left[1 - p_s(t) \right] \sum_{\substack{s' \\ s' \sim s}} \phi_{s'} \frac{A_{s'}(t^-)}{T_s(t^-)} + \frac{\Gamma \ell^2}{D} \phi_s - n_s(t^-) \phi_s \right]$$

$$= D \sum_{s \in \mathscr{S}} \left[\phi_s A_s(t^-) \sum_{\substack{s' \\ s' \sim s}} \frac{n_{s'}(t^-) \left[1 - p_{s'}(t) \right]}{T_{s'}(t^-)} + \frac{\Gamma \ell^2}{D} \phi_s - n_s(t^-) \phi_s \right], \tag{58}$$

which implies (12).

We compute the infinitesimal variance of $\langle n^\ell(t^-), \phi^\ell \rangle$

$$\mathscr{V}_2^\ell \left(\left\langle \left(B^\ell(t), n^\ell(t) \right), \phi^\ell \right\rangle \right)$$

$$= \lim_{\delta t \to 0} \frac{1}{\delta t} \mathbb{E}\left[\left(\langle n(\delta t + t^-), \phi\rangle - \langle n(t^-), \phi\rangle\right)^2 \middle| (B(t^-), n(t^-))\right]$$

$$= D\ell^2 \mathbb{E}\left[\sum_{s \in \mathscr{S}} n_s(t^-)\phi_s - \sum_{s \in \mathscr{S}} n_s(t)\phi_s \middle| (B(t^-), n(t^-))\right]^2$$

$$+ D\ell^2 \text{Var}\left(\sum_{s \in \mathscr{S}} n_s(t^-)\phi_s - \sum_{s \in \mathscr{S}} n_s(t)\phi_s \middle| (B(t^-), n(t^-))\right)$$

$$:= J_3 + J_4. \tag{59}$$

For J_3 we have

$$J_3 = \frac{\ell^2}{D}\mathscr{G}_2^2\left(\langle B(t^-), \phi\rangle, \langle n(t^-), \phi\rangle\right). \tag{60}$$

For J_4, with the independence of the related random variables, we obtain

$$J_4 = D\ell^2 \sum_{s \in \mathscr{S}} \text{Var}\left(\sum_{j=1}^{n_s(t^-)} J_{s,j}\Phi_{s,j}(t) + \xi_s(t)\phi_s - n_s(t^-)\phi_s \middle| (B(t^-), n(t^-))\right)$$

$$= D\ell^2 \sum_{s \in \mathscr{S}} \text{Var}\left(\xi_s(t)\phi_s \middle| (B(t^-), n(t^-))\right)$$

$$+ D\ell^2 \sum_{s \in \mathscr{S}} n_s(t^-)\text{Var}\left(J_{s,j}(t)\Phi_{s,j}(t) \middle| (B(t^-), n(t^-))\right)$$

$$:= J_{4,1} + J_{4,2}. \tag{61}$$

For $J_{4,1}$, by (49) we have

$$J_{4,1} = \ell^4 \Gamma\left(1 - \frac{\Gamma\ell^2}{D}\right)\sum_{s \in \mathscr{S}} \phi_s^2. \tag{62}$$

For $J_{4,2}$, by (50) and (51) we have

$$J_{4,2} = D\ell^2 \sum_{s \in \mathscr{S}} n_s(t^-)p_s(t^-)\left[1 - p_s(t^-)\right]\left[\sum_{\substack{s' \\ s' \sim s}} \phi_{s'}\frac{A_{s'}(t^-)}{T_s(t^-)}\right]^2$$

$$+ D\ell^2 \sum_{s \in \mathscr{S}} n_s(t^-) \left[1 - p_s(t^-)\right] \left[\sum_{\substack{s' \\ s' \sim s}} \phi_{s'}^2 \frac{A_{s'}(t^-)}{T_s(t^-)} - \left(\sum_{\substack{s' \\ s' \sim s}} \phi_{s'} \frac{A_{s'}(t^-)}{T_s(t^-)} \right)^2 \right]$$

$$:= J_{4,1,1} + J_{4,1,2}. \tag{63}$$

We simplify $J_{4,1,2}$ as follows

$$J_{4,1,2} = D\ell^2 \sum_{s \in \mathscr{S}} n_s(t^-) \left[1 - p_s(t^-))\right] \sum_{\substack{s' \\ s' \sim s}} \frac{A_{s'}(t^-)}{T_s(t^-)} \left[\phi_{s'} - \sum_{\substack{s' \\ s' \sim s}} \phi_{s'} \frac{A_{s'}(t^-)}{T_s(t^-)} \right]^2$$

$$= D\ell^2 \sum_{s \in \mathscr{S}} n_s(t^-) \left[1 - p_s(t^-)\right] \sum_{\substack{s' \\ s' \sim s}} \frac{A_{s'}(t^-)}{T_s(t^-)} \left[\sum_{\substack{s'' \\ s'' \neq s' \\ s'' \sim s}} (\phi_{s'} - \phi_{s''}) \frac{A_{s''}(t^-)}{T_s(t^-)} \right]^2 . \tag{64}$$

This together with (59)–(63) implies (14).

With the infinitesimal means and variances we apply Theorem (1.6), [14] or Theorem 3.32, [51], to arrive at (9), and apply Exercise 3.8.12 of [6], Lemma A 1.5.1, [44], or Proposition B.1 in [64][3] to obtain (10). To conclude the proof of Theorem 2.1 is completed.

References

1. D. Applebaum, **Lévy processes and stochastic calculus**, Cambridge University Press, Cambridge, (2009).
2. G. J. Babu and E. D. Feigelson, Spatial point processes in astronomy, *J. Statist. Plann. Inference*, **50**, 311–326, (2015).
3. N. Bellomo, F. Colasuonno, D. Knopoff, and J. Soler, From a systems theory of sociology to modeling the onset and evolution of criminality, *Netw. Het. Media*, **10**, 421–441, (2015).
4. H. Berestycki, N. Rodríguez, and L. Ryzhik, Traveling wave solutions in a reaction-diffusion model for criminal activity, *Multiscale Model. Simul.*, **11**, 1097–1126, (2013).
5. M. Bertero, P. Boccacci, G. Desiderà, and G. Vicidomini, Image deblurring with Poisson data: from cells to galaxies, *Inverse Problems*, **25**, paper n. 123006, 26 (2009).
6. K. Bichteler, **Stochastic integration with jumps**, Cambridge University Press, Cambridge, (2002).

---

[3]In [64], one can directly compute Γ_f and obtain that when $f = Id$ then the infinitesimal variance comes up.

7. D. Brockmann, L. Hufnagel, and T. Geisel, The scaling laws of human travel, *Nature*, **439**, 462–465, (2006).

8. T. Budd, *Burglary of domestic dwellings: Findings from the British Crime Survey,* Home Office Statistical Bulletin, Vol. 4 (Government Statistical Service, London, 1999).

9. L. Cao and M. Grabchak, Smoothly truncated levy walks: Toward a realistic mobility model, *2014 IEEE 33rd Int. Performance Computing and Communications Conference, (IPCCC)*, 5–7 December 2014, Austin, Texas, USA, pp. 1–8.

10. S. Chaturapruek, J. Breslau, D. Yazdi, T. Kolokolnikov, and S. G. McCalla, Crime modeling with Lévy flights, *SIAM J. Appl. Math.*, **73**, 1703–1720, (2013).

11. S. N. Chow, W. Li, and H. Zhou, Entropy dissipation of Fokker-Planck equations on graphs, *Discrete Contin. Dyn. Syst.*, **38**, 4929–4950, (2018).

12. K. L. Chung and R. J. Williams, **Introduction to stochastic integration**, Birkhäuser/Springer, New York, (2014).

13. L. Citi, D. Ba, E. N. Brown, and R. Barbieri, Likelihood methods for point processes with refractoriness, *Neural Comput.*, **26**, 237–263, (2014).

14. R. Durrett, **Stochastic calculus**, CRC Press, Boca Raton, FL, (1996).

15. R. Durrett, **Essentials of stochastic processes**, Springer-Verlag, New York, (1999).

16. R. Durrett, **Probability models for DNA sequence evolution**, Springer-Verlag, New York, (2002).

17. R. Durrett, **Probability: theory and examples**, Cambridge University Press, Cambridge, (2010).

18. P. Embrechts, R. Frey, and H. Furrer, Stochastic processes in insurance and finance, in **Stochastic processes: theory and methods**. North-Holland, Amsterdam, 365–412, (2001).

19. S. N. Ethier and T. G. Kurtz, **Markov processes**, John Wiley & Sons, Inc., New York, (1986).

20. M. Erbar and J. Maas, Ricci curvature of finite Markov chains via convexity of the entropy, *Arch. Ration. Mech. Anal*, **206**, 997–1038, (2012) .

21. M. Erbar and J. Maas, Gradient flow structures for discrete porous medium equations, *Discrete Contin. Dyn. Syst.*, **34**, 1355–1374, (2014).

22. M. Fathi and J. Maas, Entropic Ricci curvature bounds for discrete interacting systems, *Ann. Appl. Probab.*, **26**, 1774–1806, (2016).

23. G. Farrell and K. Pease, **Repeat Victimization**, Criminal Justice Press, (2001).

24. T. Franco, Interacting particle systems: hydrodynamic limit versus high density limit, in **From particle systems to partial differential equations**. Springer, Heidelberg, 179–189, (2014).

25. J. M. Gau and T. C. Pratt, Revisiting broken windows theory: Examining the sources of the discriminant validity of perceived disorder and crime, *J. Crim. Justice*, **38**, 758–766, (2010).

26. M. C. González, C. A. Hidalgo, and A.-L. Barabási, Understanding individual human mobility patterns, *Nature*, **453**, 779–782, (2008).

27. I. J. Good, Some statistical applications of Poisson's work, *Statist. Sci.*, **1**, 157–180, (1986).

28. W. Gorr and Y. Lee, Early warning system for temporary crime hot spots, *J. Quant. Criminol.*, **31**, 25–47, (2015).

29. T. Goudon, B. Nkonga, M. Rascle, and M. Ribot, Self-organized populations interacting under pursuit-evasion dynamics, *Phys. D*, **304/305**, 1–22, (2015).

30. M. Z. Guo, G. C. Papanicolaou, and S. R. S. Varadhan, Nonlinear diffusion limit for a system with nearest neighbor interactions, *Comm. Math. Phys.*, **118**, 31–59, (1988).

31. F. Hamba, Turbulent energy density in scale space for inhomogeneous turbulence, *J. Fluid Mech.*, **842**, 532–553, (2018).

32. I. Hameduddin, C. Meneveau, T A. Zaki, and D. F. Gayme, Geometric decomposition of the conformation tensor in viscoelastic turbulence, *J. Fluid Mech.*, **842**, 395–427, (2018).

33. B. E. Harcourt, Reflecting on the subject: A critique of the social influence conception of deterrence, the broken windows theory, and order-maintenance policing New York style, *Michigan Law Rev.*, **97**, 291–389, (1998).

34. S. W. He, J. G. Wang, and J. A. Yan, **Semimartingale theory and stochastic calculus**, Kexue Chubanshe (Science Press), Beijing; CRC Press, Boca Raton, FL, (1992).

35. S. J. Illingworth, J. P. Monty, and I. Marusic, Estimating large-scale structures in wall turbulence using linear models, *J. Fluid Mech.*, **842**, 146–162, (2018).
36. J. Jacod and A. N. Shiryaev, **Limit theorems for stochastic processes**, Springer-Verlag, Berlin, (2003).
37. A. James, M. J. Plank, and A. M. Edwards, Assessing Lévy walks as models of animal foraging, *J. R. Soc. Interface*, **8**, 1233–1247, (2011).
38. J. Jiménez, Coherent structures in wall-bounded turbulence, *J. Fluid Mech.*, **842**, P1, 100, (2018).
39. S. D. Johnson, W. Bernasco, K. J. Bowers, H. Elffers, J. Ratcliffe, G. Rengert, and M. Townsley, Space-time patterns of risk: A cross national assessment of residential burglary victimization, *J. Quant. Criminol.*, **23**, 201–219, (2007).
40. S. D. Johnson and K. J. Bowers, The stability of space-time clusters of burglary, *Br. J. Criminol.*, **44**, 55–65, (2004).
41. S. D. Johnson, K. Bowers, and A. Hirschfield, New insights into the spatial and temporal distribution of repeat victimization, *Br. J. Criminol.*, **37**, 224–241, (1997).
42. P. A. Jones, P. J. Brantingham, and L. R. Chayes, Statistical models of criminal behavior: the effects of law enforcement actions, *Math. Models Methods Appl. Sci.*, **20**, 1397–1423, (2010).
43. S. Karlin and H. M. Taylor, **A second course in stochastic processes**, Academic Press, Inc. [Harcourt Brace Jovanovich, Publishers], New York-London, (1981).
44. C. Kipnis, and C. Landim, **Scaling limits of interacting particle systems**, Springer-Verlag, Berlin, (1999).
45. C. Kipnis, S. Olla, and S. R. S. Varadhan, Hydrodynamics and large deviation for simple exclusion processes, *Comm. Pure Appl. Math.*, **42**, 115–137, (1989).
46. T. Kolokolnikov, M. J. Ward and J. Wei, The stability of steady-state hot-spot patterns for a reaction-diffusion model of urban crime, *Discrete Contin. Dyn. Syst. Ser. B*, **19**, 1373–1410, (2014).
47. T. M. Liggett, **Lectures on stochastic flows and applications**, Published for the Tata Institute of Fundamental Research, Bombay; by Springer-Verlag, Berlin, (1986).
48. T. Levajković, H. Mena, and M. Zarfl, Lévy processes, subordinators and crime modeling, *Novi Sad J. Math.* **46**, 65–86, (2016).
49. T. M. Liggett, Interacting Markov processes, in **Biological growth and spread (Proc. Conf., Heidelberg, 1979)**. Springer, Berlin-New York, 145–156, (1980).
50. T. M. Liggett, **Interacting particle systems**, Springer-Verlag, New York, (1985).
51. T. M. Liggett, **Continuous time Markov processes**, American Mathematical Society, Providence, RI, (2010).
52. D. J. B. Lloyd and H. O'Farrell, On localised hotspots of an urban crime model, *Phys. D*, **253**, 23–39, (2013).
53. R. N. Mantegna and H. E. Stanley, Stochastic process with ultraslow convergence to a Gaussian: the truncated Lévy flight, *Phys. Rev. Lett.*, **73**, 2946–2949, (1994).
54. M. C. Mariani and Y. Liu, Normalized truncated Levy walks applied to the study of financial indices, *Physica A, Stat. Mech. Appl.*, **377**, 590–598, (2007).
55. G. Ajmone Marsan, N. Bellomo, and L. Gibelli, Stochastic evolutionary differential games toward a systems theory of behavioral social dynamics, *Math. Models Methods Appl. Sci.*, **26**, 1051–1093, (2016).
56. A. Matacz, Financial modeling and option theory with the truncated Lévy process, *Int. J. Theor. Appl. Financ.*, **3**, 143–160, (2000).
57. S. G. McCalla, M. B. Short, and P. J. Brantingham, The effects of sacred value networks within an evolutionary, adversarial game, *J. Stat. Phys.*, **151**, 673–688, (2013).
58. M. Métivier, **Semimartingales**, Walter de Gruyter & Co., Berlin-New York, (1982).
59. M. Métivier and J. Pellaumail, **Stochastic integration**, Academic Press [Harcourt Brace Jovanovich, Publishers], New York-London-Toronto, Ont., (1980).
60. L. C. Miranda and R. Riera, Truncated Lévy walks and an emerging market economic index, *Physica A, Stat. Mech. Appl.*, **297**, 509–520, (2001) .

61. G. O. Mohler, M. B. Short, and P. J. Brantingham, The concentration-dynamics tradeoff in crime hot spotting, in **Unraveling the Crime-Place Connection** Vol. 22 (Routledge, 2017), pp. 19-40.

62. G. O. Mohler, M. B. Short, P. J. Brantingham, F. P. Schoenberg, and G. E. Tita, Self-exciting point process modeling of crime, *J. Am. Stat. Assoc.*, **106**, 100–108, (2011) .

63. G. O. Mohler, M. B. Short, S. Malinowski, M. Johnson, G. E. Tita, A. L. Bertozzi, and P. J. Brantingham, Randomized controlled field trials of predictive policing, *J. Am. Stat. Assoc.*, **110**, 1399–1411, (2015) .

64. J. Mourrat, A quantitative central limit theorem for the random walk among random conductance, *Electron. J. Probab.*, **17**, no. 97, 17, (2012).

65. H. G. Othmer, S. R. Dunbar, and W. Alt, Models of dispersal in biological systems, *J. Math. Biol.*, **26**, 263–298, (1988).

66. S. Peszat and J. Zabczyk, **Stochastic partial differential equations with Lévy noise**, Cambridge University Press, Cambridge, (2007).

67. A. B. Pitcher, Adding police to a mathematical model of burglary, *European J. Appl. Math.*, **21**, 401–419, (2010).

68. P. E. Protter, **Stochastic integration and differential equations**, Springer-Verlag, Berlin, (2005).

69. S. R. S. Varadhan, Entropy methods in hydrodynamic scaling, in **Proceedings of the International Congress of Mathematicians, Vol. 1, 2 (Zürich, 1994)**. Birkhäuser, Basel, 196–208, (1995).

70. S. R. S. Varadhan, Lectures on hydrodynamic scaling, in **Hydrodynamic limits and related topics (Toronto, ON, 1998)**. Amer. Math. Soc., Providence, RI, 3–40, (2000).

71. N. Rodríguez, On the global well-posedness theory for a class of PDE models for criminal activity, *Phys. D*, **260**, 191–200, (2013) .

72. N. Rodríguez and A. L. Bertozzi, Local existence and uniqueness of solutions to a PDE model for criminal behavior, *Math. Models Methods Appl. Sci.*, **20**, 1425–1457, (2010).

73. M. B. Short, A. L. Bertozzi, and P. J. Brantingham, Nonlinear patterns in urban crime: hotspots, bifurcations, and suppression, *SIAM J. Appl. Dyn. Syst.*, **9**, 462–483, (2010).

74. M. B. Short, P. J. Brantingham, A. L. Bertozzi, and G. E. Tita, Dissipation and displacement of hotspots in reaction-diffusion models of crime, *Proc. Natl. Acad. Sci.*, **107**, 3961–3965, (2010).

75. M. B. Short, M. R. D'Orsogna, P. J. Brantingham, and G. E. Tita, Measuring and modeling repeat and near-repeat burglary effects, *J. Quant. Criminol.*, **25**, 325–339, (2009).

76. M. B. Short, M. R. D'Orsogna, V. B. Pasour, G. E. Tita, P. J. Brantingham, A. L. Bertozzi, and L. B. Chayes, A statistical model of criminal behavior, *Math. Models Methods Appl. Sci.*, **18**, 1249–1267, (2008).

77. M. B. Short, G. O. Mohler, P. J. Brantingham, and G. E. Tita, Gang rivalry dynamics via coupled point process networks, *Discrete Contin. Dyn. Syst. Ser. B*, **19**, 1459–1477, (2014).

78. B. Snook, Individual differences in distance travelled by serial burglars, *J. Investig. Psych. Offender Profil.* **1**, 53–66, (2004).

79. D. W. Stroock and S. R. S. Varadhan, **Multidimensional diffusion processes**, Springer-Verlag, Berlin, (2006).

80. B. Tóth and B. Valkó, Onsager relations and Eulerian hydrodynamic limit for systems with several conservation laws, *J. Statist. Phys.*, **112**, 497–521, (2003).

81. W. H. Tse and M. J. Ward, Hotspot formation and dynamics for a continuum model of urban crime, *European J. Appl. Math.* **27**, 583–624, (2016).

82. P. J. van Koppen and R. W. J. Jansen, The road to the robbery: Travel patterns in commercial robberies, *Brit. J. Criminol.* **38**, 230–246, (1998).

83. J. Q. Wilson and G. L. Kelling, Broken windows: The police and neighborhood safety, *Atlantic Mon.* **249**, 29–38, (1982).

84. J. R. Zipkin, M. B. Short, and A. L. Bertozzi, Cops on the dots in a mathematical model of urban crime and police response, *Discrete Contin. Dyn. Syst. Ser. B* **19**, 1479–1506, (2014).

Correction to: A Stochastic-Statistical Residential Burglary Model with Finite Size Effects

Chuntian Wang, Yuan Zhang, Andrea L. Bertozzi, and Martin B. Short

Correction to:
Chapter 8 in: N. Bellomo et al. (eds.), *Active Particles,*
Volume 2, **Modeling and Simulation in Science, Engineering**
and Technology, https://doi.org/10.1007/978-3-030-20297-2_8

The affiliation for the corresponding author Chuntian Wang was updated incorrectly. It has now been corrected. The correct affiliation is:

The University of Alabama, Tuscaloosa, AL, USA

The updated version of this chapter can be found at
https://doi.org/10.1007/978-3-030-20297-2_8

Printed in the United States
by Baker & Taylor Publisher Services